Clinical MR Spectroscopy

Techniques and Applications

Clinical MR Spectroscopy

Techniques and Applications

Peter B. Barker
Alberto Bizzi
Nicola De Stefano
Rao P. Gullapalli
Doris D. M. Lin

CAMBRIDGE UNIVERSITY PRESS
Cambridge, New York, Melbourne, Madrid, Cape Town, Singapore,
São Paulo, Delhi

Cambridge University Press
The Edinburgh Building, Cambridge CB2 8RU, UK

Published in the United States of America by Cambridge University Press,
New York

www.cambridge.org
Information on this title: www.cambridge.org/9780521868983

© P. B. Barker, A. Bizzi, N. De Stefano, R. P. Gullapalli, D. D. M. Lin 2010

This publication is in copyright. Subject to statutory exception
and to the provisions of relevant collective licensing agreements,
no reproduction of any part may take place without the written
permission of Cambridge University Press.

First published 2010

Printed in the United Kingdom at the University Press, Cambridge

A catalog record for this publication is available from the British Library

ISBN 978-0-521-86898-3 Hardback

Cambridge University Press has no responsibility for the persistence or
accuracy of URLs for external or third-party Internet websites referred to in
this publication, and does not guarantee that any content on such websites is,
or will remain, accurate or appropriate.

Every effort has been made in preparing this publication to provide accurate and up-
to-date information which is in accord with accepted standards and practice at the
time of publication. Although case histories are drawn from actual cases, every effort
has been made to disguise the identities of the individuals involved. Nevertheless,
the authors, editors and publishers can make no warranties that the information
contained herein is totally free from error, not least because clinical standards are
constantly changing through research and regulation. The authors, editors and
publishers therefore disclaim all liability for direct or consequential damages
resulting from the use of material contained in this publication. Readers are strongly
advised to pay careful attention to information provided by the manufacturer of
any drugs or equipment that they plan to use.

Contents

Preface *page* vii
Acknowledgments viii
Abbreviations ix

1. Introduction to MR spectroscopy in vivo 1
2. Pulse sequences and protocol design 19
3. Spectral analysis methods, quantitation, and common artifacts 34
4. Normal regional variations: brain development and aging 51
5. MRS in brain tumors 61
6. MRS in stroke and hypoxic–ischemic encephalopathy 91
7. MRS in infectious, inflammatory, and demyelinating lesions 110
8. MRS in epilepsy 131
9. MRS in neurodegenerative disease 144
10. MRS in traumatic brain injury 161
11. MRS in cerebral metabolic disorders 180
12. MRS in prostate cancer 212
13. MRS in breast cancer 229
14. MRS in musculoskeletal disease 243

Index 256

Preface

Magnetic resonance spectroscopy (MRS) allows the non-invasive measurement of selected biological compounds in vivo. Feasibility was first demonstrated in humans in the mid-1980s. Since that time, much experience has been accumulated with the use of MRS in both research and clinical applications. Nearly all magnetic resonance imaging (MRI) scanners have the capability to perform MRS, and MRS techniques still continue to improve, even after two decades of development. MRS has been applied to the study of all major pathologies, particularly in the brain, but has also found application in other organ systems as well, most notably in the breast and prostate.

In spite of this considerable research effort and the unique biochemical information provided, only limited integration of MRS into clinical practice has occurred to date. There are multiple reasons for this, including non-standardization of acquisition and analysis protocols, limited vendor support, difficulties in interpretation (particularly for radiologists without a background in MRS), limited perceived "added-value" above conventional MRI, and lack of reimbursement.

This book is intended to address some of these issues. It gives the reader a solid basis for understanding both the techniques and applications of clinical MRS. Recommendations are made for MRS protocols, and information provided on normal regional- and age-related metabolic variations in the brain. Detailed information about the role of MRS in evaluating pathologies involving the central nervous system, breast, prostate, and musculoskeletal systems is provided. The book also discusses the limitations of MRS, such as its low spatial resolution (e.g. compared to MRI), common artifacts, and diagnostic pitfalls.

The aim of this book is to provide a practical reference work that covers all aspects of in vivo spectroscopy in humans for clinical purposes. As such, it should be a useful guide for radiologists, oncologists, neurologists, neurosurgeons, and other physicians who may be interested in using MRS in their practices. We hope that more widespread adoption of MRS into the clinic will lead to better diagnoses and improved outcomes for individual patients.

Peter Barker, Baltimore
Alberto Bizzi, Milan
Nicola De Stefano, Siena
Rao Gullapalli, Baltimore
Doris Lin, Baltimore

Acknowledgments

PBB	For Catherine and Stephanie
AB	For Anna, Lorenza, and Caterina, who allowed me to devote energy and time to this project
NDS	For Simona, Giorgio, and Andrea
RPG	For Asha. For her enormous patience!
DDML	In loving memory of my father, Daniel

Abbreviations

AD	Alexander disease	MCD	malformations of cortical development
AD	Alzheimer's disease	MCI	mild cognitive impairment
ADC	apparent diffusion coefficient	MEG	magnetoencephalography
ADEM	acute disseminated encephalomyelitis	mI	*myo*-inositol
AIS	Abbreviated Injury Scale	MOA	mixed oligoastrocytoma
ALS	amyotrophic lateral sclerosis	MRI	magnetic resonance imaging
ATP	adenosine triphosphate	MRS	magnetic resonance spectroscopy
CBD	corticobasal degeneration	MRSI	MR spectroscopic imaging
CBF	cerebral blood flow	MSA	multiple system atrophy
CIS	clinically isolated syndrome	MSM	methyl-sulfonyl-methane
CRB	Cramer–Rao bounds	MTR	magnetization transfer ratio
CSD	cortical spreading depression	MTS	mesial temporal sclerosis
CSF	cerebrospinal fluid	NAA	*N*-acetylaspartate
CSI	chemical shift imaging	NMR	nuclear magnetic resonance
CW	continuous wave	OVS	outer-volume suppression
DAI	diffuse axonal injury	PC	phosphocholine
DLB	dementia with Lewy bodies	PCPCS	Pediatric Cerebral Performance Category Scale
DRN	delayed radiation necrosis		
DTI	diffusion tensor imaging	PCr	phosphocreatine
DWI	diffusion-weighted imaging	PDE	phosphodiesters
ECD	Erdheim–Chester disease	PET	positron emission tomography
EPSI	echo-planar spectroscopic imaging	Pi	inorganic phosphate
FFI	fatal familial insomnia	PLIC	posterior limb of the internal capsule
FFT	fast Fourier transformation	PME	phosphomonoesters
FID	free induction decay	PML	progressive multifocal leukoencephalopathy
FOV	field of view		
FSE	fast spin echo	PNET	primitive neuroectodermal tumors
FT	Fourier transform	PRESS	Point REsolved Spectroscopy Sequence
FTD	frontotemporal dementia	PSF	point spread function
GAMT	guanidinoacetate methyl transferase	PSP	progressive supranuclear palsy
GCS	Glasgow Coma Scale	RE	Rasmussen's encephalitis
GPC	glycerophosphocholine	RF	radiofrequency
GSD	Gerstmann–Straussler disease	SAR	specific absorption rate
HE	hepatic encephalopathy	SIAM	spectroscopic imaging acquisition mode
HGG	high-grade glioma		
HIE	hypoxic–ischemic encephalopathy	SMA	supplementary motor area
HPD	human prion disease(s)	SNR	signal to noise ratio
ICA	internal carotid artery	SPECT	single photon emission computed tomography
IVS	inner volume suppression		
LGG	low-grade glioma	SSPE	subacute sclerosing panencephalitis
LOH	loss of heterozygosity	STEAM	stimulated echo acquisition mode
MCA	middle cerebral artery	SVZ	subventricular zone

Abbreviations

SWI	susceptibility-weighted imaging	VD	vascular dementia
TBI	traumatic brain injury	VOI	volume of interest
TLE	temporal lobe epilepsy	VWM	vanishing white matter
TMS	tetramethylsilane	VZ	ventricular zone

Chapter 1
Introduction to MR spectroscopy in vivo

Key points

- Magnetic resonance spectroscopy (MRS) is an analytical technique widely used in chemistry for determining the structure of compounds, and the composition of mixtures of compounds.
- MRS is an insensitive technique, because it observes the resonance signal resulting from the tiny nuclear magnetization.
- Compounds are identified by their unique spectra, based on chemical shifts and coupling constants.
- Spectra are recorded using the pulsed Fourier transform technique.
- Proton spectroscopy of the human brain is most widely used, but other organ systems (such as breast, prostate) and nuclei (particularly ^{31}P and ^{13}C) have also been studied.
- In the brain, compounds of key importance measured by MRS include N-acetyl aspartate (located predominantly in neurons), choline, myo-inositol (located predominantly in glial cells), creatine, lactate, glutamate and glutamine.

Introduction

Nuclear magnetic resonance spectroscopy

The history of magnetic resonance spectroscopy (MRS) can be traced back to the first, independent observations of a nuclear magnetic resonance (NMR) signal in bulk matter by Bloch and Purcell in 1946.[1,2] When atomic nuclei which have the property of nuclear "spin" are placed in a static, strong magnetic field, their energy levels will vary depending on their orientation within the magnetic field. Due to the properties of quantum mechanics, only limited nuclear orientations are allowed (e.g. either "up" or "down" for spin-half nuclei such as the proton (^1H)). If an oscillating radiofrequency field is then applied at the so-called "resonant frequency" corresponding to the energy difference between the different spin orientations, an absorption of power occurs which corresponds to spins being exchanged between the upper and lower states, and a radiofrequency signal is emitted by the sample. This resonant phenomenon and the resulting emitted radiofrequency signal is the fundamental principle of NMR, which is now used worldwide for both magnetic resonance imaging (MRI) and in vivo MRS.

Although NMR was originally a somewhat obscure technology of interest only to physicists for the measurement of gyromagnetic ratios (γ) of different nuclei (see below), applications of NMR to chemistry became apparent after the discovery of chemical shift and spin–spin coupling effects in 1950 and 1951, respectively.[3,4] These effects cause the resonant frequency of the NMR signal to change by small amounts (usually expressed in terms of parts per million (ppm) of the resonant frequency), because the local magnetic field surrounding each nucleus depends on both the structure of its surrounding electrons (i.e. the chemical structure of the molecule that the nuclei occur in) and also on the magnetic properties of neighboring nuclei. Thus, nuclei in different chemical environments will exhibit different resonant frequencies (or spectra in the case of molecules with multiple different nuclei), and NMR spectra can thereby be used to identify both the structure and relative concentrations of the molecules within the sample, information that can be of great value to chemists.

Major technical advances have occurred in MRS over the last several decades; two major developments in the 1960s included the introduction of superconducting magnets (1965), which were very stable and allowed higher field strengths than with conventional electromagnets to be attained, and in 1966 the use of the Fourier transform (FT) for signal processing. In nearly all contemporary spectrometers, the sample is subjected to periodic radiofrequency (RF) pulses directed perpendicular to the main magnetic field and the signal is Fourier transformed to give a spectrum in the frequency domain. FT NMR provides increased sensitivity compared to previous techniques, and also led to the development of a huge variety of pulsed NMR methods, including the methods now commonly used for MRI and in vivo MRS.

Basic theory

If the magnetic field is described by B_0, then the energy of the nuclear spin state is given by

$$E = -\mu \bullet B_0 \quad (1.1)$$

where μ is the nuclear magnetic moment. The magnetic moment is related to the spin angular momentum, P, by the gyromagnetic ratio, γ,

$$\mu = \gamma P = \gamma(Ih/2\pi) \quad (1.2)$$

where γ is a characteristic constant for each nucleus called the "gyromagnetic ratio", I is the spin quantum number and h is Planck's constant (6.626068×10^{-34} J s). By definition, the direction of B_0 is taken to specify the Z-axis, so Equation (1.1) reduces to

$$E = -\mu_Z B_0 \quad (1.3)$$

where μ_Z is the component of μ along the Z-axis. For a nucleus with spin quantum number I, there are $(2I+1)$ different possible orientations of μ in the field, each with a component m_I in the Z direction. For example, for a spin 1/2 nucleus ($I = 1/2$), m_I can take the values $+1/2$ and $-1/2$ (Figure 1.1).

Applying the magnetic dipole allowed selection rule $\Delta m_I = \pm 1$, the resonance frequency is given by

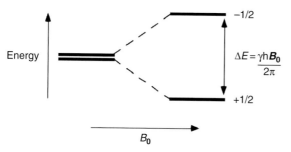

Figure 1.1. Energy levels for a nucleus with spin quantum number $I = \pm 1/2$. In the absence of an externally applied field, B_0, the two energy levels are degenerate. Nuclei with spin 1/2 have two energy levels corresponding to the two discrete values of I. The spin with $I = +1/2$ is aligned with the external magnetic field and thus is lower in energy.

$$\nu_0 = \Delta E/h = \gamma h B_0/(2\pi h) = \gamma B_0/2\pi \quad (1.4)$$

which can be expressed in angular frequency units (radians/sec) as

$$\omega_0 = \gamma B_0 \quad (1.5)$$

ω is called the Larmor frequency. With typical magnetic field strengths currently available for MRS in humans (≈ 1 to 7 Tesla), most magnetic nuclei resonate in the very high frequency (VHF) region of the electromagnetic spectrum (e.g. ≈ 42–300 MHz for the protons). Magnets (1.5 and 3.0 T) are most commonly used for clinical MRS studies, corresponding to resonant frequencies of 64 and 128 MHz, respectively.

Signal and signal-to-noise ratio

In NMR, the difference between energy levels is very small and this results in very small population differences between the upper and lower energy levels. The excess population of the lower level compared to the upper level can be calculated from the Boltzmann factor

$$\frac{n(upper)}{n(lower)} = \exp(-\Delta E/kT) \quad (1.6)$$

where k is the Boltzmann constant (1.380650×10^{-23} J K^{-1}) and T is the absolute temperature (measured in Kelvin (K)). For protons at body temperature (310 K) at 128 MHz, $n(upper)/n(lower) \approx 0.9999801$. The excess population of the lower level creates the macroscopic nuclear magnetization that is observed in the NMR experiment – note that this magnetization is very small, since, for instance in the example above, only 0.002% of the total number of spins contribute to the net nuclear magnetization, M_0, which can be expressed as

$$M = NB_0\gamma^2\hbar^2 I(I+1)/3kT \quad (1.7)$$

N is the total number of spins in the sample, and $\hbar$ is Planck's constant divided by 2π. The signal (S) detected in the receiver coil is proportional to the magnetization times the resonant frequency

$$S \propto M_0\omega_0 = NB_0^2\gamma^3\hbar^2 I(I+1)/3kT \quad (1.8)$$

giving a final dependence of the signal on B_0^2 for a given nuclei (γ fixed), or alternatively a γ^3 dependence for different nuclei at a fixed B_0. Notice also that the signal increases with decreasing temperature. Therefore, the best signal is obtained at high fields, from high γ nuclei, and at low temperature. While the temperature and nuclei to be observed are often fixed for in vivo experiments, it is clear that the use of high magnetic fields results in larger signals.

In an NMR experiment, the ability to detect a signal depends not only on the signal amplitude, but also on the amount of noise in the spectrum (i.e. the signal-to-noise ratio (SNR)). Noise voltages arise from the random, thermal motion of electrons in the radiofrequency (RF) coil used to detect the signal, and depend on the resistance of the coil (including any effects that the sample may have on the coil). Resistance typically increases with increasing RF frequency, with the exact dependence depending on both the coil and the sample properties. For biological samples (which have appreciable electrical conductivity), it is generally thought that the noise voltage increases approximately linearly with frequency, so that there is expected to be a linear increase in SNR with B_0.

The rotating frame: simple pulse sequences for spectroscopy

The simplest NMR pulse sequence involves applying a radiofrequency pulse (B_1) for a short period of time (τ), followed by collection of the signal without any field gradients applied (the so-called "free induction decay", or FID). The way this experiment is best understood is to use the "rotating frame" reference; in this frame, the B_0 field is taken to define the direction of the Z-axis, while the B_1-field is static (so long as the applied RF frequency pulse is exactly on-resonance with the Larmor frequency) and defines either the X- or Y-axes. Initially, when the spin system is at equilibrium, all the magnetization is aligned along the Z-axis; application of the RF pulse results in the rotation of the Z-magnetization about the axis of the applied RF field at a nutation rate of γB_1. At the end of the RF pulse (of length τ), the magnetization will have rotated through the angle (α, the "flip angle")

$$\alpha = \gamma B_1 \tau \tag{1.9}$$

The RF receiver coil only detects magnetization in the XY plane, so the largest signal occurs (with long repetition times, TR) when all the magnetization is tipped from the Z-axis into the transverse plane – this is called a 90 degree pulse ($\alpha = 90°$) (Figure 1.2A). In general, the signal amplitude will vary as $\sin(\alpha)$ for a single RF pulse.

In addition to detecting FIDs, much of MRI and in vivo MRS relies on the detection of signals from *spin echoes*. Spin echoes, first discovered by Erwin Hahn in 1950, [5] are signals that occur after the application of two or more RF pulses. Echoes generally can occur with pulses of any flip angle, but are conceptually easiest to understand when considering the 90°–180° echo sequence first introduced by Carr and Purcell in 1954.[6] After the initial 90° pulse, spins precess around the Z-axis at a rate determined by the strength of the main magnetic field B_0. Since the B_0 field is never completely uniform (homogeneous) throughout the different parts of the sample, different regions of the sample precess at different speeds, leading to a loss of phase-coherence between regions, and the loss of signal. By applying an 180° pulse at time TE/2 (where TE is echo time), the positions of the slow and fast precessing components are alternated in the transverse plane, and after another period of time TE/2 the magnetization vectors from all parts of the sample are in-phase again, leading to the creation of the echo signal (Figure 1.2B). If additional RF pulses are applied, additional echoes are formed. Commonly used pulse sequences for localized spectroscopy (see Chapter 2 for details) employ 3 pulses, such as a 90°–180°–180° (called the "PRESS" sequence), or 90°–90°–90° (called the "STEAM" sequence).

Relaxation times

After an RF pulse, the magnetization is tipped away from its equilibrium position aligned along the Z-axis of the rotating frame, and as a result it will start recovering back to its equilibrium position. The time-constant for this process (recovery of longitudinal magnetization) is known as the spin–lattice relaxation time constant, or T_1. T_1 is usually measured using pulse sequences known as inversion-recovery (180°–TI–90°) or saturation recovery (90°–TI–90°), where the experiment is repeated with several different TI delay times in order to map out the signal vs. time curve (Figure 1.3). For the inversion recovery experiment, this curve takes the form $S(TI) = S(0)(1 - 2^* \exp(-TI/T_1))$, while for saturation recovery it becomes $S(TI) = S(0)(1 - \exp(-TI/T_1))$. Note that it is important to wait a sufficiently long repetition (TR, typically $> 5^*T_1$) between experiments for magnetization to fully recover before the next measurement or time average.

In addition to T_1 relaxation times, there is an additional time constant (T_2, the transverse relaxation time) which describes how fast the transverse magnetization decays in a spin–echo experiment. The signal decay is

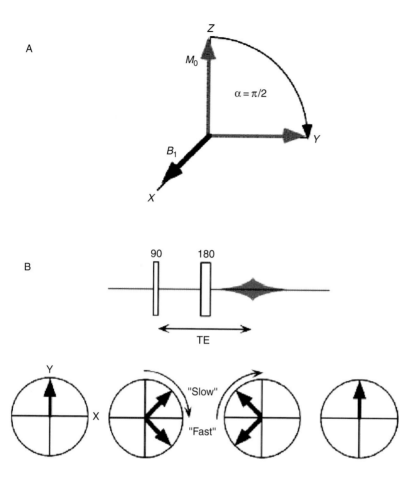

Figure 1.2. (A) Effects of a 90° pulse in the rotating frame coordinate system. (B) Formation of a spin-echo using a 90°–180°–acquire sequence.

Figure 1.3. (A) Saturation and (B) inversion recovery experiments for the measurement of T_1 relaxation times. (A) The simplest pulsed experiment consists of a single RF pulse of flip angle, a, followed by detection of the FID, and can be repeated n times to improve the signal-to-noise ratio. (B) Diagram of the inversion recovery experiment that can be used to determine the T_1 relaxation time by varying the delay, τ, provided the recovery delay, RD, is $\sim 5 > T_1$.

described by the expression $SE(TE) = S(0)\exp(-TE/T_2)$, so T_2 is estimated from measurements of S performed as a function of TE. This can be done by repeating the measurement several times with different TE values (Carr–Purcell "Method A") or by performing a single-shot multi-spin–echo experiment (i.e. multiple refocusing 180° pulses played out in a single experiment – Carr–Purcell "Method B") (Figure 1.4).

Knowledge of relaxation times in vivo is important for designing experimental parameters for optimal sensitivity, as well as for correcting metabolite concentration measurements for effects of variable relaxation times.

Chemical shifts

Early in the development of NMR, it was discovered that nuclei in different molecular environments resonated at slightly different frequencies.[7] The origin of this effect lies in the response of the molecule's electrons to the applied magnetic field; in simple terms, a rotating current is induced which generates a small magnetic field that (usually) opposes the external field (a *diamagnetic* effect). The nucleus thereby experiences a smaller net field than that which is actually applied – the effective field at the nucleus, B_{0eff}, can be expressed in terms of a shielding parameter, σ,

$$B_{0eff} = B_0(1 - \sigma) \quad (1.10)$$

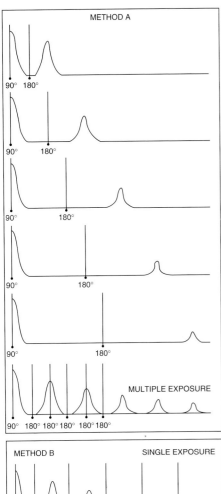

Figure 1.4. Measurement of T_2 relaxation times using either the Carr–Purcell Method "A" (a single 180° pulse) or Method "B" (multiple 180° pulses).

so that the nuclear resonance frequency becomes

$$\nu = \gamma B_{0\text{eff}}/2\pi = \nu_0(1-\sigma) \qquad (1.11)$$

This effect is known as the chemical shift; it is of prime importance, since nuclei in different chemical environments can be distinguished on the basis of their resonant frequencies. The shielding parameter is usually defined in parts per million (ppm) of the resonance frequency, measured relative to a reference compound. While the shielding parameter is a constant, the chemical shift (measured in Hz) increases linearly with field strength (Equation (1.10)). Thus the resolution of the NMR spectrum increases linearly with increasing field strength, provided that the linewidths do not change. To compare chemical shifts measured at different field strengths, it is standard to report chemical shift values (δ) in ppm relative to a standard reference compound; for ^{1}H and ^{13}C spectra (in vitro), tetramethylsilane (TMS) is often used as the reference,

$$\delta = (\nu - \nu_{\text{ref}})^* 10^6 / \nu_{\text{ref}} \qquad (1.12)$$

where ν and ν_{ref} are the frequencies of the signal of interest and the reference signal, respectively. In vivo, reference signals from compounds such as TMS are not available, so usually one of the indigenous spectral signals is used as a spectral reference (e.g. for ^{1}H spectra in the brain, the N-acetyl resonance of N-acetylaspartate (NAA), set to 2.02 ppm, is often used as a chemical shift reference).

Figure 1.5 illustrates the typical ^{1}H chemical shift range for various functional groups. While the typical chemical shift range for ^{1}H is quite small ($\approx$ 10 ppm), much larger chemical shift ranges (e.g. up to several 100 ppm) exist for other nuclei such ^{19}F, ^{31}P, ^{13}C and ^{15}N.

Spin–spin (scalar) coupling constants

In addition to chemical shift effects, it was also discovered early in the development of NMR spectroscopy that spectra from liquid samples often exhibited further fine structure in the forms of splittings (or multiplets).[4] These arise from the electron-coupled spin–spin interaction, also known as J- or scalar-coupling. Spin–spin coupling results from nuclei experiencing the magnetic fields of their neighboring nuclei through polarization of the electrons in the molecular bonds between them. The effective magnetic field experienced by one nucleus depends on the spin state of a neighboring, coupled nucleus. The nomenclature to describe spin systems assigns letters to each individual spin, which are close together in the alphabet if the spins are strongly coupled or far from one another in the alphabet if the spins are weakly coupled ($|\delta_A - \delta_X| \gg J_{AX}$). The simplest multiplet pattern that can be observed is in a 2-spin, "AX" spin system (Figure 1.6).

Spin–spin couplings have the following properties:

1. They act through the bonding electrons and are therefore intramolecular.
2. They are independent of the strength of the applied magnetic field.
3. Spin multiplet structure reflects the states of neighbor nuclei.

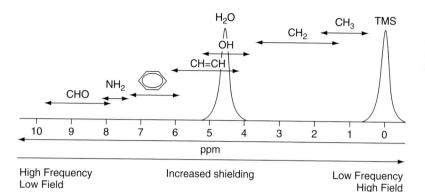

Figure 1.5. Typical ¹H chemical shift range (10 ppm) for various functional groups. The resonances of water and the standard reference compound TMS are shown at 4.7 and 0.0 ppm, respectively.

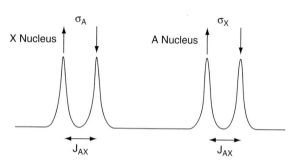

Figure 1.6. Multiplet pattern of a weakly coupled two-spin (AX) system. The splitting is equal to J_{AX}. σ represents the shielding constant of the A or X spin. The arrows indicate alignment with (↓) or against (↑) the static magnetic field.

4. The interaction is reciprocal; if A splits X, then X splits A.
5. Splitting cannot be observed due to coupling between equivalent nuclei.
6. Coupling can be homonuclear or heteronuclear.

For spins which have widely different chemical shifts, the multiplet pattern is symmetrical around the chemical shift frequency, with relative intensities given by the binomial coefficients $(1+x)^N$ for $I = 1/2$ nuclei.

The simple treatment of multiplet patterns above is only valid if $|\delta_A - \delta_X| \gg J_{AX}$, which is known as the weak coupling, or first-order approximation. When this is not true, the spectra are strongly coupled or second-order, the symmetry of the multiplet is destroyed, and the splittings are no longer equal to the coupling constant. New lines may appear (combination lines), and lines are no longer assignable to a single nucleus. In general, the spectrum can only be assigned through the use of computer simulation. At the relatively low field strengths commonly used for in vivo MRS in humans, strong coupling is quite commonly encountered.

Nuclei with $I > 1/2$ produce more complicated multiplet patterns. The deuterium nucleus ($I = 1$) produces a triplet in neighboring atoms with an intensity ratio 1:1:1, corresponding to $m_I = -1$, 0 and +1, respectively. However, some quadrupolar nuclei relax too rapidly to generate observable splittings. It should also be noted that spin–spin splitting could be modified or eliminated by chemical exchange or double resonance experiments (decoupling).

Fourier transform spectroscopy

As mentioned above, virtually all MRS studies are performed by collecting time domain data after application of either a 90° pulse, or an echo-type of sequence. All resonances from the different molecules are collected simultaneously in the time domain, and the time domain signal (FID) is largely uninterpretable to the human eye. In order for a spectrum to be generated, it is necessary to perform Fourier transformation (FT), which allows the viewing of the signal intensity as a function of frequency (i.e. in the frequency domain) (Figure 1.7). Various filtering and other manipulations are often performed on the data both before and after fast Fourier transformation (FFT), which may have quite profound effects on the quality of the final spectrum; these are discussed in more detail in Chapter 3.

One advantage of pulsed FT is that all signals are being recorded at once, so it has a sensitivity advantage over alternative acquisition methods ("continuous-wave", or CW), which recorded each part of the spectrum separately. In order to accumulate sufficient SNR with the pulsed FT method, the scan can be repeated many (N) times and averaged together ("time-averaging") to improve SNR ($\propto \sqrt{N}$). The scan time will be $N*TR$; it is important to choose the correct TR and the flip angle for optimum SNR. The seminal

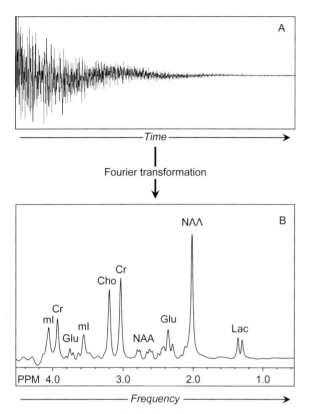

Figure 1.7. (A) An example of a free induction decay (FID, recorded as a function of time) and (B) the corresponding frequency domain spectrum obtained by Fourier transformation. The sample is a phantom containing N-acetyl aspartate (NAA, 2.01 and 2.6 ppm), creatine (Cr, 3.02 and 3.91 ppm), choline (Cho, 3.21 ppm), myo-inositol (ml, 3.56 and 4.05 ppm), glutamate (Glu, 2.34 and 3.74 ppm), and lactate (Lac, 1.31 ppm (doublet)), recorded at 3 T with an echo time of 30 msec.

work by Ernst indicates that for optimum SNR the minimum TR should be chosen consistent with the pulse sequence being used and the desired spectral resolution (in Hz, equal to the inverse of the data readout window (acquisition time)), and then the flip angle set (the "Ernst Angle") according to the expression, $\alpha = \cos^{-1}(\exp(-TR/T_1))$. For example, if $TR = 1.5$ sec and $T_1 = 1$ sec, $\alpha = 77.15°$.[8,9]

In vivo MR spectroscopy

In vivo magnetic resonance spectroscopy (MRS) of the brain was first reported in the late 1970s in animal models.[10] Previous studies of biological tissues by NMR spectroscopy had focused on isolated, perfused organ systems or cell suspensions, and had indicated the feasibility of obtaining biochemical information non-invasively using NMR. Most of these studies used the phosphorus-31 nucleus (^{31}P), since there was interest in measuring metabolism relating to bioenergetics, which involved compounds such as adenosine triphosphate (ATP), phosphocreatine (PCr) and inorganic phosphate (Pi). ^{31}P spectroscopy was relatively straightforward to perform, since the ^{31}P nucleus is spin-1/2 and has a reasonably high gyromagnetic (γ) ratio and chemical shift range (~40 ppm), and does not require any water suppression. It was also found that the resonance frequency of Pi was sensitive to pH,[11] and could be used to determine brain pH non-invasively, [12, 13] As interest in in vivo MRS and MRI increased, larger bore horizontal superconducting magnet systems were developed for this purpose, for larger animals and humans, although at lower field strengths than used for high-resolution NMR (e.g. 1.5–2.0 Tesla (T), 64–85 MHz for ^{1}H). An important technical advance was the introduction of local, surface RF coils that had high sensitivity,[14] and also limited signal reception to only tissues that are proximal to the coil, thereby eliminating signal from unwanted regions or other organs. These advances enabled the first observation of in vivo MRS in humans,[15] and the detection of birth asphyxia in the brain of infants,[16] using ^{31}P MRS. In addition to ^{31}P, there was also interest in the carbon-13 (^{13}C) nucleus, which, like ^{31}P, has some technical advantages, such as a wide chemical shift range (~200 ppm), spin-1/2, and no need for solvent suppression.[17] However, both ^{31}P and ^{13}C suffer from relatively low sensitivity. In the case of ^{13}C, sensitivity is very low because of its low natural abundance (1%), although exogenously introduced isotopic enrichment can be used, which is in fact an important method for studying kinetics of metabolism.[17]

It was recognized that proton (^{1}H) spectroscopy would offer a large sensitivity advantage over these other nuclei, because the proton has the highest gyromagnetic ratio γ of non-radioactive nuclei, as well as a high natural abundance. Sensitivity is also enhanced compared to other nuclei because of favorable metabolite relaxation times, and also because several important brain metabolites have resonances resulting from functional groups with multiple protons (methyl groups with three protons). In order for proton MRS to be successful, however, water suppression techniques had to be developed for in vivo MRS, so as to remove the much larger water signal (compared to the metabolite signals), and magnetic field homogeneity and field strengths had to be sufficient to allow one to resolve the smaller chemical shifts of protons (range ~ 10 ppm). In 1983, Behar et al. reported the first in vivo MRS of the brain,[18] in rats at 360 MHz (~8.5 T) using a surface RF coil and a continuous, pre-saturation pulse for water suppression. Resonances were assigned (by comparison with high-resolution NMR spectra of perchloric acid extracts of brain tissue) to phosphocholine (PCho), phosphocreatine (PCr), creatine (Cr), aspartate (Asp), glutamate (Glu), N-acetyl aspartate (NAA), γ-amino butyric acid (GABA), alanine (Ala) and lactate (Lac). It was also demonstrated in the same paper that induction of hypoxia (by lowering the inspired oxygen fraction from 25% to 4% for 15 min) caused an elevation of the brain lactate signal, which could be reversed by restoration of 25% oxygen. In 1984, the same group demonstrated that proton spectra, somewhat less well-resolved, could also be obtained at the more clinically relevant field strength of 1.9 T, and that the use of a spin–echo pulse sequence provided improved water suppression, and removed lipids and broad baseline components, all of which have relatively shorter transverse (T_2) relaxation times than the small molecular weight metabolites.[19]

In 1985, Bottomley et al. reported the first spatially localized human brain spectrum, at 1.5 T using a slice-selective spin–echo excitation technique, and frequency selective water suppression.[20] At an echo time of 80 ms, signals were observed from NAA, Asp, creatine and choline-containing (Cho) compounds, as well as from lipids and residual water. Although this paper demonstrated the feasibility of human brain spectroscopy in vivo on a 1.5 T MRI system, the spatial localization and spectral resolution were limited. Similar approaches were used by Luyten and den Hollander [21] and Hanstock et al.[22] to record spectra from the human brain, using a spin–echo depth pulse localization scheme and surface coil reception. Using a 2.1 T magnet, Hanstock et al. demonstrated well-resolved signals for choline, creatine, NAA, as well as a combined peak of GABA, glutamate (Glu) and glutamine (Gln), which could be recorded from an approximate volume of 14 ml in a 4-min scan time. T_2 values of Cr and NAA were also estimated to be in the range 140–530 ms (with Cr appreciably shorter than NAA), and it was further demonstrated that a normal human brain lactate concentration of the order of 0.5 mM may be detectable using modulation of the spin–echo by J-coupling (a TE of 150 ms was used[22]).

However, spatial localization provided by depth pulses and surface coils is relatively ill-defined, and therefore improved spatial localization techniques (in particular, definition of localization in all three dimensions) were required. Spatial localization allows signals to be recorded from well-defined structures or lesions within the brain, and by recording signals from smaller volume elements, improved field homogeneity can be obtained.[23] In the 1980s, a wide range of spatial localization techniques were developed for in vivo spectroscopy;[24] however, many were either difficult to implement, involved too many RF pulses, or were inefficient (i.e. involved too much signal loss, or did not fully suppress out-of-voxel magnetization). Out of this plethora of sequences, two emerged as simple and robust enough for routine use, each based on three slice-selective pulses applied in orthogonal directions. The STEAM sequence (Stimulated Echo Acquisition Mode)[25,26,27,28] uses three 90° pulses and detects the resulting stimulated echo from the volume intersected by all three pulses, while the PRESS sequence uses one 90° pulse and two 180° pulses to detect a spin echo from the localized volume.[29,30] A detailed description of these techniques is provided in the next chapter. The demonstration in 1989 of high-resolution human proton brain spectra from relatively small, well-defined regions of interest in short scan times (generally less than 5 min) led to a rapid increase in the use of this methodology, and allowed for the non-invasive study of human brain metabolism in neuropathology by proton MRS in the early 1990s.[28]

Commercial availability and automation[31] of proton MRS on clinical MRI systems expedited these studies, as well as the transition of this method to the clinical examination. While human brain MRS in the 1980s was mainly focused on ^{31}P MRS,[32] the realization of the higher signal-to-noise ratios available with ^{1}H MRS, and the consequent improvements in spatial resolution, led to the adoption of ^{1}H MRS at the expense of ^{31}P. ^{1}H MRS also has the highly significant advantage over all other nuclei that it uses exactly the same hardware (RF coils, amplifiers, preamplifiers, receivers, etc.) as used in conventional MRI, thereby allowing it to be performed on most commercial MRI scanners without significant hardware modifications. At 1.5 T, proton MRS can typically be performed on voxel volumes of the order of 1–8 cm^3, while ^{31}P MRS usually requires voxel volumes greater than 30 cm^3. Therefore, the clinical applications of ^{31}P MRS are limited in the brain, because of low spatial resolution and signal-to-noise ratios (SNR). As mentioned above, ^{13}C MRS in the brain has even lower sensitivity (and hence spatial resolution) than ^{31}P, and therefore has also remained solely a research tool, rather than becoming a clinical technique for use in radiology.

The remainder of this handbook is focused on methods for proton MRS of the human brain, since this is the technique that is overwhelmingly used to study human brain metabolism at present. Proton MRS also shows promise for the evaluation of other organ systems, particularly the prostate and breast. [33,34] In general, MRS outside of the brain presents a number of additional technical challenges that make it much harder to perform than in the central nervous system. For this reason, there has been greater emphasis on CNS applications so far; however, in the future this may change. The last few chapters of the handbook deal with techniques and application of MRS in the body.

Nuclei for in vivo MRS

By far the most in vivo MRS studies have been performed using the proton (^{1}H) nucleus, because of several reasons; the proton has high sensitivity because of its high γ and high natural abundance, as well as quite favorable relaxation times and spin half. In addition, the proton is also the nucleus used for conventional MRI, so proton MRS can be performed usually with exactly the same hardware as is used for conventional MRI. However, there are other nuclei which can be used for in vivo MRS if appropriate RF coils, amplifiers, and electronics are available; some examples include carbon-13 (^{13}C), nitrogen-15 (^{15}N), or phosphorus-31 (^{31}P): generally, these have lower sensitivity and natural abundance, which results in longer scan times and lower spatial resolution (increased voxel size). The properties of these nuclei are listed in Table 1.1. Since clinical applications of these nuclei are not yet routinely available, the majority of this book will focus on the use of the proton nucleus for in vivo MRS. However, it should be mentioned that the use of these nuclei with hyperpolarization and/or isotropic enrichment can give large sensitivity increases, and are the subject of active research efforts at present (2008). If technical challenges and cost issues can be overcome, methods based on these nuclei may offer unprecedented opportunities to study dynamic metabolic processes in humans non-invasively.

Information content of proton MR spectra of the brain

Because of the relatively low sensitivity of in vivo MRS, in order for a compound to be detectable, generally its concentration must be in the millimolar range, and it must be a small, mobile molecule. Large and/or membrane-associated molecules are not usually detected, although they may exhibit broad resonances that contribute to the baseline of the spectrum.[35]

The information content of a proton brain spectrum depends on quite a few factors, such as the field strength used, echo time, and type of pulse sequence. At the commonly used 1.5 T field strength, at long echo times (e.g. 140 or 280 ms are often used; Figure 1.8) only signals from Cho, Cr, and NAA are observed in normal brain, while compounds such as lactate, alanine, or others may be detectable if their concentrations are elevated above normal levels due to pathological processes.[36,37,38] At short echo times (e.g. 35 ms or less) compounds with shorter T_2 relaxation times (or multiplet resonances which become dephased at longer echo times) also become detectable. These include resonances from glutamate, glutamine, and GABA, which are not resolved from each other at 1.5 T, *myo*-inositol, as well as lipids and macromolecular resonances (Figure 1.8). Spectral appearance at 3.0 T is generally similar to that at 1.5 T (Figures 1.8C and D), although the coupling patterns of the multiplet resonances are somewhat different. Most of the multiplets (e.g. Glu, Gln, mI, taurine) are strongly coupled at these field strengths, and Glu and Gln overlap slightly less at 3 T than at 1.5 T. As field strengths increase further, to 4.0 and 7.0 T, spectral resolution progressively increases (provided that magnetic field homogeneity can be maintained) and more compounds can be assigned with confidence, including separating *N*-acetyl aspartyl glutamate (NAAG) from NAA, separation of Glu from Gln, and the detection of up to 14 different compounds at short echo times at 7 T (Figure 1.8E).[39] A complete

Table 1.1. Properties of nuclei for in vivo MRS.

Nucleus	Frequency (MHz @1.5 T)	Spin	Natural abundance (%)
Proton (^{1}H)	63.9	1/2	99.98
Phosphorus (^{31}P)	25.9	1/2	100.00
Sodium (^{23}Na)	16.9	3/2	100.00
Carbon (^{13}C)	16.1	1/2	1.10
Deuterium (^{2}D)	9.8	1	0.02
Nitrogen (^{15}N)	6.5	1/2	0.37
Oxygen (^{17}O)	8.7	5/2	0.04
Fluorine (^{19}F)	59.8	1/2	100.00
Lithium (^{7}Li)	24.9	3/2	92.50

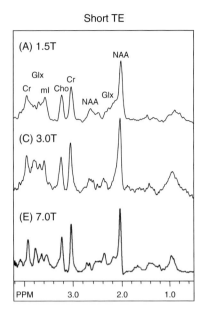
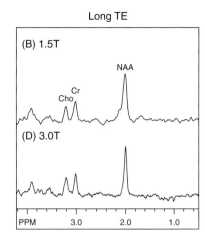

Figure 1.8. Single-voxel proton MRS of normal human brain white matter at different field strengths and echo times recorded using the STEAM pulse sequence (2 × 2 × 2 cm voxel size). 1.5 T: (A) TE 20 ms, (B) TE 136 ms; 3.0 T (C) TE 20 ms, (D) TE 136 ms; and 7.0 T (E) TE 18 ms. Spectrum (E) provided by Dr James Murdoch, Philips Medical Systems. As field strength increases, spectral resolution improves, particularly for the strongly coupled resonances such as glutamate, glutamine, and *myo*-inositol.

list of metabolite structures, chemical shift, coupling constants, and multiplet patterns can be found in reference [40], and a summary of all compounds that have been detected in the human brain by proton MRS is given in Table 1.2.

NAA

The largest signal in the normal adult brain spectrum, the acetyl group of *N*-acetyl aspartate resonates at 2.01 ppm, with a usually unresolved (except at very high fields) contribution from *N*-acetyl aspartyl glutamate (NAAG) at 2.04 ppm.[41,42] The aspartyl group also exhibits a pH-sensitive, strongly coupled resonance at approximately 2.6 ppm. Despite being one of the most abundant amino acids in the central nervous system, NAA was not discovered until 1956, and its function has been the subject of considerable debate.[43] It has been speculated to be a source of acetyl groups for lipid synthesis, a regulator of protein synthesis, a storage form of acetyl-CoA or aspartate, a breakdown product of NAAG (which, unlike NAA, is a neurotransmitter), or an osmolyte.[44] NAA is believed to be synthesized in neuronal mitochondria, from aspartate and acetyl-CoA (Figure 1.9A). NAA is often referred to as a neuronal marker, based on several lines of evidence. For instance, immunocytochemical staining techniques have indicated that NAA is predominantly restricted to neurons, axons, and dendrites within the central nervous system;[45] and studies of diseases known to involve neuronal and/or axonal loss (for instance, infarcts, brain tumors, or multiple sclerosis (MS) plaques) have without exception found NAA to be decreased. In pathologies such as MS, good correlations between brain NAA levels and clinical measures of disability have been found, suggesting that higher NAA levels may be associated with better neuronal function.[46] Animal models of chronic neuronal injury have also been shown to give good correlations between NAA levels (as measured by MRS) and in vitro measures of neuronal survival. [47] All of these studies therefore suggest that MRS measurements of NAA may be useful for assessment of neuronal health or integrity in the central nervous system.

However, other experiments suggest that caution should be used in interpreting NAA solely as a neuronal marker. For instance, it has also been reported that NAA may be found in non-neuronal cells, such as mast cells or isolated oligodendrocyte preparations, suggesting that NAA may not be specific for neuronal processes.[48,49,50] It is unclear if these cells are present at high enough concentrations in the normal human brain to contribute significantly to the NAA signal, however. There are also some rare cases where NAA metabolism is perturbed, almost certainly independently of neuronal density or function. One example is the leukodystrophy, Canavan's disease, which is associated with a large elevation of intracellular NAA, owing to deficiency of aspartoacylase, the enzyme that

Table 1.2. Compounds detected by proton MRS in the human brain.

Compounds normally present	Compounds which may be detected under pathological or other abnormal conditions
Large signals at long TE	*Long TE*
N-acetylaspartate (NAA)	Lactate (Lac)
Creatine (Cr) and Phosphocreatine (PCr)	β-Hydroxy-butyrate, acetone
Cholines (Cho):	Succinate, pyruvate
Glycerophosphocholine (GPC)	Alanine
Phosphocholine (PC), free choline (Cho)	Glycine
Large signals at short TE	*Short TE*
Glutamate (Glu)	Lipids
Glutamine (Gln)	Macromolecules
myo-Inositol (mI)	Phenylalanine
	Galactitol
Small signals (short or long TE)	*Exogenous compounds (short or long TE)*
N-acetyl-aspartyl-glutamate (NAAG), aspartate	Propan-1,2-diol
Taurine, betaine, scyllo-inositol, ethanolamine	Mannitol
Threonine	Ethanol
Glucose, glycogen	Methylsulfonylmethane (MSM)
Purine nucleotides	
Histidine	
Small signals that can be detected with the use of two-dimensional and/or spectral editing techniques	
γ-Amino-butyric acid (GABA)	
Homocarnosine, pyrrolidinone	
Glutathione	
Threonine	
Vitamin C (ascorbic acid)	

degrades NAA to acetate and aspartate (Figure 1.9B).[51] In addition, there has been a case report of a child, with mental retardation, with a complete absence of NAA (Figure 1.9C).[52] This case suggests that neurons can exist without the presence of NAA, and indeed that NAA is not necessary for neuronal function.

While these observations indicate that there is evidence both for and against NAA as a measure of neuronal density and function, on balance, NAA does appear to be one of the better *surrogate* neuronal markers that can be measured non-invasively in humans. Like all surrogate markers, there will be occasions when it does not reflect the *true* neuronal status.

Decreases in NAA in some diseases have been shown to be reversible, suggesting that low NAA does not always indicate permanent neuronal damage.[53] Reversible NAA deficits (either spontaneous, or in response to treatment) have been observed in diseases such as multiple sclerosis, mitochondrial diseases, AIDS, temporal lobe epilepsy, amyotrophic lateral sclerosis, or acute disseminated encephalomyelitis (ADEM).[44,54] Therefore, in individual patients, while a low NAA signal in some pathologies may indicate irreversible neuroaxonal damage (e.g. strokes, brain tumors), in others it may be due to dysfunction (perturbed NAA synthesis or degradation) that may be reversible with either treatment-related or spontaneous recovery.

Choline

The choline signal (Cho, 3.20 ppm) is a composite peak consisting of contributions from the trimethyl

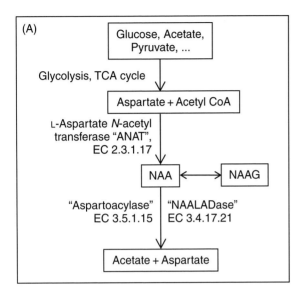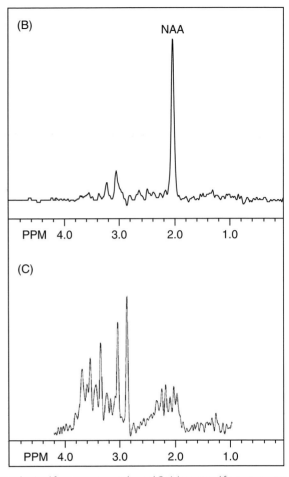

Figure 1.9. (A) Schematic of metabolism of NAA within the brain. NAA is synthesized from aspartate and acetyl CoA (generated from precursors such as glucose, acetate or pyruvate) and degraded to acetate and aspartate. There is also a pathway relating NAA and NAAG catalyzed by the enzyme NAALADase. (B) In Canavan's Disease, aspartoacylase is deficient, leading to large accumulations of NAA in the brain. (C) A case of NAA-deficiency, reproduced with permission from [52], resulting from a presumed deficit in the synthetic pathway of NAA. Normal signals from Cho, Cr, Glx, and ml are observed; however, NAA is absent. Spectra recorded at 1.5 T.

amine (–N(CH3)3) groups of glycerophosphocholine (GPC), phosphocholine (PC), and a small amount of free choline itself.[55]

These compounds are involved in membrane synthesis and degradation, and it has often been suggested that they are elevated in disease states where increased membrane turnover is involved (e.g. tumors). Glial cells have also been reported to have high levels of Cho.[56,57] Other pathological processes which lead to Cho elevation include active demyelination,[58] either resulting from the degradation of myelin phospholipids primarily to GPC, or perhaps due to inflammation.[59] Elevated Cho levels seem to be a characteristic of many types of neoplasms, including high-grade brain tumors (provided that they are not necrotic), prostate, breast, head and neck, and other tumors. In particular, it would appear that malignant transformation of tumors involves an increase in PC relative to GPC.[60]

Low brain Cho has been observed in hepatic encephalopathy,[61] and there is also some evidence to suggest that dietary intake of choline can modulate cerebral Cho levels.[62] In both cases, this may be due to altered (decreased or increased) systemic transport of Cho to the brain.

Cho also shows quite strong regional variations in the brain, usually with somewhat higher levels in white matter than gray, although the thalamus, hypothalamus, and insular cortex also show high levels in the normal brain.

Creatine

The creatine methyl resonance (Cr, 3.03 ppm) is a composite peak consisting of both creatine and phosphocreatine, compounds that are involved in energy metabolism via the creatine kinase reaction, generating ATP. In many spectra, a second resonance from the CH_2 of creatine is also observed at 3.91 ppm (provided that it is not saturated by the water-suppression pulses). In vitro, glial cells contain a two- to fourfold higher concentration of creatine than do neurons.[63] Creatine also shows quite large regional variations, with lower levels in white matter than gray matter in normal brain, as well as very high levels of Cr in the cerebellum compared to supratentorial regions.[64]

Since creatine is synthesized in the liver and transported to the brain, chronic liver disease leads to lower cerebral creatine concentration.[65] There is also a rare group of diseases which involve total Cr deficiency in the brain, resulting from either a lack of synthesis in the liver (GAMT, guanidinoacetate methyl transferase deficiency) or defective transport to the brain.[66,67,68]

Lactate

In the human brain, the lactate methyl resonance (1.31 ppm) is below (or at the very limit of) detectability in most studies, due to the low concentration of lactate within the brain under normal conditions. A small lactate signal may sometimes be observed in ventricular cerebrospinal fluid (CSF), where it is more visible due to either being present in higher concentration (than brain), or because it has a longer T_2 relaxation time.[69]

Lactate is often increased and detected by MRS in pathological conditions; lack of oxygen (due to either hypoxia or ischemia) will cause an increase in lactate when metabolism of glucose through the Krebs cycle can no longer be sustained.[70] Therefore, increased levels of brain lactate have been observed using MRS in a variety of conditions, including both acute and chronic ischemia,[36,71] and hypoxia (where it is a poor prognostic indicator).[72] Also, defects in the Krebs cycle (even in the presence of oxygen) can cause lactate to become elevated. Some examples of pathologies where this may occur include brain tumors,[73] mitochondrial diseases,[38,74] and other conditions.[75]

Small elevations of lactate have also been reported in the visual cortex during photic stimulation,[76] believed to be due to increased non-oxidative glycolysis, but this effect does not appear to be particularly reproducible.[77]

Lactate may also be difficult to distinguish from overlapping lipid resonances, either originating from the brain itself, or spatial contamination from the very strong lipid signals in the scalp. Several approaches can be used to distinguish lactate from lipid, including the use of spectral editing techniques,[78] although one of the simplest ways is to use an echo time of approximately 140 ms ($1/J$, where $J \approx 7$ Hz) where the lactate methyl resonance should be inverted.[79]

Myo-inositol

One of the larger signals in short echo time spectra occurs from *myo*-inositol (mI) at 3.5–3.6 ppm. mI is a pentose sugar, which is part of the inositol triphosphate intracellular second messenger system. Levels have been found to be reduced in hepatic encephalopathy,[65] and increased in Alzheimer's dementia[80] and demyelinating diseases.[81] The exact pathophysiological significance of alterations in mI is uncertain, but a leading hypothesis is that elevated mI reflects increased populations of glial cell, which are known to express higher levels of mI than neurons;[82,83] this may be related to differences in *myo*-inositol/sodium co-transporter activity that appears to play a key role in astrocyte osmoregulation.[84] This would explain chronic disturbance in mI both in degenerative and inflammatory disease, and transiently in hypo- and hyperosmolar states.

Myo-inositol resonates at almost the same frequency in the spectrum as glycine; however, glycine is a singlet, while mI is a strongly coupled multiplet, so the two can usually be distinguished by using different echo times (glycine should be the predominant signal at long echo times), or field strengths. Glycine is usually at low concentration in the normal brain, but can increase to detectable levels in some diseases, such as nonketotic hyperglycinemia.[85]

Glutamate and glutamine

Glutamate (Glu) and glutamine (Gln) are key compounds in brain metabolism. Glutamate is the most abundant amino acid in the brain, and is the dominant neurotransmitter.[86] During neuronal excitation, glutamate is released and diffuses across the synapse, where it is rapidly taken up by astrocytes (along with sodium ions (Na^+)). The astrocyte converts the glutamate to glutamine, which is then released and reuptaken by neurons. In the neuron, glutamine is converted back to glutamate, and the process repeated. This glutamate–glutamine cycling is an energy-demanding process, which has been speculated to consume as much as 80–90% of the total cortical glucose usage.[87]

Since at a field strength of 1.5 T there is almost complete overlap of Glu and Gln, they are usually labeled as a composite peak Glx, and are very difficult to separate, although some authors have attempted to distinguish them.[61] The ^{2}CH protons of both Glu and Gln resonate around 3.7 ppm, while the ^{3}CH$_2$ and ^{4}CH$_2$ multiplets occur between 2.1 and 2.4 ppm. At 3 T, Glu and Gln may be determined quite reliably with an appropriate pulse sequence and/or curve-fitting methods.[88,89] At higher fields (at 4 T or above), the ^{4}CH$_2$ resonances of Glu and Gln start to become well resolved, and hence more reliably determined.[39]

Because of the difficulty of measuring Glu and Gln at 1.5 T, relatively few studies have looked at pathology-related changes in these compounds. However, recently Glu was found to be elevated in MS plaques at 3 T,[90] and previous studies at 1.5 T found elevated cerebral Gln in patients with liver failure (for example, hepatic encephalopathy[65] and Reye's syndrome[91]), most likely as the result of increased blood ammonia levels, which increases glutamine synthesis.

Less commonly detected compounds

A survey of the literature reveals more than 25 additional compounds that have been assigned in proton spectra of the human brain (Table 1.2). Some of these compounds are present in the normal human brain, but are difficult to detect routinely because they are very small and/or have overlapping peaks. Some examples of these compounds include NAAG, aspartate, taurine, scyllo-inositol, betaine, ethanolamine, purine nucleotides, histidine, glucose and glycogen.[92] Other compounds are yet more difficult to detect, and require the use of spectral editing pulse sequences in order to be detected, because their resonances overlap almost completely with those of other, more abundant, compounds. Examples of these include γ-amino-butyric acid (GABA) and glutathione.[93,94]

Under disease conditions, certain compounds may become visible as their concentration increases sufficiently high to be detected. Examples of compounds that have been detected under pathological conditions include the ketone bodies β-hydroxy-butyrate and acetone,[95,96] and other compounds such as phenylalanine (in phenylketonurea[97]), galactitol, ribitol, arabitol in polyol disease,[98] and succinate, pyruvate, alanine, glycine, and threonine in various disorders.

Exogenous compounds which are able to cross the blood brain barrier may also reach sufficiently high concentrations to be detected by proton MRS. Examples of exogenous compounds, sometimes termed "xenobiotics", include the drug delivery vehicle propan-1,2-diol,[99] mannitol (used to reduce swelling and edema in neurosurgical procedures and intensive care),[100] ethanol,[101] and the health food supplement methyl-sulfonyl-methane (MSM).[102]

In addition to metabolite concentrations, other information may also be measured from brain proton spectra. For instance, measurements of absolute (as opposed to relative, as can be measured by MRI) brain temperature have been made using the water–NAA chemical shift difference (the water chemical shift has a 0.01 ppm/°C temperature dependence, whereas that of NAA is temperature-independent).[103]

In addition, the exchangeable protons of metabolites resonating downfield of water may be used to estimate brain pH. These compounds (histidine, homocarnosine, and the amide resonance of NAA) generally have low signal intensity, but are detectable by the use of short echo times, appropriate water suppression methods, and high magnetic field strengths. Using oral loading of histidine (to increase its detectability), Vermathen *et al.* were able to estimate brain pH from the chemical shift difference of the C2 and C4 resonances of the imidazole ring;[104] similarly, Rothman *et al.* were able to use the same resonances of homocarnosine to estimate brain pH in epilepsy patients who were receiving therapy which caused increased brain homocarnosine concentrations.[105] The rate of exchange of the NAA amide protons with water is also pH sensitive, and can be used to estimate brain pH.[106]

Compounds detected by proton MRS outside the CNS

The discussion so far has focused entirely on the information content of proton spectra of the human brain; however, when going to other organ systems, different compounds are detected in the spectra – for instance, in normal prostate tissue, a signal from citrate at 2.6 ppm is typically detected, while normal breast tissue usually only contains visible water and fat signals. In muscle, signals may be detected from intra- and extramyocellular lipids, acetylcarnotine, creatines, cholines, taurine and carnosine.[107] Spectra from these other organ systems are discussed in more detail in later chapters.

References

[1] Bloch F. Nuclear induction. *Phys Rev* 1946; **70**: 460–74.

[2] Purcell EM, Torrey HC, Pound RV. Resonance absorption by nuclear magnetic moments in a solid. *Phys Rev* 1946; **69**: 37–8.

[3] Proctor WG, Yu FC. The dependence of a nuclear magnetic resonance frequency. *Phys Rev* 1950; **77**: 717.

[4] Gutowsky HS, McCall DW. Nuclear magnetic resonance fine structure in liquids. *Phys Rev* 1951; **82**: 748–9.

[5] Hahn EL. Spin echoes. *Phys Rev* 1950; **80**: 580–94.

[6] Carr HY, Purcell EM. Effects of diffusion on free precession in nuclear magnetic resonance experiments. *Phys Rev* 1954; **94**: 630–8.

[7] Proctor WG, Yu FC. The dependence of a nuclear magnetic resonance frequency upon chemical compound. *Phys Rev* 1950; **77**: 717.

[8] Ernst RR, Anderson WA. Application of Fourier transform spectroscopy to magnetic resonance. *Rev Sci Instr* 1966; **37**: 93–102.

[9] Ernst RR, Bodenhausen G, Wokaun A. *Principles of Nuclear Magnetic Resonance in One and Two Dimensions*. New York, NY: Oxford University Press, 1990.

[10] Chance B, Nakase Y, Bond M, Leigh JS, Jr., McDonald G. Detection of 31P nuclear magnetic resonance signals in brain by in vivo and freeze-trapped assays. *Proc Natl Acad Sci USA* 1978; **75**: 4925–9.

[11] Moon RB, Richards JH. Determination of intracellular pH by 31P magnetic resonance. *J Biol Chem* 1973; **248**: 7276–8.

[12] Petroff OA, Prichard JW. Cerebral pH by NMR. *Lancet* 1983; **2**: 105–06.

[13] Petroff OA, Prichard JW, Behar KL, Alger JR, den Hollander JA, Shulman RG. Cerebral intracellular pH by 31P nuclear magnetic resonance spectroscopy. *Neurology* 1985; **35**: 781–8.

[14] Ackerman JJ, Grove TH, Wong GG, Gadian DG, Radda GK. Mapping of metabolites in whole animals by 31P NMR using surface coils. *Nature* 1980; **283**: 167–70.

[15] Ross BD, Radda GK, Gadian DG, Rocker G, Esiri M, Falconer-Smith J. Examination of a case of suspected McArdle's syndrome by 31P nuclear magnetic resonance. *N Engl J Med* 1981; **304**: 1338–42.

[16] Cady EB, Costello AM, Dawson MJ, Delpy DT, Hope PL, Reynolds EO, et al. Non-invasive investigation of cerebral metabolism in newborn infants by phosphorus nuclear magnetic resonance spectroscopy. *Lancet* 1983; **1**: 1059–62.

[17] Alger JR, Sillerud LO, Behar KL, Gillies RJ, Shulman RG, Gordon RE, et al. In vivo carbon-13 nuclear magnetic resonance studies of mammals. *Science* 1981; **214**: 660–2.

[18] Behar KL, den Hollander JA, Stromski ME, Ogino T, Shulman RG, Petroff OA, et al. High-resolution 1H nuclear magnetic resonance study of cerebral hypoxia in vivo. *Proc Natl Acad Sci USA* 1983; **80**: 4945–8.

[19] Behar KL, Rothman DL, Shulman RG, Petroff OA, Prichard JW. Detection of cerebral lactate in vivo during hypoxemia by 1H NMR at relatively low field strengths (1.9 T). *Proc Natl Acad Sci USA* 1984; **81**: 2517–9.

[20] Bottomley PA, Edelstein WA, Foster TH, Adams WA. In vivo solvent-suppressed localized hydrogen nuclear magnetic resonance spectroscopy: a window to metabolism? *Proc Natl Acad Sci USA* 1985; **82**: 2148–52.

[21] Luyten PR, den Hollander JA. Observation of metabolites in the human brain by MR spectroscopy. *Radiology* 1986; **161**: 795–8.

[22] Hanstock CC, Rothman DL, Prichard JW, Jue T, Shulman RG. Spatially localized 1H NMR spectra of metabolites in the human brain. *Proc Natl Acad Sci USA* 1988; **85**: 1821–5.

[23] Bax A, Freeman R. Enhanced NMR resolution by restricting the effective sample volume. *J Magn Reson* 1980; **37**: 177–81.

[24] Aue WP. Localization methods for in vivo NMR spectroscopy. *Rev Magn Reson Med* 1986; **1**: 21–72.

[25] Frahm J. Localized proton spectroscopy using stimulated echoes. *J Magn Reson* 1987; **72**: 502–08.

[26] Granot J. Selected Volume Excitation Using Stimulated Echoes (VEST). Applications to spatially localized spectroscopy and imaging. *J Magn Reson* 1986; **70**: 488–92.

[27] Kimmich R, Hoepfel D. Volume-selective multipulse spin-echo spectroscopy. *J Magn Reson* 1987; **72**: 379–84.

[28] Frahm J, Bruhn H, Gyngell ML, Merboldt KD, Hanicke W, Sauter R. Localized high-resolution proton NMR spectroscopy using stimulated echoes: initial applications to human brain in vivo. *Magn Reson Med* 1989; **9**: 79–93.

[29] Bottomley PA, inventor General Electric Company, assignee. Selective volume method for performing localized NMR spectroscopy. USA patent 4480228. October 30th 1984.

[30] Ordidge RJ, Gordon RE, inventors; Oxford Research Systems Limited, assignee. Methods and apparatus of obtaining NMR spectra. United States patent 4531094. 1983.

[31] Webb PG, Sailasuta N, Kohler SJ, Raidy T, Moats RA, Hurd RE. Automated single-voxel proton MRS: technical development and multisite verification. *Magn Reson Med* 1994; **31**: 365–73.

[32] Radda GK. The use of NMR spectroscopy for the understanding of disease. *Science* 1986; **233**: 640–5.

[33] Kurhanewicz J, Vigneron DB, Males RG, Swanson MG, Yu KK, Hricak H. The prostate: MR imaging and spectroscopy. Present and future. *Radiol Clin North Am* 2000; **38**: 115–38, viii–ix.

[34] Bolan PJ, Nelson MT, Yee D, Garwood M. Imaging in breast cancer: magnetic resonance spectroscopy. *Breast Cancer Res* 2005; **7**: 149–52.

[35] Behar KL, Ogino T. Characterization of macromolecule resonances in the 1H NMR spectrum of rat brain. *Magn Reson Med* 1993; **30**: 38–44.

[36] Barker PB, Gillard JH, van Zijl PC, Soher BJ, Hanley DF, Agildere AM, et al. Acute stroke: evaluation with serial proton MR spectroscopic imaging. *Radiology* 1994; **192**: 723–32.

[37] Remy C, Grand S, Lai ES, Belle V, Hoffmann D, Berger F, et al. 1H MRS of human brain abscesses in vivo and in vitro. *Magn Reson Med* 1995; **34**: 508–14.

[38] Lin DD, Crawford TO, Barker PB. Proton MR spectroscopy in the diagnostic evaluation of suspected mitochondrial disease. *Am J Neuroradiol* 2003; **24**: 33–41.

[39] Tkac I, Andersen P, Adriany G, Merkle H, Ugurbil K, Gruetter R. In vivo 1H NMR spectroscopy of the human brain at 7 T. *Magn Reson Med* 2001; **46**: 451–6.

[40] Govindaraju V, Young K, Maudsley AA. Proton NMR chemical shifts and coupling constants for brain metabolites. *NMR Biomed* 2000; **13**: 129–53.

[41] Frahm J, Michaelis T, Merboldt KD, Hanicke W, Gyngell ML, Bruhn H. On the N-acetyl methyl resonance in localized 1H NMR spectra of human brain in vivo. *NMR Biomed* 1991; **4**: 201–04.

[42] Pouwels PJ, Frahm J. Differential distribution of NAA and NAAG in human brain as determined by quantitative localized proton MRS. *NMR Biomed* 1997; **10**: 73–8.

[43] Moffett JR, Ross B, Arun P, Madhavarao CN, Namboodiri AM. N-Acetylaspartate in the CNS: from neurodiagnostics to neurobiology. *Prog Neurobiol* 2007; **81**: 89–131.

[44] Barker PB. N-acetyl aspartate – a neuronal marker? *Ann Neurol* 2001; **49**: 423–4.

[45] Simmons ML, Frondoza CG, Coyle JT. Immunocytochemical localization of N-acetyl-aspartate with monoclonal antibodies. *Neuroscience* 1991; **45**: 37–45.

[46] De Stefano N, Narayanan S, Francis GS, Arnaoutelis R, Tartaglia MC, Antel JP, et al. Evidence of axonal damage in the early stages of multiple sclerosis and its relevance to disability. *Arch Neurol* 2001; **58**: 65–70.

[47] Guimaraes A, Schwartz P, Prakash MR, Carr CA, Berger UV, Jenkins BG, et al. Quantitative in vivo 1H nuclear magnetic resonance spectroscopic imaging of neuronal loss in rat brain. *Neuroscience* 1995; **69**: 1095.

[48] Bhakoo KK, Pearce D. In vitro expression of N-acetyl aspartate by oligodendrocytes: implications for proton magnetic resonance spectroscopy signal in vivo. *J Neurochem* 2000; **74**: 254–62.

[49] Burlina AP, Ferrari V, Facci L, Skaper SD, Burlina AB. Mast cells contain large quantities of secretagogue-sensitive N-acetylaspartate. *J Neurochem* 1997; **69**: 1314–7.

[50] Urenjak J, Williams SR, Gadian DG, Noble M. Specific expression of N-acetylaspartate in neurons, oligodendrocyte-type-2 astrocyte progenitors, and immature oligodendrocytes in vitro. *J Neurochem* 1992; **59**: 55–61.

[51] Barker PB, Bryan RN, Kumar AJ, Naidu S. Proton NMR spectroscopy of Canavan's Disease. *Neuropediatrics* 1992; **23**: 263–7.

[52] Martin E, Capone A, Schneider J, Hennig J, Thiel T. Absence of N-acetylaspartate in the human brain: impact on neurospectroscopy? *Ann Neurol* 2001; **49**: 518–21.

[53] De Stefano N, Matthews PM, Arnold DL. Reversible decreases in N-acetylaspartate after acute brain injury. *Magn Reson Med* 1995; **34**: 721–7.

[54] Bizzi A, Ulug AM, Crawford TO, Passe T, Bugiani M, Bryan RN, et al. Quantitative proton MR spectroscopic imaging in acute disseminated encephalomyelitis. *Am J Neuroradiol* 2001; **22**: 1125–30.

[55] Barker P, Breiter S, Soher B, Chatham J, Forder J, Samphilipo M, et al. Quantitative proton spectroscopy of canine brain: in vivo and in vitro correlations. *Magn Reson Med* 1994; **32**: 157–63.

[56] Gill SS, Small RK, Thomas DG, Patel P, Porteous R, van Bruggen N, et al. Brain metabolites as 1H NMR markers of neuronal and glial disorders. *NMR Biomed* 1989; **2**: 196–200.

[57] Gill SS, Thomas DG, Van BN, Gadian DG, Peden CJ, Bell JD, et al. Proton MR spectroscopy of intracranial

tumours: in vivo and in vitro studies. *J Comput Assist Tomogr* 1990; **14**: 497–504.

[58] Davie CA, Hawkins CP, Barker GJ, Brennan A, Tofts PS, Miller DH, *et al.* Detection of myelin breakdown products by proton magnetic resonance spectroscopy. *Lancet* 1993; **341**: 630–1.

[59] Brenner RE, Munro PM, Williams SC, Bell JD, Barker GJ, Hawkins CP, *et al.* The proton NMR spectrum in acute EAE: the significance of the change in the Cho:Cr ratio. *Magn Reson Med* 1993; **29**: 737–45.

[60] Aboagye EO, Bhujwalla ZM. Malignant transformation alters membrane choline phospholipid metabolism of human mammary epithelial cells. *Cancer Res* 1999; **59**: 80–4.

[61] Kreis R, Ross BD, Farrow NA, Ackerman Z. Metabolic disorders of the brain in chronic hepatic encephalopathy detected with H-1 MR spectroscopy. *Radiology* 1992; **182**: 19–27.

[62] Stoll AL, Renshaw PF, De Micheli E, Wurtman R, Pillay SS, Cohen BM. Choline ingestion increases the resonance of choline-containing compounds in human brain: an in vivo proton magnetic resonance study. *Biol Psychiatry* 1995; **37**: 170–4.

[63] Urenjak J, Williams SR, Gadian DG, Noble M. Proton nuclear magnetic resonance spectroscopy unambiguously identifies different neural cell types. *J Neuroscience* 1993; **13**: 981.

[64] Jacobs MA, Horska A, van Zijl PC, Barker PB. Quantitative proton MR spectroscopic imaging of normal human cerebellum and brain stem. *Magn Reson Med* 2001; **46**: 699–705.

[65] Ross BD, Michaelis T. Clinical applications of magnetic resonance spectroscopy. *Magn Reson Q* 1994; **10**: 191–247.

[66] Bizzi A, Bugiani M, Salomons GS, Hunneman DH, Moroni I, Estienne M, *et al.* X-linked creatine deficiency syndrome: a novel mutation in creatine transporter gene SLC6A8. *Ann Neurol* 2002; **52**: 227–31.

[67] Cecil KM, Salomons GS, Ball WS, Jr., Wong B, Chuck G, Verhoeven NM, *et al.* Irreversible brain creatine deficiency with elevated serum and urine creatine: a creatine transporter defect? *Ann Neurol* 2001; **49**: 401–04.

[68] Stockler S, Holzbach U, Hanefeld F, Marquardt I, Helms G, Requart M, *et al.* Creatine deficiency in the brain: a new, treatable inborn error of metabolism. *Pediatr Res* 1994; **36**: 409–13.

[69] Nagae-Poetscher LM, McMahon M, Braverman N, Lawrie WT, Jr., Fatemi A, Degaonkar M, *et al.* Metabolites in ventricular cerebrospinal fluid detected by proton magnetic resonance spectroscopic imaging. *J Magn Reson Imaging* 2004; **20**: 496–500.

[70] Veech RL. The metabolism of lactate. *NMR Biomed* 1991; **4**: 53–8.

[71] Petroff OA, Graham GD, Blamire AM, al-Rayess M, Rothman DL, Fayad PB, *et al.* Spectroscopic imaging of stroke in humans: histopathology correlates of spectral changes. *Neurology* 1992; **42**: 1349–54.

[72] Penrice J, Cady EB, Lorek A, Wylezinska M, Amess PN, Aldridge RF, *et al.* Proton magnetic resonance spectroscopy of the brain in normal preterm and term infants, and early changes after perinatal hypoxia-ischemia. *Pediatr Res* 1996; **40**: 6–14.

[73] Alger JR, Frank JA, Bizzi A, Fulham MJ, DeSouza BX, Duhaney MO, *et al.* Metabolism of human gliomas: assessment with H-1 MR spectroscopy and F-18 fluorodeoxyglucose PET. *Radiology* 1990; **177**: 633–41.

[74] Mathews PM, Andermann F, Silver K, Karpati G, Arnold DL. Proton MR spectroscopic characterization of differences in regional brain metabolic abnormalities in mitochondrial encephalomyopathies. *Neurology* 1993; **43**: 2484–90.

[75] Sutton LN, Wang Z, Duhaime AC, Costarino D, Sauter R, Zimmerman R. Tissue lactate in pediatric head trauma: a clinical study using 1H NMR spectroscopy. *Pediatr Neurosurg* 1995; **22**: 81–7.

[76] Prichard J, Rothman D, Novotny E, Petroff O, Kuwabara T, Avison M, *et al.* Lactate rise detected by 1H NMR in human visual cortex during physiologic stimulation. *Proc Natl Acad Sci USA* 1991; **88**: 5829–31.

[77] Merboldt K-D, Bruhn H, Hanicke W, Michaelis T, Frahm J. Decrease of glucose in the human visual cortex during photic stimulation. *Magn Reson Med* 1992; **25**: 187–94.

[78] Hurd RE, Freeman D. Proton editing and imaging of lactate. *NMR Biomed* 1991; **4**: 73–80.

[79] Kelley DA, Wald LL, Star-Lack JM. Lactate detection at 3 T: compensating J coupling effects with BASING. *J Magn Reson Imaging* 1999; **9**: 732–7.

[80] Shonk TK, Moats RA, Gifford P, Michaelis T, Mandigo JC, Izumi J, *et al.* Probable Alzheimer disease: diagnosis with proton MR spectroscopy. *Radiology* 1995; **195**: 65–72.

[81] Kruse B, Hanefeld F, Christen HJ, Bruhn H, Michaelis T, Hanicke W, *et al.* Alterations of brain metabolites in metachromatic leukodystrophy as detected by localized proton magnetic resonance spectroscopy in vivo. *J Neurol* 1993; **241**: 68–74.

[82] Brand A, Richter-Landsberg C, Leibfritz D. Multinuclear NMR studies on the energy metabolism of glial and neuronal cells. *Dev Neurosci* 1993; **15**: 289–98.

[83] Flogel U, Willker W, Leibfritz D. Regulation of intracellular pH in neuronal and glial tumour cells, studied by multinuclear NMR spectroscopy. *NMR Biomed* 1994; **7**: 157–66.

[84] Strange K, Emma F, Paredes A, Morrison R. Osmoregulatory changes in *myo*-inositol content and Na+/myo-inositol cotransport in rat cortical astrocytes. *Glia* 1994; **12**: 35–43.

[85] Heindel W, Kugel H, Roth B. Noninvasive detection of increased glycine content by proton MR spectroscopy in the brains of two infants with nonketotic hyperglycinemia. *Am J Neuroradiol* 1993; **14**: 629–35.

[86] Magistretti PJ, Pellerin L, Rothman DL, Shulman RG. Energy on demand. *Science* 1999; **283**: 496–7.

[87] Sibson NR, Dhankhar A, Mason GF, Rothman DL, Behar KL, Shulman RG. Stoichiometric coupling of brain glucose metabolism and glutamatergic neuronal activity. *Proc Natl Acad Sci USA* 1998; **95**: 316–21.

[88] Hurd R, Sailasuta N, Srinivasan R, Vigneron DB, Pelletier D, Nelson SJ. Measurement of brain glutamate using TE-averaged PRESS at 3 T. *Magn Reson Med* 2004; **51**: 435–40.

[89] Provencher SW. Estimation of metabolite concentrations from localized in vivo proton NMR spectra. *Magn Reson Med* 1993; **30**: 672–9.

[90] Srinivasan R, Sailasuta N, Hurd R, Nelson S, Pelletier D. Evidence of elevated glutamate in multiple sclerosis using magnetic resonance spectroscopy at 3 T. *Brain* 2005; **128**: 1016–25.

[91] Kreis R, Pfenninger J, Herschkowitz N, Boesch C. In vivo proton magnetic resonance spectroscopy in a case of Reye's syndrome. *Intensive Care Med* 1995; **21**: 266–9.

[92] van Zijl PCM, Barker PB. Magnetic resonance spectroscopy and spectroscopic imaging for the study of brain metabolism. In *Imaging Brain Structure and Function*. New York, NY: Proceedings of the New York Academy of Sciences; 1997: 75–96.

[93] Rothman DL, Petroff OA, Behar KL, Mattson RH. Localized 1H NMR measurements of gamma-aminobutyric acid in human brain in vivo. *Proc Natl Acad Sci USA* 1993; **90**: 5662–6.

[94] Terpstra M, Henry PG, Gruetter R. Measurement of reduced glutathione (GSH) in human brain using LCModel analysis of difference-edited spectra. *Magn Reson Med* 2003; **50**: 19–23.

[95] Pan JW, Telang FW, Lee JH, de Graaf RA, Rothman DL, Stein DT, et al. Measurement of beta-hydroxybutyrate in acute hyperketonemia in human brain. *J Neurochem* 2001; **79**: 539–44.

[96] Seymour KJ, Bluml S, Sutherling J, Sutherling W, Ross BD. Identification of cerebral acetone by 1H-MRS in patients with epilepsy controlled by ketogenic diet. *Magma* 1999; **8**: 33–42.

[97] Kreis R, Pietz J, Penzien J, Herschkowitz N, Boesch C. Identification and quantitation of phenylalanine in the brain of patients with phenylketonuria by means of localized in vivo 1H magnetic-resonance spectroscopy. *J Magn Reson B* 1995; **107**: 242–51.

[98] van der Knaap MS, Wevers RA, Struys EA, Verhoeven NM, Pouwels PJ, Engelke UF, et al. Leukoencephalopathy associated with a disturbance in the metabolism of polyols. *Ann Neurol* 1999; **46**: 925–8.

[99] Cady EB, Lorek A, Penrice J, Reynolds EO, Iles RA, Burns SP, et al. Detection of propan-1,2-diol in neonatal brain by in vivo proton magnetic resonance spectroscopy. *Magn Reson Med* 1994; **32**: 764–7.

[100] Maioriello AV, Chaljub G, Nauta HJ, Lacroix M. Chemical shift imaging of mannitol in acute cerebral ischemia. Case report. *J Neurosurg* 2002; **97**: 687–91.

[101] Hanstock CC, Rothman DL, Shulman RG, Novotny EJ, Jr., Petroff OA, Prichard JW. Measurement of ethanol in the human brain using NMR spectroscopy. *J Stud Alcohol* 1990; **51**: 104–07.

[102] Rose SE, Chalk JB, Galloway GJ, Doddrell DM. Detection of dimethyl sulfone in the human brain by in vivo proton magnetic resonance spectroscopy. *Magn Reson Imaging* 2000; **18**: 95–8.

[103] Cady EB, D'Souza PC, Penrice J, Lorek A. The estimation of local brain temperature by in vivo 1H magnetic resonance spectroscopy. *Magn Reson Med* 1995; **33**: 862–7.

[104] Vermathen P, Capizzano AA, Maudsley AA. Administration and (1)H MRS detection of histidine in human brain: application to in vivo pH measurement. *Magn Reson Med* 2000; **43**: 665–75.

[105] Rothman DL, Behar KL, Prichard JW, Petroff OA. Homocarnosine and the measurement of neuronal pH in patients with epilepsy. *Magn Reson Med* 1997; **38**: 924–9.

[106] Mori S, Eleff SM, Pilatus U, Mori N, van Zijl PC. Proton NMR spectroscopy of solvent-saturable resonances: a new approach to study pH effects in situ. *Magn Reson Med* 1998; **40**: 36–42.

[107] Boesch C. Musculoskeletal spectroscopy. *J Magn Reson Imaging* 2007; **25**: 321–38.

Chapter 2

Pulse sequences and protocol design

Key points

- There are two classes of spatial localization techniques: single voxel (SV) and multi-voxel (MRSI).
- SV-MRS is usually performed using "STEAM" or "PRESS" sequences.
- Multi-voxel acquisitions are usually 2D (or sometimes 3D) MRSI with partial brain coverage.
- Correct adjustment of field homogeneity ("shimming"), water and lipid suppression are vital.
- Spectral editing ("MEGA-PRESS") can be used to detect low-concentration, coupled spin systems such as GABA, glutathione
- MRS is better at high field.
- Correct choice of protocol (SV or MRSI, short or long TE) depends on many factors.

Introduction

The correct choice of pulse sequence, and its associated acquisition parameters, are often the key factors in determining whether an MRS study is successful or not. As described in the previous chapter, early attempts at spatially localized MRS were relatively crude and not particularly accurate or efficient. However, rapid development in the 1980s and 1990s led to more sophisticated and reliable approaches. Spatial localization techniques fall into two general categories: either single-voxel techniques (i.e. where a spectrum is recorded from a single brain region) or multi-voxel techniques (where multiple regions are acquired simultaneously – also called MR spectroscopic imaging (MRSI) or chemical shift imaging (CSI)). Currently, the two most commonly used single voxel sequences are STEAM (Stimulated Echo Acquisition Mode),[1,2,3] and PRESS (Point REsolved Spectroscopy Sequence).[4,5] For MRSI, a variety of different approaches are used, but most commonly two- or three-dimensional MRSI is performed, most often (but not always) used in combination with PRESS excitation.[6,7,8]

Single-voxel techniques

The basic principle underlying nearly all single-voxel localization techniques is to apply three mutually orthogonal slice-selective pulses and design the pulse sequence to only collect the echo signal from the volume in space (voxel) where all three slices intersect (Figure 2.1). The sequence is designed so that signals from other regions outside the desired voxel are eliminated (usually by using "crusher gradients").[1,9] Since the volume of the human brain is of the order of $1500\,cm^3$ (and the whole head is larger), and a typical voxel size used for brain spectroscopy might be $8\,cm^3$, it is apparent that the spatial localization has to be very efficient – e.g. if the sequence excited 1% of the magnetization outside of the voxel, and 100% within the voxel, there would be almost twice as much signal coming from outside the voxel as inside. Therefore, it is very important that out-of-voxel magnetization is suppressed as efficiently as possible, by the use of slice selection pulses with good excitation profiles, and sufficiently large crusher gradients. In addition, "out-of-voxel" artifacts can be further reduced by the use of outer-volume suppression (OVS) pulses;[10,11] it is recommended that these are routinely used with STEAM or PRESS localization.

In the STEAM sequence (Figure 2.1B), three 90° pulses are used, and the "stimulated echo" is collected. All other signals (e.g. spin-echoes from the second or third pulse) should be dephased by a large crusher gradient applied during the so-called mixing time (TM) between the second and third pulses. Crusher gradients applied during the echo time (TE) on selected gradient channels are also necessary for consistent formation of the stimulated echo. In PRESS, the second and third pulses are refocusing (180°) pulses, and crusher gradients are applied around these pulses

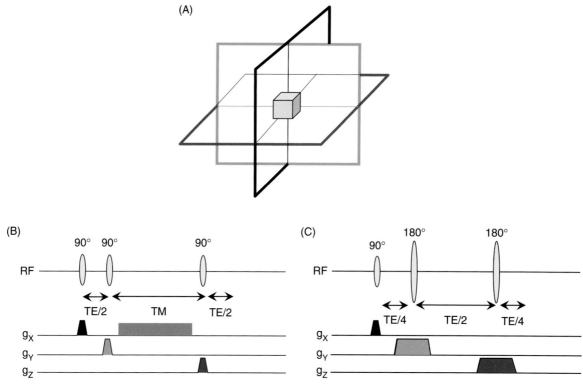

Figure 2.1. Common pulse sequences for single-voxel localization techniques. (A) Spatial localization is achieved by collecting signals from the intersection of three slice-selective RF pulses applied in orthogonal directions. (B) The STEAM sequence, consisting of three 90° slice selective pulses. (C) The PRESS sequence, consisting of a slice selective 90° excitation pulse and two 180° refocusing pulses. For simplicity, only slice selection gradients are shown (except for the mixing time (TM) crusher gradient in the STEAM sequence depicted in 1.1B).

in order to select the desired spin echo signal arising from all three RF pulses (Figure 2.1C). Care should be taken to vary the crusher direction or strength for different crusher pairs, so as to avoid accidental refocusing of unwanted coherences.[7,9]

STEAM and PRESS have been the subject of a detailed comparison;[12] they are generally similar, but differ in a few key respects.

(a) Slice profile (i.e. sharpness of edges of voxel). STEAM is somewhat better because it is usually easier to produce a 90° pulse with a sharp slice profile than a 180° pulse. High bandwidth 90° excitation pulses are also more available than high bandwidth 180° pulses, and require less power. Therefore STEAM may be particularly useful for high-field brain MRS.[13]

(b) Signal-to-noise ratio (SNR). Provided that equal volumes of tissue are observed and using the same parameters (Repetition Time (TR), TE, number of averages, etc.), PRESS should have approximately a factor of 2 better SNR than STEAM, because the stimulated echo in STEAM is formed from only half the available equilibrium magnetization, whereas PRESS refocuses the full magnetization. In practice, because of the better slice profiles usually obtained with the STEAM sequence, the experimental SNR improvement with PRESS may be somewhat less than a factor of 2.

(c) Minimum TE. STEAM should have a shorter minimum TE than PRESS, since it uses a TM time period during which T_1 (rather than T_2) relaxation occurs, and shorter 90° than 180° pulses may be possible.

(d) Other minor differences between STEAM and PRESS may be water suppression (sometimes better with STEAM), and spectral appearance (e.g. coupled spin systems such as glutamate, lactate, etc., may evolve slightly differently as a function of TE).

The differences listed above are fairly subtle, so that the results obtained by STEAM or PRESS are usually similar in clinical brain spectroscopy, but most commonly PRESS is used because of its higher SNR.

While single-voxel techniques are popular in clinical practice for several reasons (they are widely available, have short scan times, usually good field homogeneity, can be readily performed at short echo times, and are relatively easy to process and interpret), they do also suffer some limitations. Probably their greatest single limitation is the lack of ability to determine spatial heterogeneity of spectral patterns (often very important in brain tumors, for instance), and the fact that only a small number of brain regions can be covered within the time constraints of a normal clinical MR exam. In the next section, the alternative, multi-voxel techniques are considered.

Multiple-voxel (spectroscopic imaging) techniques

The concept of MRSI was first demonstrated in phantoms in 1982,[14] combining phase-encoding in a single direction with an FID readout in the absence of any field gradients. Early attempts at MRSI in the human brain also used 1D-MRSI (i.e. phase-encoding in a single direction), using a line excitation technique consisting of slice selective pulses applied in two directions to form an echo.[15] While these studies demonstrated proof-of-principle, generally one-dimensional localization is insufficient for detailed studies of focal brain pathology. Therefore, MRSI techniques were extended to two dimensions by using phase-encoding gradients in two directions,[7,8,10,16] or, subsequently, with full three-dimensional encoding.[6]

A problem with MRSI in the brain is that, usually, it is necessary to restrict excitation to only a fairly small region, for instance using the PRESS sequence (advanced methods for whole-brain MRSI will be discussed below). Restricted excitation is performed for several reasons: (a) it allows B_0 field homogeneity to be optimized on the region of interest (it can be difficult to obtain sufficient field homogeneity for larger volumes); (b) it reduces the number of phase-encoding steps required (i.e. the larger the field of view, the greater the number of phase-encode steps required to achieve a given spatial resolution, and hence the longer the scan time); and (c) it eliminates signals from outside the brain, in particular the very strong lipid signals from the scalp and retro-orbital fat. Figure 2.2 shows an example of a 2D-PRESS-MRSI sequence (adapted from [7]), and Figure 2.3 shows example data using this pulse sequence in a patient with a lesion in the left hippocampus.

However, the PRESS-MRSI approach also has some problems. Because conventional slice-selective 180° refocusing pulses do not have particularly good slice profiles, metabolite excitation across the PRESS box is not particularly uniform, and some signal from outside the PRESS box will generally be excited also. The pulses may also have relatively low bandwidths, leading to different excitation profiles for different metabolites, such that the ratio of NAA/Cho, for instance, may be different in the left and right, or anterior and posterior, edges of the PRESS box.[17] These problems may be partially overcome by using slightly larger PRESS volume excitation, and then applying high-bandwidth spatial saturation pulses to sharpen the edges of the excitation profile.[18]

Another problem with PRESS-MRSI is that, by definition, it restricts excitation to a rectangular volume. Because the brain has an elliptical shape, it is difficult to apply the PRESS excitation so that a signal is obtained from cortical regions close to the skull without also exciting a considerable amount of lipid signal. An alternative approach is to use a spin-echo sequence which excites a whole transverse slice, and then use multiple, carefully placed OVS pulses to suppress the lipid signals from the scalp.[10] This allows spectra from all regions within the chosen slice to be evaluated,

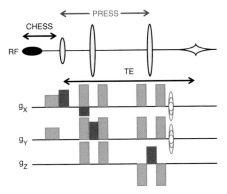 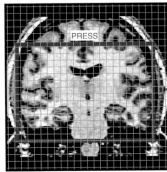

Figure 2.2. 2D-PRESS-MRSI pulse sequence: a PRESS sequence is used to excite a large volume of brain tissue while excluding signal from lipid in the scalp and/or regions of poor field homogeneity, and then phase-encoding gradients (g_X and g_Y) are used to localize spectra from regions within the excited region. A CHESS pulse and crusher gradient is applied prior to the PRESS sequence for water suppression. Crusher gradients applied around the 180° refocusing pulses are also shown.

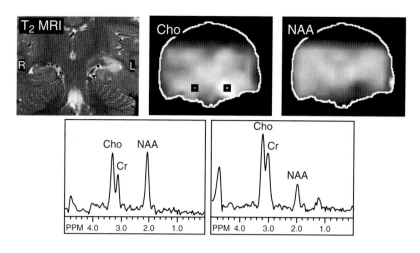

Figure 2.3. A 2D-PRESS-MRSI scan in the coronal plane of a patient with a lesion in the left hippocampus (visualized on T_2 MRI) recorded at 1.5 T. The lesion can be seen to have a low NAA signal and elevated Cho (and Cr) on both metabolic images and selected MRSI spectra from regions of interest in the left and right hippocampi, most likely consistent with a neoplastic process. The alternative diagnosis, mesial temporal sclerosis, rarely shows increased Cho levels.

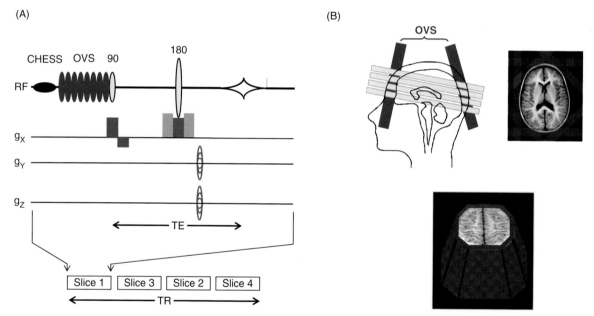

Figure 2.4. (A) Multi-slice 2D MRSI pulse sequence; a slice-selective spin echo is preceded by 8 outer-volume suppression pulses (OVS, arranged to saturate the lipid signals in the scalp) and a CHESS water suppression pulse. Two dimensional phase-encoding gradients are applied on g_Y and g_Z. In this example, four slices are collected within one repetition time (TR). Gradients associated with the OVS and CHESS pulses are omitted for clarity. (B) Schematic representation of the location of the OVS pulses, forming an octagonal cone in order to conform to the contours of the skull. The four oblique axial MRSI slices are also represented on the sagittal schematic.

and also can be straightforwardly extended to a multi-slice approach (Figure 2.4).[10] An example of one slice from a multi-slice 2D MRSI data set from a normal human volunteer is shown in Figure 2.5.

The multi-slice technique allows for greater brain coverage, although with conventional phase-encoding techniques it has proven difficult to record enough slices within a reasonable scan time to provide whole-brain coverage. Brain coverage can also be quite large with the alternative 3D-PRESS-MRSI sequence, although again scan time becomes very long if high spatial resolution in all three directions is required. Another problem when large spatial coverage is required is the difficulty of obtaining sufficient magnetic field homogeneity over the volume of the brain. For this reason, MRSI with large volume coverage is usually performed at long echo time (e.g. 140 or 280 ms). Generally, field homogeneity requirements are less stringent for long

Chapter 2: Pulse sequences and protocol design

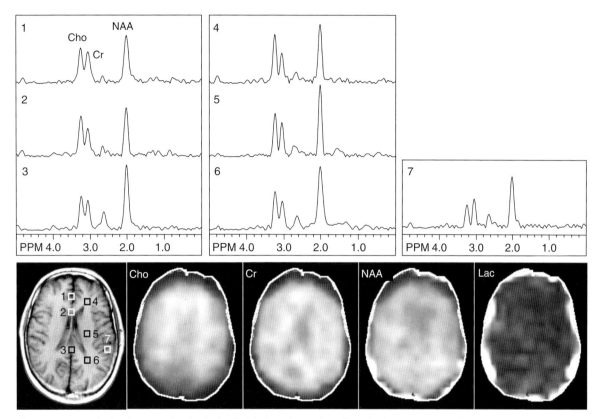

Figure 2.5. Example data from one slice (at the level of the lateral ventricles) of a multi-slice 2D-MRSI data from a normal human subject recorded at 1.5 T. In addition to the anatomical MRI scan, spectroscopic images of choline, creatine, N-acetylaspartate, and lactate are shown, and selected spectra (showing regional variations) from various white and gray matter regions within the brain.

TE spectra than for short TE, because the spectra are simpler, with flatter baselines and less overlapping resonances. Note that echo times of 140 or 280 msec are used to be optimal for lactate detection; at TE 140 msec lactate should be inverted, and at 280 msec it will be a positive, in-phase signal (see below).

Spatial resolution and scan time in MRSI

The nominal spatial resolution in MRSI is calculated as the field-of-view divided by the number of phase-encoding steps, so an MRSI scan with a field-of-view of 24 cm and 32 phase-encoding steps would have a nominal, linear voxel size of 0.75 cm. However, this is only an approximation, since the actual pixel dimensions (and shape) will be given by a "point-spread function" (PSF) which depends in detail on the k-space sampling, and any spatial filtering that may be used in processing the data. In particular, limited k-space sampling (as is commonly used in MRSI) results in truncation artifacts, the familiar sinc(x) Gibbs Ringing pattern (Figure 2.6), in which the intensity of the sidebands of the sinc(x) function decrease slowly (as $1/x$), and hence allows much signal from outside the nominal voxel to contribute to its spectrum.

Scan time in MRSI experiments is usually given by the product of the TR times the number of phase encoding steps; e.g. for a 3D-MRSI experiment, scan time = $TR \times (N_X \times N_Y \times N_Z)$, so for typical parameters (e.g. TR 1 s, $N_X = N_Y = 16$, $N_Z = 8$, $TR = 1$ s), the scan time is 34 min. Scan time can be shortened using reduced (such as circular) phase-encoding schemes.[19] Also, the number of phase-encoding steps is related to the field of view (FOV), and the desired spatial resolution Δ (e.g. $N_X = FOV_X/\Delta_X$). Therefore, in order to minimize the scan time (i.e. minimize N) for a given spatial resolution, it is important to use as small as possible FOV consistent with the dimensions of the object to be imaged. So, for the brain, the left–right FOV should be smaller than the anterior–posterior, since the brain is smaller in this dimension.[20] Typically, scan savings of 25–30% can be achieved by using a

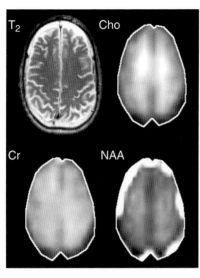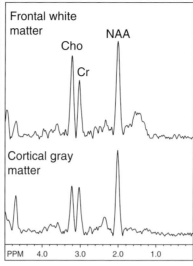

Figure 2.6. SENSE-MRSI, example data in a patient with HIV infection and dementia recorded at 3.0 T. The conventional T_2-weighted MRI scan is unremarkable except for some volume loss, while MRSI shows an elevated choline signal bilaterally in the frontal lobe white matter, believed to be due to gliosis. The rapid data acquisition using SENSE-MRSI allowed this relatively uncooperative patient to be scanned without using sedation or anesthesia.

reduced FOV in the left–right direction, compared to a square FOV.

Fast MRSI techniques

Even with circular k-space sampling, and optimal FOVs, 2D or 3D-MRSI experiments are still relatively time-consuming because there is still a large number of phase-encoding gradient steps to be collected. This is particularly true in MRSI experiments that require both high spatial resolution and extended brain coverage. Therefore, a number of different approaches for fast MRSI have been developed, and have been reviewed previously in reference [21]; however, so far they have had limited clinical impact, mainly because many of them are not (at the time of writing) commercially available; a brief summary of these techniques is presented below.

Fast MRSI techniques usually involve the use of techniques that have also been used to "speed-up" MRI scans; for instance, methods are based on the use of multiple spin echoes ("turbo"-MRSI, analogous to fast-spin-echo MRI), echo-planar techniques (echo-planar spectroscopic imaging, or EPSI)[22,23,24], spiral-MRSI,[25] or parallel imaging techniques (e.g. sensitivity-encoded (SENSE)-MRSI). Each generally has their own set of advantages and disadvantages; for instance, turbo-MRSI generally involves long echo times because each echo readout is quite long (depending on the required spectral resolution), while EPSI is very fast but may incur artifacts and/or somewhat reduced SNR due to the oscillating read gradient that it employs. Also, at high field, it may be difficult to switch the EPSI read gradient fast enough to get the desired spectral width. However, because of its speed, EPSI (also called proton EPSI, or PEPSI) has been used for whole-brain 3D MRSI,[26] and has also been used to study various brain pathologies.[27,28]

Spiral MRSI is a fast-MRSI technique that has some similarities to EPSI, in that an oscillating read gradient is applied during data acquisition. In EPSI, k-space is traversed in a rectilinear manner by the oscillating read gradient that is applied at the same time as the readout time is evolving. In spiral MRSI, a gradient waveform in two-dimensions is applied which traverses a spiral trajectory in two directions of k-space, again concurrent with the evolution of the readout time. A regridding algorithm[29] is used on the raw data to interpolate the collected data onto a Cartesian k-space grid, at which point it can be processed conventionally by Fourier transformation. While spiral MRSI has several theoretical advantages over EPSI, there have been relatively few clinical applications, presumably because of lack of widespread availability and the need for sophisticated reconstruction software; however, spiral MRSI has been used to detect metabolic abnormalities in gray matter in patients with secondary progressive multiple sclerosis.[30]

Another approach for scan time reduction in 2D or 3D MRSI is to apply "parallel imaging" methods,[31] similar to those that have been developed for MR imaging.[32,33] The basic principle is to use the inhomogeneous B_1 fields of multiple, phased-array coils to encode some of the spatial information, thereby allowing fewer conventional phase-encoding steps to be used, hence reducing scan time. In parallel MRI, reconstruction algorithms have been developed which either interpolate the missing k-space data (e.g. SMASH, GRAPPA[34,35,36]), or that unfold the images produced by Fourier transformation of the incomplete k-space data (SENSE[32]).

An attractive feature of SENSE is that it can be combined with any existing MRSI pulse sequence, so that there are no *pulse sequence*-related SNR losses, unlike those which can potentially occur with other fast MRSI methods.[21] Scan time reductions can be increased by using larger R factors, or by extending SENSE into the second or third dimensions for 2D or 3D MRSI.[31] However, insufficient lipid suppression is a problem for SENSE-MRSI, since residual scalp lipid signals often contaminate brain spectra, due to incomplete "unfolding" by the SENSE algorithm.

An example of a multi-slice MRSI scan with a SENSE factor of 2, applied in a patient with human immunodeficiency virus (HIV), is shown in Figure 2.6.

Water suppression

Brain metabolite concentrations are on the order of 10 mM or less, whereas protons in brain water are approximately 80 M. Therefore, water suppression techniques are critical for proton spectroscopy in order to reliably observe the much smaller metabolite signals.

The most common approach is to pre-saturate the water signal using frequency selective, 90° pulses (CHESS pulses)[37] prior to the localization pulse sequence. By using more than one pulse, and with correct choices of pulse flip angles,[38,39] very good suppression factors can be attained (>1000). Since some T_1 relaxation will occur between the suppression pulse and the read pulse of the localization sequence, a flip angle somewhat greater than 90° is required in order to catch the water longitudinal magnetization at its null point. By using more than one pulse, the sequence can be designed to give good suppression factors over a range of T_1 relaxation times and B_1 field strengths (B_1 inhomogeneity).[38,39] Some examples include the WET scheme, which usually employs three or four CHESS pulses of optimized flip angle,[39] or the VAPOR scheme that uses seven CHESS pulses of variable flip angle and timing.[40]

Water suppression can also be achieved during the localization sequence (as opposed to the use of saturation pulses applied prior to the start of the localization sequence). For instance, in the STEAM pulse sequence, since the magnetization for the stimulated echo is stored along the Z-axis during the mixing time (TM), it is possible to place additional saturation pulses during this time period in order to improve suppression. It is also possible to suppress the water signal through the use of selective 180° pulses and crusher gradients (e.g. the MEGA or BASING sequences[41,42]). For instance, a bipolar gradient pair (i.e. gradients with opposite polarities), separated by a selective 180° pulse applied only to the water signal, will refocus the metabolite signals while simultaneously dephasing the water (and/or any other signals affected by the 180°). Alternatively, a balanced gradient crusher pair (of the same polarity) can be used to crush the water signal (and perhaps also lipids) while rephasing the metabolites, if the 180° pulse is applied selectively to the metabolite resonances. An elegant way to achieve this is to use spectral-spatial 180° pulses in the PRESS sequence, which refocus magnetization only within a slice over a limited, controllable bandwidth.[8] One disadvantage of using selective 180° pulses within the localization sequence is that it will lengthen the minimum echo time achievable, since low bandwidth selective pulses are of longer duration than higher bandwidth, slice-selective pulses.

Lipid suppression

Lipids in peri-cranial fat are present in very high (molar) concentrations. Lipid signal suppression can be performed in several different ways. One approach is to avoid exciting the lipid signal, using e.g. STEAM or PRESS localization to avoid exciting lipid-containing regions outside the brain. Alternatively (or in addition), outer-volume suppression pulses can be used to pre-saturate the lipid signals (Figure 2.4).[10] An inversion pulse can also be used for lipid signal suppression, exploiting the difference in T_1 values between lipid (typically 300 ms) and metabolites (typically 1000–2000 ms)[43] at 1.5 T. At this field strength, an inversion time of around 200 ms (= T_1 * ln[2]) will selectively null the lipid signal, while most of the metabolite signal remains inverted. This method has the advantage that no assumption is made about the spatial distribution of the lipid, but may somewhat reduce metabolite SNR. Interestingly, with a correct choice of TR and TI, the metabolite signals can be made invariant to changes in T_1 over quite a wide range of T_1 values.[44,45] It is also possible to reduce lipid artifacts by post-processing methods,[46] by making use of the fact that the spatial distribution of peri-cranial lipid is known, for instance, from T_1-weighted MRI scans.

Other approaches to lipid suppression include the use of frequency selective saturation pulses,[47] although these will preclude the observation of compounds such as lactate or alanine, which resonate in the same region of the spectrum. Similar to water suppression, lipid suppression can also be improved by incorporating spectral-spatial pulses into the localization sequence (such as PRESS[48]). For instance, spectral-spatial refocusing pulses can be designed which avoid exciting the lipid signal, which is then dephased by the application of crusher gradients.[42]

Spectral editing sequences using selective pulses

An interesting property of selective excitation and/or refocusing pulses is that they open up the possibility of selectively detecting specific compounds in the spectrum. This may be particularly useful for detecting signals from relatively lower concentration compounds that overlap with larger signals in the conventional spectrum. For instance, the MEGA-PRESS sequence has been used for the detection of the inhibitory neurotransmitter γ-aminobutyric acid (GABA). GABA is a relatively simple molecule with three coupled resonances at 1.9, 2.1 and 3.0 ppm (Figure 2.7A); unfortunately, all of these resonances overlap larger signals from more abundant compounds. Nevertheless, it is possible to detect GABA using the MEGA-PRESS editing sequence (Figure 2.7B). At a TE of 68 msec, the outer lines of the GABA 3.0 ppm triplet are inverted due to modulation by the scalar coupling to the 1.9 ppm methylene group. The application of frequency-selective 180°

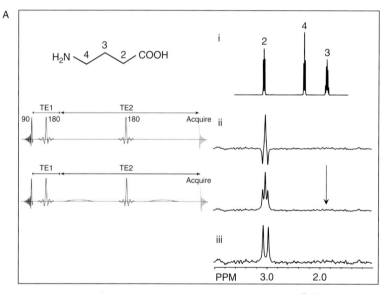

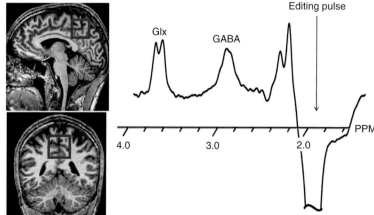

Figure 2.7. (A) Spectral editing of GABA using the MEGA-PRESS pulse sequence. (i) Structure of GABA and its conventional spectrum (equivalent to TE = 0 msec). (ii) The MEGA-PRESS pulse sequence and simulated spectra from the GABA 3.0 ppm triplet at TE 68 msec with (arrow) and without editing pulses applied. (iii) Difference spectrum, subtraction of spectrum without editing pulse from spectrum with editing pulse. (B) Example of MEGA-PRESS in vivo in a normal human volunteer at 3 T. A 3 × 3 × 3 cm voxel was placed in the medial posterior gray matter, TE 68 msec and 15 min scan time. The GABA+ signal is clearly seen at 3.0 ppm, as well as the co-edited Glx resonance at 3.7 ppm.

pulses at 1.9 ppm during the TE time period of the MEGA-PRESS sequence has the effect of removing the modulation of the 3.0 ppm resonance. By performing two scans, one with and one without the editing pulses – or better, by performing two scans with editing pulse frequencies of 1.9 and 7.5 ppm (i.e. symmetric about the water frequency (4.7 ppm)) – subtraction of the two scans results in the selective detection of the 3.0 ppm resonance, and removal of all resonances not affected by the editing pulses (Figure 2.7).

Other molecules may be edited using the MEGA-PRESS sequence (also known as BASING, which can also be used for water suppression); some of these molecules include lactate (to distinguish from lipid), glutathione, ascorbic acid (vitamin C), or N-acetyl aspartyl glutamate.[49,50,51] However, spectral-editing sequences are currently not commercially available, so these methodologies at present remain largely used for research purposes only. It should also be recognized that appreciable "co-editing" of unwanted molecules can occur with MEGA-PRESS; for instance, when observing GABA by using editing pulses applied at 1.9 ppm (and control "off" irradiation the other side of the water peak at 7.5 ppm), homocarnosine (a dipeptide of GABA and histidine) and macromolecules will also contribute to the edited signal seen at 3.0 ppm, in addition to GABA. For this reason, this signal is sometimes referred to as "GABA+"; with further refinement of the editing sequence (e.g. more selective editing pulses, or T_1-based discrimination of macromolecules) it may be possible to improve editing selectivity. Glutamate (the ^2CH resonance at ~3.75 ppm) will also co-edit, although it does not overlap with the GABA 3.0 ppm signal. Advanced curve-fitting of MEGA-PRESS spectra (e.g. using the "LCModel" – see Chapter 3) is recommended to quantify and separate the different contributions of co-edited molecules.

MRS at different field strengths

As discussed in Chapter 1, SNR and chemical shift dispersion are expected to increase approximately linearly with increasing magnetic field strength, and so in vivo MRS is expected to be superior at higher field strengths. Over the last few years, therefore, there has been a gradual progression for human brain MRS studies to be performed at 3.0, rather than 1.5, Tesla. In practice, gains in SNR and spectral resolution have been found to be somewhat less than linear,[52,53] so

that at first glance, spectra at 1.5 and 3.0 T may actually appear quite similar. This is because, in addition to field strength, a number of other factors affect in vivo spectroscopy, including metabolite relaxation times and magnetic field homogeneity.

A major factor that affects spectral appearance is linewidth, which depends on both metabolite T_2 and field homogeneity. Typical metabolite (e.g. for Cr or NAA) linewidths in the human brain are 3.5 Hz at 1.5 T, 5.5 Hz at 4.0 T, and 9.5 Hz at 7.0 T.[13] This increased linewidth would appear to be largely due to increased macro- and microscopic B_0 field homogeneity, which also affects metabolite apparent T_2 relaxation times, which are found to decrease with increasing B_0. As a result, high field MRS is best performed at short echo times (such at 35 msec or less). Also, while MRS at 1.5 T can be quite successfully performed at TE 280 msec, at 3.0 T the recommended echo time to be considered for long TE scans is 140 msec (unless an in-phase lactate signal is desired, in which case 280 msec will be needed).

Despite increasing linewidth, spectra at 7 and 3 T nevertheless demonstrate improved SNR and resolution compared to those at 1.5 T,[13,52] and in particular resonances from coupled spin systems such as glutamate, glutamine, and *myo*-inositol are better visualized at high field (e.g. see Figure 1.8 in Chapter 1). High-order shimming is particularly important at higher fields as magnetic susceptibility effects increase linearly with increasing field.

Other factors which become important at high fields include the use of high bandwidth radiofrequency slice selective pulses, in order to minimize chemical shift displacement artifacts which increase linearly with increasing magnetic field strength.

MRS with multiple RF receiver coils

Since their invention in 1990,[54] so-called phased-arrays of receiver coils have been increasingly used for MRI. However, until comparatively recently, they were little used for MRS, despite their advantage (compared to larger, volume coils) of higher local sensitivity, as well as extended coverage due to multiple elements. [55,56,57] More recently, particularly with the widespread commercial availability of phased-array head coils and multiple receiver channel MRI systems, their use for MRS and MRSI has increased. Usually, a large body RF coil is used for transmitting RF pulses, while the phased-array is used for signal reception. SNR improvements for phased-array coils depend on their geometry, the number of coils, and the voxel position (e.g. deep vs. superficial), but factors of more than 2 can usually be obtained, and coils with large numbers of elements (e.g. 32) may offer increases of 4 or more, at least in cortical regions.[58] The main challenge with multi-channel MRS and MRSI is the increased complexity of data handling, and the appropriate means of combining channels to produce uniform sensitivity and optimal SNR. Methods for channel combination are relatively straightforward, [54,55] and generally involve a weighting factor based on the channel sensitivity (i.e. a more sensitive channel, closest to the region-of-interest, should have a greater weight than a less sensitive channel) and a phase-correction to account for phase differences between channels. These factors can be determined by calibrations of coil sensitivity, [47] or from the MRS(I) data itself.[55] For instance, for a two-dimensional MRSI data set collected with $n = 1$ to N channels, the uniform, optimum sensitivity MRSI data S_u can be reconstructed

$$S_u(x,y) = \frac{\sum_{n=1}^{N} \exp(-i\arg(A_n(x,y))) \times S_n(x,y)}{\sum_{n=1}^{N} A_n(x,y)}$$

(2.1)

where $A_n(x,y)$ is the complex sensitivity (usually measured relative to the body coil), and $S_n(x,y)$ is the complex MRSI data of the nth coil, at point (x,y), respectively. The use of multiple, phased-array coils also opens up the possibility of reducing scan time for MRSI by using parallel acquisition techniques such as SENSE,[31] as discussed above.

Shimming and other "prescan" functions

The "quality" of a spectrum is critically dependent on the success of the prescan procedures that calibrate various aspects of the scanner function prior to data collection. Typically, a prescan will consist of adjusting transmitter and receiver gains, adjusting the field homogeneity ("shimming"), setting the scanner center frequency (on-resonance with water), and adjusting the flip angle(s) of the water suppression pulses (and often a final adjustment of receiver gain). The failure of any one of these processes can lead to an uninterpretable spectrum. In the early days of MRS, many of these steps were performed manually by the operator, but currently these steps are usually performed in an automated procedure by the scanner.

Shimming methods may be either iterative trial and error, or field map-based to calculate optimal shim currents.[59,60] If appropriately designed, field map methods should be quicker and more accurate. At lower fields (e.g. 1.5 T), often only adjusting the linear shim coils is sufficient, but at higher fields, since susceptibility affects scale linearly with field strength, it is important that high-order shim coils are available.[61] Proper shielding and pre-emphasis (i.e. adjustment of the gradient waveforms to compensate for eddy current effects) of magnetic field gradients is also important, since eddy currents induced in the magnetic bore/cryostat can have a serious effect on field homogeneity and spectral quality.

Water suppression flip angles are usually adjusted on an iterative, trial and error basis, although with the design of B_1 and T_1-insensitive suppression schemes, [40] this step may not be necessary.

Recommended acquisition protocols

In choosing a particular MRS protocol to use in an individual patient, a number of different factors need to be considered. First of all – what is the diagnostic question? If the observation of lactate is required (e.g. looking for ischemic injury, or mitochondrial disease) then a long TE is preferable (140 or 280 msec), since lactate is more cleanly detected at long TE with less lipid contamination. Note that lactate should be inverted at TE 140 msec, and in-phase at 280 msec; at high fields, however, depending on the RF pulses used, inversion at 3 T may be incomplete due to chemical shift displacement effects, resulting in reduced detection. Conversely, if compounds such as mI or Glx are of interest (e.g. looking for gliosis, or hepatic encephalopathy (Gln)), then short TE (< 35 msec) will be needed.

A second question will be – should the protocol be single voxel or multi-voxel? This will depend on whether it is valuable to probe the spatial distribution of metabolites – e.g. many brain tumors are heterogeneous, and the spectrum will depend where in the lesion the voxel is sampled. Also, in some cases, the location of the metabolic abnormality may not be known a priori, which would also favor the multi-voxel approach. From a spatial coverage viewpoint, multi-voxel would always appear to be favorable; however, this is not always the case; for instance, some brain regions are not very amenable to multi-voxel spectroscopy (e.g. posterior fossa, anterior temporal or frontal lobes, or when the patient has metal artifacts (e.g. brace, post-surgical, etc.)), and MRSI with large coverage tends not to work well at short TE. In addition, scan time is longer with MRSI, and therefore MRSI may not be feasible in uncooperative patients, or when time is limited. MRSI is also somewhat more sensitive to head motion than SV-MRS, although neither technique will work well when motion is appreciable.

Some recommended protocols for different clinical scenarios are described below, but it is also worth remembering some of the basic limitations of MRS.

(1) Single-voxel, standard voxel size $2 \times 2 \times 2$ cm (8 cm^3). Minimum voxel size: approximately $1.5 \times 1.5 \times .1.5$ cm (~3.38 cm^3).
(2) Multi-voxel MRSI, standard voxel size ~ $1-2$ cm^3, minimum ~ 0.5 cm^3.
(3) Protocol setup: avoid lipid signals from scalp, regions of bad field homogeneity (anterior temporal lobe, inferior frontal lobe, anywhere adjacent to sinuses or other air spaces (e.g. auditory canal)).

When performing SV-MRS, it is useful to also collect a non-water suppressed spectrum from the same ROI (some manufacturers do this automatically) which can be used to check linewidth, do an eddy current and phase correction, and for quantitation purposes. In addition, it is often very helpful to record a spectrum from the contralateral hemisphere, since this will aid interpretation. It is also handy to save a screenshot of the voxel location for future reference.

In the next sections, specific recommended protocols for particular clinical conditions are described.

Temporal lobe epilepsy (TLE)

While 2D-PRESS-MRSI has been successfully used to study TLE (covering both temporal lobe simultaneously (Figure 2.8)) it is more common to apply single-voxel MRS in the left and right temporal lobes sequentially, usually looking for lower NAA in the epileptogenic hippocampus (Figure 2.9). Since the main compounds of interest in seizure disorders are NAA and lactate (as well as Cr and Cho), these protocols work best at long TE (e.g. TE 140 msec at 3.0 T, 140 or 280 msec at 1.5 T). Figure 2.9 shows an example of the SV-MRS protocol applied to a patient with an abnormal left hippocampus. Both SV and MRSI

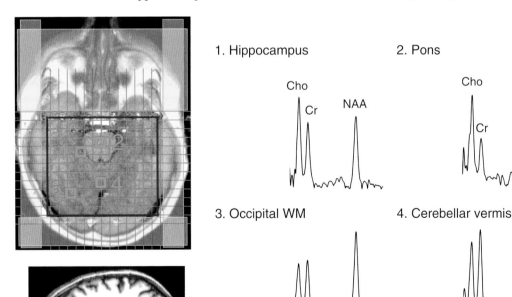

Figure 2.8. 2D-PRESS MRSI for the bilateral evaluation of temporal lobe metabolism. A PRESS box is chosen that is angulated parallel to the long-axis of the temporal lobe, at the level of the hippocampus. The box location is chosen to avoid regions of lipid in the scalp or poor field homogeneity. Four saturation bands (indicated in blue) are used to suppress out of voxel signal. Phase-encoding is applied in the transverse plane (voxel size ~1 cm^3). Spectra are recorded at 3 T (TR 2000, TE 140 msec). Note strong regional variations in pons, hippocampus, cerebellar vermis, and occipital white matter. This protocol works best when high order shimming, and high bandwidth slice selective pulses, are available.

protocols are angulated along the long-axis of the hippocampus.

Hepatic encephalopathy (HE)

The triad of metabolic changes associated with hepatic encephalopathy (HE) are increased glutamine, decreased choline, and decreased *myo*-inositol. Therefore MRS in patients with suspected HE should be examined with short TE MRS; while the metabolic changes are most likely global in these patients, following from the original paper most studies have used standardized single voxel locations in the mesial occipital gray matter, and parietal white matter (Figure 2.10; $2 \times 2 \times 2$ cm, TR 1500, TE 35 msec, 128 averages).[62]

Supra- and infra-tentorial multi-slice MRSI

Multi-slice or 3D MRSI can often be used, with 3 or 4 slices in the axial plane, to provide coverage of the larger part of the supratentorial brain. Representative

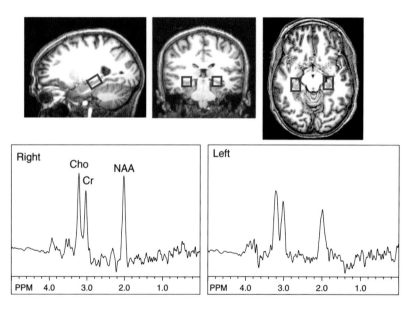

Figure 2.9. Single voxel protocol for evaluating temporal lobe epilepsy (1.5 T, TR 1500, TE 140 msec, 128 averages). (A) Voxel locations are best prescribed from high-resolution 3D T_1-weighted images that have been reconstructed in axial, sagittal, and coronal orientations. The voxel is anisotropic ($2 \times 1.5 \times 1.5$ cm) with the longest dimension applied along the long-axis of the hippocampus. Since hippocampal spectra vary as a function of position in the A–P direction, the voxel position should be carefully positioned in the mid-hippocampus on both sides. The anterior portion of the hippocampus usually cannot be studied in most adults because of susceptibility artifacts from the sphenoid sinus. (B) Example spectra from a 51-year-old subject with nocturnal seizures. Note the lower ratios of NAA/Cho and NAA/Cr in the abnormal left hippocampus; all metabolite peak heights appear slightly lower in the left hippocampus because of its worse field homogeneity (broader linewidth) than on the right.

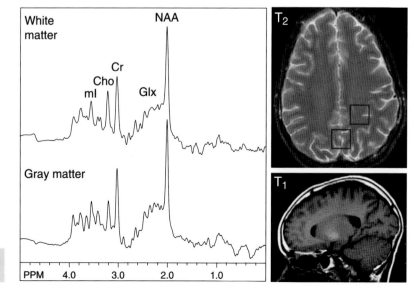

Figure 2.10. Voxel locations commonly used to study hepatic encephalopathy in mesial occipital gray matter and parietal white matter. Spectra recorded at short TE in this patient with mild hepatic encephalopathy showed increased Glx signal (due to increased glutamine), and decreased Cho and mI signals. Note also basal ganglia hyperintensity in sagittal T_1-weighted MRI often seen in patients with liver failure.

Chapter 2: Pulse sequences and protocol design

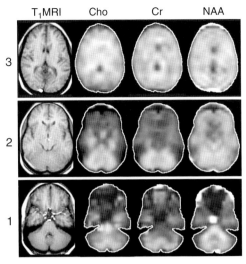

Figure 2.11. Multi-slice MRSI protocol covering the cerebellum and brain stem. Slice locations are chosen to avoid exciting unwanted signal in the clivus, oral cavity, and sinuses. Metabolic images show high levels of Cr and Cho in the cerebellum.

slice locations are shown in Figure 2.4, and data from a sequence of this type are presented in Figure 2.5 (1.5 T, TR 2300 TE 280 msec, nominal voxel size 0.8 cm^3). Regional variations are discussed in more detail in Chapter 4.

For the posterior fossa, Figure 2.11 shows slice locations and data from a 3-slice MRSI covering the cerebellum, brain stem, and inferior frontal lobe (1.5 T, TR 1700 TE 280 msec, nominal voxel size 0.8 cm^3). The slices are angulated to avoid exciting water or lipid signal in the vicinity of the clivus, oral cavity and sinuses. OVS pulses should also be placed in these regions to help suppress any residual unwanted signals. It can be seen that the cerebellum exhibits high levels of Cho and Cr; this is discussed further in Chapter 4. An alternative protocol for the posterior fossa is to use 3D-PRESS-MRSI, again avoiding exciting any signals from regions of poor field homogeneity and high lipid content.

Finally, it should be mentioned that there is currently development underway of fast MRSI sequences that provide whole-brain coverage.[63] While these are not yet commercially available, it is hoped that in the future these will become routinely available and standardized for both research investigation and clinical studies.

References

[1] Frahm J. Localized proton spectroscopy using stimulated echoes. *J Magn Reson* 1987; **72**(3): 502–08.

[2] Frahm J, Bruhn H, Gyngell ML, Merboldt KD, Hanicke W, Sauter R. Localized proton NMR spectroscopy in different regions of the human brain in vivo. Relaxation times and concentrations of cerebral metabolites. *Magn Reson Med* 1989; **11**: 47–63.

[3] Frahm J, Bruhn H, Gyngell ML, Merboldt KD, Hanicke W, Sauter R. Localized high-resolution proton NMR spectroscopy using stimulated echoes: initial applications to human brain in vivo. *Magn Reson Med* 1989; **9**: 79–93.

[4] Bottomley PA, inventor General Electric Company, assignee. Selective volume method for performing localized NMR spectroscopy. USA patent 4480228. 1984 October 30th 1984.

[5] Ordidge RJ, Gordon RE, inventors; Oxford Research Systems Limited, assignee. Methods and apparatus of obtaining NMR spectra. United States patent 4531094. 1983.

[6] Duijn JH, Matson GB, Maudsley AA, Weiner MW. 3D phase encoding 1 H spectroscopic imaging of human brain. *Magn Reson Imaging* 1992; **10**: 315–9.

[7] Moonen CT W, Sobering G, van Zijl PC M, Gillen J, von Kienlin M, Bizzi A. Proton spectroscopic imaging of human brain. *J Magn Reson* 1992; **98**: 556–75.

[8] Spielman D, Meyer C, Macovski A, Enzmann D. 1 H spectroscopic imaging using a spectral-spatial excitation pulse. *Magn Reson Med* 1991; **18**: 269–79.

[9] van Zijl PC, Moonen CT, Alger JR, Cohen JS, Chesnick SA. High field localized proton spectroscopy in small volumes: greatly improved localization and shimming using shielded strong gradients. *Magn Reson Med* 1989; **10**: 256–65.

[10] Duyn JH, Gillen J, Sobering G, van Zijl PC, Moonen CT. Multisection proton MR spectroscopic imaging of the brain. *Radiology* 1993; **188**: 277–82.

[11] Ordidge RJ. Random noise selective excitation pulses. *Magn Reson Med* 1987; **5**: 93–8.

[12] Moonen CT, von Kienlin M, van Zijl PC, Cohen J, Gillen J, Daly P, et al. Comparison of single-shot localization methods (STEAM and PRESS) for in vivo proton NMR spectroscopy. *NMR Biomed* 1989; **2**: 201–08.

[13] Tkac I, Andersen P, Adriany G, Merkle H, Ugurbil K, Gruetter R. In vivo 1 H NMR spectroscopy of the human brain at 7 T. *Magn Reson Med* 2001; **46**: 451–6.

[14] Brown TR, Kincaid BM, Ugurbil K. NMR chemical shift imaging in three dimensions. *Proc Natl Acad Sci USA* 1982; **79**: 3523–6.

[15] Petroff OA, Graham GD, Blamire AM, al-Rayess M, Rothman DL, Fayad PB, et al. Spectroscopic imaging of stroke in humans: histopathology correlates of spectral changes. *Neurology* 1992; **42**: 1349–54.

[16] Luyten PR, Marien AJ, Heindel W, van Gerwen PH, Herholz K, den Hollander JA, et al. Metabolic imaging of patients with intracranial tumors: H-1 MR spectroscopic imaging and PET. *Radiology* 1990; **176**: 791–9.

[17] Nelson SJ. Analysis of volume MRI and MR spectroscopic imaging data for the evaluation of patients with brain tumors. *Magn Reson Med* 2001; **46**: 228–39.

[18] Tran TK, Vigneron DB, Sailasuta N, Tropp J, Le Roux P, Kurhanewicz J, et al. Very selective suppression pulses for clinical MRSI studies of brain and prostate cancer. *Magn Reson Med* 2000; **43**: 23–33.

[19] Maudsley AA, Matson GB, Hugg JW, Weiner MW. Reduced phase encoding in spectroscopic imaging. *Magn Reson Med* 1994; **31**: 645–51.

[20] Golay X, Gillen J, van Zijl PC, Barker PB. Scan time reduction in proton magnetic resonance spectroscopic imaging of the human brain. *Magn Reson Med* 2002; **47**: 384–7.

[21] Pohmann R, von Kienlin M, Haase A. Theoretical evaluation and comparison of fast chemical shift imaging methods. *J Magn Reson* 1997; **129**: 145–60.

[22] Mansfield P. Spatial mapping of chemical shift in NMR. *Magn Reson Med* 1984; **1**: 370–86.

[23] Posse S, Tedeschi G, Risinger R, Ogg R, Le Bihan D. High speed 1 H spectroscopic imaging in human brain by echo planar spatial-spectral encoding. *Magn Reson Med* 1995; **33**: 34–40.

[24] Ebel A, Maudsley AA. Improved spectral quality for 3D MR spectroscopic imaging using a high spatial resolution acquisition strategy. *Magn Reson Imaging* 2003; **21**: 113–20.

[25] Adalsteinsson E, Irarrazabal P, Topp S, Meyer C, Macovski A, Spielman DM. Volumetric spectroscopic imaging with spiral-based k-space trajectories. *Magn Reson Med* 1998; **39**: 889–98.

[26] Ebel A, Soher BJ, Maudsley AA. Assessment of 3D proton MR echo-planar spectroscopic imaging using automated spectral analysis. *Magn Reson Med* 2001; **46**: 1072–8.

[27] Pelletier D, Nelson SJ, Grenier D, Lu Y, Genain C, Goodkin DE. 3-D echo planar (1)HMRS imaging in MS: metabolite comparison from supratentorial vs. central brain. *Magn Reson Imaging* 2002; **20**: 599–606.

[28] Govindaraju V, Gauger GE, Manley GT, Ebel A, Meeker M, Maudsley AA. Volumetric proton spectroscopic imaging of mild traumatic brain injury. *Am J Neuroradiol* 2004; **25**: 730–7.

[29] Block KT, Frahm J. Spiral imaging: a critical appraisal. *J Magn Reson Imaging* 2005; **21**: 657–68.

[30] Adalsteinsson E, Langer-Gould A, Homer RJ, Rao A, Sullivan EV, Lima CA, et al. Gray matter N-acetyl aspartate deficits in secondary progressive but not relapsing-remitting multiple sclerosis. *Am J Neuroradiol* 2003; **24**: 1941–5.

[31] Dydak U, Weiger M, Pruessmann KP, Meier D, Boesiger P. Sensitivity-encoded spectroscopic imaging. *Magn Reson Med* 2001; **46**: 713–22.

[32] Pruessmann KP, Weiger M, Scheidegger MB, Boesiger P. SENSE: Sensitivity Encoding for fast MRI. *Magn Reson Med* 1999; **42**: 952–62.

[33] Sodickson DK, Manning WJ. Simultaneous acquisition of spatial harmonics (SMASH): fast imaging with radiofrequency coil arrays. *Magn Reson Med* 1997; **38**: 591–603.

[34] Jakob PM, Griswold MA, Edelman RR, Sodickson DK. AUTO-SMASH: a self-calibrating technique for SMASH imaging. SiMultaneous Acquisition of Spatial Harmonics. *Magma* 1998; **7**: 42–54.

[35] McKenzie CA, Yeh EN, Ohliger MA, Price MD, Sodickson DK. Self-calibrating parallel imaging with automatic coil sensitivity extraction. *Magn Reson Med* 2002; **47**: 529–38.

[36] Griswold MA, Jakob PM, Heidemann RM, Nittka M, Jellus V, Wang J, et al. Generalized autocalibrating partially parallel acquisitions (GRAPPA). *Magn Reson Med* 2002; **47**: 1202–10.

[37] Haase A, Frahm J, Hanicke W, Matthei D. ^{1}H NMR chemical shift selective imaging. *Phys Med Biol* 1985; **30**: 341–4.

[38] Moonen CTW, van Zijl PCM. Highly efficient water suppression for in vivo proton NMR spectroscopy. *J Magn Reson* 1990; **88**: 28–41.

[39] Ogg RJ. WET, a T1- and B1-insensitive water-suppression method for in vivo localized 1 H NMR spectroscopy. *J Magn Reson B* 1994; **104**: 1–10.

[40] Tkac I, Starcuk Z, Choi IY, Gruetter R. In vivo 1 H NMR spectroscopy of rat brain at 1 ms echo time. *Magn Reson Med* 1999; **41**: 649–56.

[41] Mescher M, Merkle H, Kirsch J, Garwood M, Gruetter R. Simultaneous in vivo spectral editing and water suppression. *NMR Biomed* 1998; **11**: 266–72.

[42] Star-Lack J, Nelson SJ, Kurhanewicz J, Huang LR, Vigneron DB. Improved water and lipid suppression for 3D PRESS CSI using RF band selective inversion with gradient dephasing (BASING). *Magn Reson Med* 1997; **38**: 311–21.

[43] Spielman DM, Pauly JM, Macovski A, Glover GH, Enzmann DR. Lipid-suppressed single- and multisection proton spectroscopic imaging of the human brain. *J Magn Reson Imaging* 1992; **2**: 253–62.

[44] Hetherington HP, Mason GF, Pan JW, Ponder SL, Vaughan JT, Twieg DB, *et al*. Evaluation of cerebral gray and white matter metabolite differences by spectroscopic imaging at 4.1 T. *Magn Reson Med* 1994; **32**: 565–71.

[45] Hetherington HP, Pan JW, Mason GF, Ponder SL, Twieg DB, Deutsch G, *et al*. 2D 1 H spectroscopic imaging of the human brain at 4.1 T. *Magn Reson Med* 1994; **32**: 530–4.

[46] Haupt CI, Schuff N, Weiner MW, Maudsley AA. Removal of lipid artifacts in 1 H spectroscopic imaging by data extrapolation. *Magn Reson Med* 1996; **35**: 678–87.

[47] Smith MA, Gillen J, McMahon MT, Barker PB, Golay X. Simultaneous water and lipid suppression for in vivo brain spectroscopy in humans. *Magn Reson Med* 2005; **54**: 691–6.

[48] Star-Lack J, Vigneron DB, Pauly J, Kurhanewicz J, Nelson SJ. Improved solvent suppression and increased spatial excitation bandwidths for three-dimensional PRESS CSI using phase-compensating spectral/spatial spin-echo pulses. *J Magn Reson Imaging* 1997; **7**: 745–57.

[49] Rothman DL, Petroff OA, Behar KL, Mattson RH. Localized 1 H NMR measurements of gamma-aminobutyric acid in human brain in vivo. *Proc Natl Acad Sci USA* 1993; **90**: 5662–6.

[50] Trabesinger AH, Boesiger P. Improved selectivity of double quantum coherence filtering for the detection of glutathione in the human brain in vivo. *Magn Reson Med* 2001; **45**: 708–10.

[51] Freeman DM, Sotak CH, Muller HH, Young SW, Hurd RE. A double quantum coherence transfer proton NMR spectroscopy technique for monitoring steady-state tumor lactic acid levels in vivo. *Magn Reson Med* 1990; **14**: 321–9.

[52] Barker PB, Hearshen DO, Boska MD. Single-voxel proton MRS of the human brain at 1.5 T and 3.0 T. *Magn Reson Med* 2001; **45**: 765–9.

[53] Gonen O, Gruber S, Li BS, Mlynarik V, Moser E. Multivoxel 3D proton spectroscopy in the brain at 1.5 versus 3.0 T: signal-to-noise ratio and resolution comparison. *Am J Neuroradiol* 2001; **22**: 1727–31.

[54] Roemer PB, Edelstein WA, Hayes CE, Souza SP, Mueller OM. The NMR phased array. *Magn Reson Med* 1990; **16**: 192–225.

[55] Brown MA. Time-domain combination of MR spectroscopy data acquired using phased-array coils. *Magn Reson Med* 2004; **52**: 1207–13.

[56] Natt O, Bezkorovaynyy V, Michaelis T, Frahm J. Use of phased array coils for a determination of absolute metabolite concentrations. *Magn Reson Med* 2005; **53**: 3–8.

[57] Wald LL, Moyher SE, Day MR, Nelson SJ, Vigneron DB. Proton spectroscopic imaging of the human brain using phased array detectors. *Magn Reson Med* 1995; **34**: 440–5.

[58] Wiggins GC, Triantafyllou C, Potthast A, Reykowski A, Nittka M, Wald LL. 32-channel 3 Tesla receive-only phased-array head coil with soccer-ball element geometry. *Magn Reson Med* 2006; **56**: 216–23.

[59] Sukumar S, Johnson MO, Hurd RE, van Zijl PC. Automated shimming for deuterated solvents using field profiling. *J Magn Reson* 1997; **125**: 159–62.

[60] Gruetter R. Automatic, localized in vivo adjustment of all first- and second-order shim coils. *Magn Reson Med* 1993; **29**: 804–11.

[61] Spielman DM, Adalsteinsson E, Lim KO. Quantitative assessment of improved homogeneity using higher-order shims for spectroscopic imaging of the brain. *Magn Reson Med* 1998; **40**: 376–82.

[62] Kreis R, Ross BD, Farrow NA, Ackerman Z. Metabolic disorders of the brain in chronic hepatic encephalopathy detected with H-1 MR spectroscopy. *Radiology* 1992; **182**: 19–27.

[63] Maudsley AA, Darkazanli A, Alger JR, Hall LO, Schuff N, Studholme C, *et al*. Comprehensive processing, display and analysis for in vivo MR spectroscopic imaging. *NMR Biomed* 2006; **19**: 492–503.

Chapter 3
Spectral analysis methods, quantitation, and common artifacts

Key points

- Correct post-processing and quantitation are key aspects of in vivo MRS.
- Filtering, phase-correction and baseline correction improve MRS data.
- Peak area estimation can be done using parametric or non-parametric routines in either the time domain or frequency domains.
- "LCModel" software is becoming widely used and accepted, particularly for single-voxel MRS data.
- MRSI processing requires additional steps; k-space filtering and other manipulations can improve MRSI data quality.
- A variety of strategies are available for quantitation, based on either internal or external reference standards, or phantom replacement methodology.
- Quantitation routines should take into account voxel composition, particularly the amount of CSF partial volume present.
- MRS is sensitive to field inhomogeneity and other artifacts.

Introduction

Methods for spectral analysis and the quantitative analysis of spectral data are arguably as important as the techniques used to collect the data; the use of incorrect analysis methods can lead to systematic errors or misinterpretation of spectra. In general, the ultimate goal of spectral analysis is to determine the concentrations of the compounds present in the spectra. In MRS, the area under the spectral peak is proportional to the metabolite concentration; however, determining the proportionality constant can be challenging. In addition, peak area measurements in in-vivo spectroscopy are complicated by resonance overlap, baseline distortions, and lineshapes that often only poorly approximate conventional models such as Gaussian or Lorentzian functions. Therefore, quantitative analysis of in vivo MRS data is challenging. This chapter reviews basic spectral processing techniques, methods for determining peak areas, and strategies for calculating metabolite concentrations.

Basic MRS processing techniques

Time domain processing

As discussed in Chapter 1, MRS involves the collection of an FID or echo signal in the time domain following pulsed excitation. In order for the data to be interpretable to the human eye, it is necessary to Fourier transform (FT) the time domain data in order to represent it as spectral intensity as a function of frequency (Figure 1.7). The properties of the spectrum (for instance, SNR and resolution) can be manipulated by applying digital filters to the time-domain data prior to FT. One or more of the following filters are commonly applied to in vivo MRS data.

1. Multiplication by negative exponential or Gaussian (i.e. decaying) functions – i.e. functions of the form $\exp(-\lambda t)$ or $\exp(-\lambda^2 t^2)$, which cause the spectral linewidth (i.e. width at half-height, or $\Delta\nu_{1/2}$) to increase by λ/π and $1.67\lambda/\pi$, respectively. While the increase in linewidth is generally undesirable, the filter also has the effect of smoothing out noise in the spectrum, so applying filters of this type improve the SNR of the spectrum, which is often critically important. The best SNR is obtained when the decay constant of the filter, λ, equals the rate of decay of the signal in the FID ($1/T_2^*$) – this is called a "matched filter". Occasionally, it is desirable to apply a filter to increase resolution rather than SNR; in these cases, exponentially increasing filters may be applied (at the expense of decreased SNR). One commonly used filter

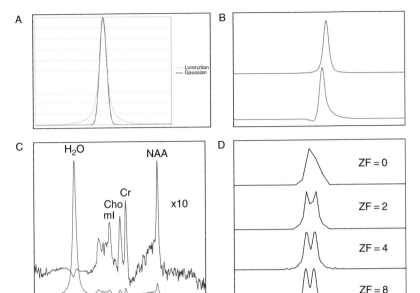

Figure 3.1. (A) An example of Gaussian and Lorentzian lineshapes with the same width at half height (0.6 Hz; displayed spectral width is 10 Hz). (B) Bottom, a localized (2 × 2 × 2 cm) single voxel water signal (1.5 T, TE 35 msec) from a phantom without and (top) with eddy current correction. (C) Localized proton spectra of the normal brain with (red) and without (blue) residual water-suppression using a high-pass band-reject filter. (D) A simulated lactate doublet (7 Hz coupling, T_2^* 100 ms, linewidth ~ 3 Hz) digitized with 512 data points at a rate of 2 kHz. Without zero-filling, the spectral digital resolution is approximately 4 Hz per point, and the doublet structure is not visible. Zero-filling by a factor of 2 allows the doublet to be visualized, and by a factor of 4 or 8 gives the doublet a smoother appearance.

is the so-called Lorentzian-to-Gaussian transformation, $\exp(+\lambda_1 t)\exp(-\lambda_2^2 t^2)$, which converts the "natural" Lorentzian lineshape into a Gaussian function. This is often considered preferable, since the Gaussian lineshape has less intensity in the "wings" of the resonance compared to a Lorentzian (Figure 3.1A).

2. Correction of DC offsets in the FID. Occasionally, there may be a DC offset (i.e. static, non-zero signal) in the FID; when Fourier transformed, this gives rise to a zero-frequency glitch in the spectrum. This can be removed by first subtracting the DC offset from the FID prior to Fourier transformation (usually estimated from a portion of the FID at the end of acquisition, where little real signal is expected). With modern spectrometers, with well-adjusted receivers, this step is usually not necessary.

3. Eddy current correction. An eddy current is a small, transient current which is induced in the magnet structure after the application of a pulsed field gradient. The current produces a small, time-varying change in the magnetic field strength which causes artifacts in the spectral lineshape. With careful gradient design, including the use of active screening and pre-emphasis, these currents can be minimized, but generally not eliminated. However, so long as an appropriate reference signal is available (e.g. the unsuppressed water signal from the localized voxel), it is possible to apply a time-dependent phase correction which corrects for this effect; as the water signal is on-resonance, it can be described by $A_{ec}(t)^*\exp(-\lambda t)^*\exp(i\varphi_{ec})$, where A_{ec} and φ_{ec} are amplitude and phase variations due to eddy currents. The eddy current corrected spectrum is obtained by applying the phase-correction $\exp(-i\varphi_{ec})$ to the water suppressed FID prior to Fourier transformation. [1] In principle, the factor $1/A_{ec}$ could also be applied,[2] although in practice this is usually not done because it can cause increased noise (and artifacts), particularly when SNR is low. Note that eddy current correction requires the acquisition of a non-water suppressed scan in addition to the normal water suppressed scan. An example of a localized brain water signal with and without eddy current correction is shown in Figure 3.1B.

4. Band-reject filters. Often, in proton MRS, the signals of interest (metabolites) are much smaller amplitude than residual signals from high concentration species such as water and fat. This can make visualization and quantitation of the metabolite signals more difficult; therefore, time-domain convolution filters (e.g. high-pass band-reject) filters are often used to remove these signals (particularly water) before Fourier transformation. [3] Commonly, a Gaussian high-pass filter is used with a bandwidth of 50–100 Hz for spectra recorded

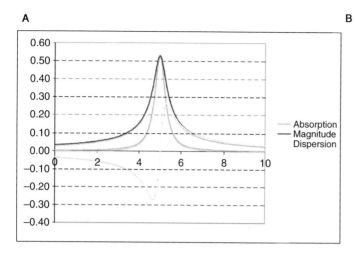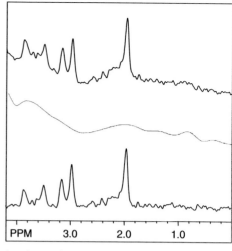

Figure 3.2. (A) Absorption and dispersion Lorentzian lineshapes ($\Delta\nu_{1/2}$ = 0.6 Hz; displayed spectral width is 10 Hz), as well as magnitude. (B) (Top) Example of a short echo time proton brain spectrum without baseline correction. (Middle) Baseline contribution estimated using a cubic spline routine. (Bottom) baseline corrected spectrum.

at 1.5 or 3.0 T. An example of a localized brain spectrum with and without residual water suppression using a band-reject filter is shown in Figure 3.1C.

5. Zero-filling. If the FID is digitized at a sampling interval Δt (equal to the inverse of the spectral width (SW), Δt = 1/SW), then the total length of the data readout is $N^*\Delta t$, where N is the number of data points. The digital resolution in the spectrum, Δf = SW/N = $1/N^*\Delta t$. If the readout window is relatively short, this will lead to coarse digital resolution in the spectrum and poorly defined peaks. One solution to this is so-called "zero-filling", the addition of extra data points of zero intensity at the end of the FID before FT. This results in higher digital resolution in the spectrum and also provides some interpolation between adjacent data points (at least for zero-filling by a factor of 2 (i.e. 2^*N final data points). Figure 3.1D shows examples of simulated spectra processed with and without zero-filling.

Frequency domain processing

After Fourier transformation, additional processing steps are also usually performed. It should be remembered that MR data are always complex (real and imaginary), since quadrature detection is used. If the time-domain MRS signal is expressed, an exponentially decaying oscillator ($f(t) = \exp(i\Omega t)\exp(-\lambda t)$), the FT of this signal has real and imaginary parts as follows:

$$f(\omega) = F.T.[f(t)] = \int_{-\infty}^{+\infty} \exp(-i\omega t) f(t) dt \quad (3.1)$$

$$f(\omega) = \frac{\lambda}{\lambda^2 + (\Omega - \omega)^2} + \frac{i(\Omega - \omega)}{\lambda^2 + (\Omega - \omega)^2}$$
$$= A(\omega) + iD(\omega) \quad (3.2)$$

where the real ($A(\omega)$) and imaginary ($D(\omega)$) parts of this expression are called the "absorption" and "dispersion" lineshapes, respectively (Figure 3.2). The absorption lineshape is preferable, since it has only positive intensity, and narrower wings compared to the dispersion lineshape.

In practice, there is nearly always a phase-error associated with the scanner receiver pathway and electronics (ϕ_0), which can be written as follows:

$$f(t) = \exp(i\Omega t)\exp(-\lambda t)\exp(i\phi_0) \quad (3.3)$$

and there may also be a small timing error between the start of data collection and the "true" $t = 0$ (i.e. the start of the FID, or top of the echo). In this case, the f(t) becomes:

$$f(t) = \exp(i\Omega(t + t_0))\exp(-\lambda t)\exp(i\phi_0)$$
$$\approx \exp(i\Omega t)\exp(-\lambda t)^*\exp(i(\phi_0 + \Omega t_0)) \quad (3.4)$$

where t_0 is the timing error. After Fourier transformation, this has the effect of mixing the absorption and dispersion lineshapes:

$$f(\omega) = \cos(\phi_0 + \omega t_0) \times A(\omega) \\ + i\sin(\phi_0 + \omega t_0) \times D(\omega) \quad (3.5)$$

Therefore, in order to obtain pure absorption lineshapes, it is necessary to apply the phase correction $\exp(-i(\phi_0 + \omega t_0))$ to the frequency domain data. Note that the phase correction may be the same across the entire spectrum (i.e. "zero order", $\exp(-i(\phi_0))$, and/or it may vary linearly with frequency across the spectrum ("first order", $\exp(-i\omega t_0)$). Since the phase errors are unknown, traditionally phase correction has been applied manually by the operator, although more recently various algorithms have been developed for automated phase correction. Unfortunately, the automated routines are not always completely reliable, and manual intervention is still sometimes required. It is usually easier to phase-correct a high SNR spectrum (such as the unsuppressed water signal), and then apply this phase-correction to the water-suppressed spectrum (so long as it is recorded under the same pulse sequence timings, etc., and therefore has the same phase error as the unsuppressed spectrum).

Another alternative is to simply calculate the magnitude of the complex signal, i.e.:

$$S(\omega) = \sqrt{A(\omega)^2 + iD(\omega)^2} \quad (3.6)$$

which removes any phase error from the signal, but has two undesirable properties: (a) it increases the width at half height because both the absorption and dispersion lineshapes are combined, and (b) it decreases SNR because noise from both the real and imaginary channels is combined (Figure 3.2A). Note that this may not always be apparent, since the noise is forced to be positive in the magnitude spectrum, whereas it is both positive and negative in the phase-sensitive spectra, so the apparent peak-to-peak noise voltage may appear less in the magnitude spectra in regions where there is no signal. There is one scenario where calculating the magnitude has no effect on linewidth; this occurs when a whole echo is sampled; the FT of a symmetric function (such as an echo) integrates to zero for the dispersion lineshape, so magnitude calculation of echo signals should have similar lineshapes to phase-corrected spectra. This can be useful, for instance, in large data sets (such as high resolution 2D or 3D-MRSI data) where individual, manual phase-correction of spectra is not feasible.

Generally, MRS data also contain broad signals (from large molecules such as proteins, other macromolecules, and molecules whose motion is restricted (solids, membranes, etc.)) that give the spectra an uneven, non-zero baseline. The baseline may also be distorted for instrumental reasons (e.g. if the first point of the FID is not correctly digitized). In order to accurately estimate metabolite peak areas, it is usually necessary to apply a baseline correction. Some spectral fitting programs directly estimate the baseline resonances, although the more common approach is to fit the baseline to a function such as a cubic spline (choosing data points where there is no metabolite signal intensity, and then interpolating) and then subtract the fitted function from the spectrum before peak area estimation. An example of baseline correction using a cubic spline routine is given in Figure 3.2B.

MRSI processing techniques

In addition to the processing steps listed above for single voxel spectra, additional processing steps are involved in the processing of MRSI data. MRSI data are usually collected using phase-encoding in two or three spatial dimensions, and therefore requires Fourier transformation in these spatial ("k-space") domains. The resolution and SNR of the data can be manipulated using digital filters prior to Fourier transformation; since MRSI is usually done at low spatial resolution, only a limited range of k-space is sampled, and this can lead to "truncation artifacts" manifesting as "wiggles" ($sinc(x)$) in the reconstructed spectroscopic images. These can be reduced by applying filters (e.g. "Hanning" or "Hamming filters" are often used) in k-space prior to acquisition; this also has the effect of increasing SNR, but does also increase the effective voxel size (i.e. decrease spatial resolution). In MRSI (and MRI), the shape of the voxel is described by the "point-spread function" (PSF); some two-dimensional PSFs and the effects of various different filters are shown in Figure 3.3. A "Fermi" filter[4] is particularly useful, since it makes the PSF spherically symmetric (i.e. same resolution in any direction) and with two- or three-dimensional phase-encoding can be used to save scan time (since phase-encoding steps with zero-weighting do not need to be collected).

The exact location of the center of the MRSI pixel can also be altered ("voxel-shifting") by applying linear phase corrections on the k-space data prior to Fourier transformation;[5] this can sometimes be useful, particularly since MRSI voxels are quite large and of similar dimensions of the structure or lesion to be examined. Pixels shifts ($\Delta_x, \Delta_y, \Delta_z$) can be achieved using:

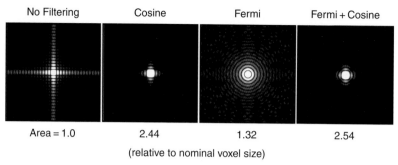

Figure 3.3. 2D-MRSI point spread functions (PSF), and the effects of digital filtering. The spatial response function of a 32 × 32 acquisition zero-filled to 256 × 256 is presented. The gray scale is adjusted so that maximum brightness (white) corresponds to 10% of the maximum signal. The area (integral) of the PSF over the full FOV provides an index of the MRSI voxel size. When no filtering is applied, the slowly (1/x) decaying wiggles of the sinc function are very apparent, and the voxel size equals the MRSI nominal voxel size. When a cosine filter is applied, the sinc-wiggles are much reduced, but at the expense of decreased resolution (voxel size = 2.44 times the nominal). When a Fermi filter is applied (circular k-space sampling), the sinc wiggles are marginally reduced, and now show a circularly symmetric pattern (voxel size = 1.32 times nominal). Finally, when both the Fermi and cosine filters are applied, a nearly symmetric PSF with 2.54-times the nominal voxel size is obtained.

$$s_{shifted}(t, k_x k_y, k_z) = s_{original}(t, k_x, k_y, k_z) \exp(i \Delta x k_x) \\ \exp(i \Delta y k_y) \exp(i \Delta z k_z) \quad (3.7)$$

When using voxel-shifting, however, it should be remembered that the "nominal voxel size" displayed by many software packages (= field-of-view (FOV)/number of phase-encoding steps) is usually significantly smaller than the true voxel size based on the integration of the PSF; therefore, voxel-shifting may have a smaller influence on the spectral appearance than might be anticipated.

After FT, additional processing steps may be taken, including phase-correction (which may vary from voxel-to-voxel; in particular, depending on the phase-encoding gradients used, there may be linearly dependent zero-order phase corrections in each direction), susceptibility correction (i.e. correction for field inhomogeneity by left or right shifting the spectrum, usually based on measuring the frequency of the residual water, or NAA, peak). Low-resolution metabolic images may be created by integrating peak areas in each voxel, and are often interpolated up to a higher matrix size for display.

Spectral fitting

The signal voltage induced in the receiver coil is directly proportional to the metabolite concentration. From the Fourier transformation, it can be shown that the area under the curve in the spectrum (frequency domain) is equal to the amplitude of the first point of the time domain signal (free induction decay, or FID). Therefore, quantitative analysis either requires the determination of peak areas in the frequency domain, or direct time-domain analysis to estimate the amplitude of the first data point of the different frequency components that comprise the FID.

A variety of methods have been developed for *frequency domain* analysis.[6,7,8]. The simplest approach is simply to use numerical integration, although this method will work poorly when spectral overlap occurs. More sophisticated analysis methods include parametric curve-fitting routines, using various model functions (e.g. Lorentzian, Gaussian, Voigt, others[9,10]) and fitting algorithms (simplex, non-linear least squares, etc.). The most sophisticated method, and one which is becoming widely used, is the so-called linear combination model (LCModel) that fits the spectrum as a linear combination of the pure compound spectra known to exist in the spectrum.[11] The LCModel is particularly attractive for several reasons: (a) it makes full use of all the resonances in the molecule; (b) it is fully automated and user independent, including both baseline and phase correction; and (c) with appropriate calibration data, it can give absolute metabolite concentrations, and an estimate of the uncertainty (e.g. Cramer–Rao lower bounds (CRB)). An example of a LCModel analysis of a short echo time spectrum recorded at 3 T in the normal human brain is shown in Figure 3.4. It is generally recommended that metabolite concentrations should not be used for further analysis if the CRB exceeds 20%; it is also important to check the quality of the spectra (and the fitting results) even if the CRB <20%, since sometimes low uncertainty estimates occur even when spectra are of poor quality, or contain artifacts.

Chapter 3: Spectral analysis methods, quantitation, and common artifacts

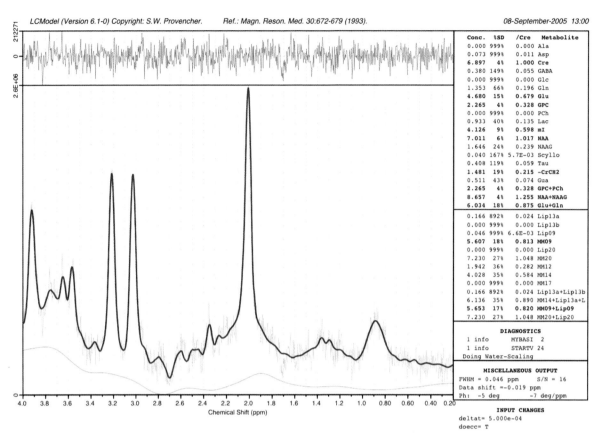

Figure 3.4. An example of the LCModel analysis method. The experimental data is fit as a linear combination of spectra of pure compounds recorded under the same experimental conditions as the in vivo spectrum. Automated baseline and phase-correction is performed, and an estimate of metabolite concentrations provided if a water reference signal is available for quantitation. In this example of a 2×2×2 cm PRESS spectrum recorded at 3.0 T from a normal control subject (*TR/TE*/number of averages = 2000/35/128), the difference between the original experimental data and the results of the curve fit is shown in the top trace. Metabolite concentrations highlighted in blue correspond to those with an estimated uncertainty of less than 20%.

A variety of *time-domain* fitting methods have been proposed, usually using parametric models based on the exponentially decaying oscillations.[12,13] One of the main perceived advantages of time-domain fitting is that it can avoid artifacts that may be induced by Fourier transformation of incomplete or partially corrupted time-domain data (e.g. missing or incorrect data points at the beginning of the FID). While this is true, since the properties of the Fourier transform are well known, it is usually possible to account for these problems using appropriate functions in frequency domain fitting as well, so that the choice of method may depend as much on numerical convenience, rather than any fundamental difference in approach.

Quantitation methods

The voltage detected by the spectrometer from a sample containing a metabolite (M) can be expressed as:

$$S_M = \beta \times [M] \times v \times f(T_1^M, T_2^M, TR, TE, B_1) \quad (3.8)$$

where v is the MRS voxel size (in cm^3), [M] is the millimolar concentration of the metabolite, f() is a pulse sequence modulation factor which will depend on the pulse sequence used, its repetition and echo times (*TR* and *TE*), the T_1 and T_2 relaxation times of M, and B_1 is the strength of the radiofrequency field within the MRS voxel. β is a scaling factor that can be expressed as:

$$\beta \propto NS \times G \times \omega_0 \times Q \times n\xi V_c/a \quad (3.9)$$

where *NS* is the number of scans (averages) performed, *G* is the receiver gain, ω_0 is the spectrometer operating frequency (e.g. 64 MHz for a 1.5 T magnet) and *Q*, n, ξ, V_c, and a are all factors related to the geometry and quality of the radiofrequency receiver coil.[14] The proportionality constant for Equation (3.9) is unknown, prohibiting the calculation of [M] without additional calibration measurements. However, Equations (3.8) and (3.9) do indicate how the signal (and hence sensitivity) can be enhanced, by using large numbers of scans, larger voxel size, high field systems, sensitive RF coils (high B_1), and maximizing the modulation factor f (by choosing optimum TR and flip angles, and short TE). Sensitivity is usually largely independent of the receiver gain, G, since this usually increases both signal and noise equally.

In order to calculate metabolite concentrations, a reference signal must be acquired, ideally under identical conditions to those used to record the NAA signal, so that all proportionality factors in Equations (3.1) and (3.2) are identical, allowing the ratio equation to be written:

$$[M] = [\text{Reference}] \times \frac{S_M}{S_{reference}} \quad (3.10)$$

where we assume from hereon that S_M and $S_{reference}$ have already been corrected for possible differences in T_1 and T_2 relation times according to standard Equations (3.7).

The choice of the reference compound is of key importance for the accuracy of the quantitation procedure. The most commonly used method is to select some other compound in the brain spectrum (e.g. most often creatine (Cr)) and report ratios of M/Cr. This is an example of an *internal* intensity reference, namely one that comes from the same voxel as the signal to be measured. Internal references have the advantage that many of the factors in Equations (3.8) and (3.9) are virtually identical for both signals (e.g. volume of tissue, B_1 (and B_0) field strength, flip angle and other pulse sequence-related factors), and so are insensitive to systematic errors associated with these parameters. Using a reference signal from the same spectrum as the compound to be measured also has the advantage that no additional scan time is required. However, while initially it was hoped that Cr levels might be relatively constant throughout the brain and invariant with pathology, subsequent studies have shown substantial regional [7,15] and pathology-related changes in Cr,[16] so that in general it is somewhat unsafe to infer changes in, for instance, NAA from the measurement of only the NAA/Cr ratio. However, ratios may be useful if the constituent metabolites change in different directions; for instance, a common neuropathological pattern is for NAA to decrease and Cho to increase; in this situation, larger changes will be seen in the NAA/Cho ratio, than in either metabolite alone.

An alternative, and widely used, internal intensity reference is the unsuppressed tissue water signal, [13,17,18] which can be easily and quickly (at least for single-voxel spectroscopy) recorded by turning off the water suppression pulses. Brain water content is relatively well known, and pathology-associated changes are relatively small. Furthermore, it is possible to estimate voxel water content from appropriate MRI sequences.[19] Finally, the unsuppressed water signal may also be helpful for phase- and eddy current-correction of the water-suppressed spectrum.[20,21] For all these reasons, quantitation of single-voxel spectra using the internal tissue water signal has become a popular technique over the last few years. Studies have demonstrated that this is a reliable method, e.g. for multi-site trials of brain spectroscopy.[22]

However, there may be situations where water referencing is not optimal; for instance, where brain water content is variable or not well known (e.g. in neonatal studies, or pathologies involving major changes in brain water content). Also, water referencing may not be convenient in certain MR spectroscopic imaging (MRSI) studies. For instance, it may be prohibitively time-consuming to record both water suppressed and non-suppressed MRSI data sets. In these instances, other approaches to quantitation should be considered. These approaches can be considered to fall into two classes, either *external* references, or the *phantom replacement* technique. External references involve the recording of a reference signal from a region outside of the primary region of interest. Often, a vial of known concentration compound (or simply water) is placed next to the head, and the signal from this region measured, either before or after the brain spectrum is recorded. While external standards have the advantage that the reference concentration is exactly known, their use is complicated by the fact that some of the factors in Equation (3.8) may no longer be constant. The phantom T_1 and T_2 relaxation times will almost certainly be different from those in vivo, while the biggest source of error is likely differences in B_1 field strength (and probably also RF pulse flip angles) due to inhomogeneities in the RF coils used for reception and/or transmission. The external standard may also induce magnetic susceptibility effects that degrade the B_0 field homogeneity, again leading to systematic

Table 3.1. Summary of relative advantages and disadvantages of commonly used quantitation techniques for proton MR spectroscopy of the human brain.

Internal	Advantages	Disadvantages
Creatine	Simple, no extra scan time, no B_0 or B_1 errors	Pathological and regional variations common
Water	Minimal extra scan time, simple, no B_0 or B_1 errors	May change up to ≈20% in pathological conditions
External		
Contralateral hemisphere	Simple, no coil loading errors	Extra scan time required, sensitive to B_0 or B_1 errors, cannot be used in global/diffuse diseases, or in midline structures
External Standard	Reference concentration exactly known, no coil loading errors	Extra scan time required, sensitive to B_0 or B_1 errors, may cause susceptibility effects, cannot be used with all pulse sequences
Phantom Replacement	Reference concentration exactly known, no extra *patient* scan time required	Coil loading correction required, requires stable system

errors because of suboptimal spectral quality. Also, care has to be taken to ensure that the spatial localization sequence interrogates the same volume of tissue and reference sample. Sometimes a spectrum from a different (e.g. contralateral) brain region can be used as a reference, e.g. in patients with focal brain disease, if the contralateral hemisphere is known to be normal. This approach avoids the susceptibility problems associated with the placement of an external vial, but shares the other potential problems of external references. Also, in many patients, it may be unsafe to assume that metabolite concentrations in apparently uninvolved brain regions (i.e. with normal brain MRI) are in fact the same as in the normal populations.[23]

The phantom replacement technique can be regarded as a hybrid of external and internal intensity references. The basic idea is to record a spectrum from a patient, remove the patient from the magnet, insert a standard sample, and then record its spectrum using as closely matched experimental conditions as possible. The method has the advantage that the reference concentration is exactly known, and most of the factors in Equations (3.8) and (3.9) are also known and can be controlled for. One major difference between the two acquisitions, however, is that the RF coil quality factor (Q) will almost certainly be different, because the electrical properties (impedance) of the phantom will be different from the human head. Fortunately, this loading factor (F) can be readily determined from the RF power or voltage calibration required to obtain a 90° pulse, and applied as a correction factor:

$$[M] = [\text{Reference}] \times F \times \frac{S_M}{S_{reference}} \quad (3.11)$$

It may also sometimes be necessary to include correction factors if the brain and reference scans are collected with different numbers of scans and/or different receiver gains. The phantom replacement technique is convenient to use for MRSI scans, since it does not require any additional *patient* scan time, and, in fact, on stable clinical scanners, the reference scan may change little from one day to the next, so that it does not need to be recorded for every patient. As with the external standard method, correction factors have to be applied to account for differences in T_1 and T_2 relaxation times between the phantom and in vivo. Potential errors can occur due to B_1 inhomogeneity if the brain and reference scans are from different locations. The method therefore works best with highly homogeneous transmit and/or receive coils (such as quadrature birdcage head coils found on most 1.5 T scanners), otherwise care has to be taken to attempt to match the reference scan location as close as possible to those in the brain. Alternatively, approaches for correcting MRSI scans for B_1 inhomogeneity prior to quantification (e.g. for use with phased-array receiver coils) may be required.

Table 3.1 contains a summary of the different quantitation methods available for MR spectroscopy and spectroscopic imaging, and lists their relative advantages and disadvantages.

Units and tissue compartmentalization

There are several different units which can be used to express in vivo tissue metabolite concentrations. From Equation (3.9), it can be seen that the in vivo concentration will be expressed in the same units as the

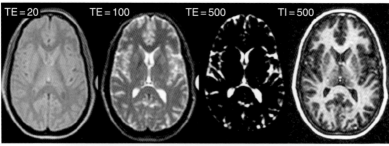
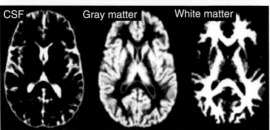

Figure 3.5. An example of multi-spectral brain and CSF segmentation for use with MRSI. Top row: rapid fast-spin echo sequences are acquired with different degrees of T_2-weighting (proton density (TE 20 msec), T_2-weighting (TE 100 msec), CSF only (TE 500 msec), and a T_1-weighted (TI 500, TE 20 msec). Through the use of region of interest measurements of individual tissue types and subsequent Eigenimage filtering,[36] these images can be processed to generate maps (bottom row) corresponding to pure CSF, gray matter, and white matter. This information can be used to correct metabolite concentrations determined by MRS(I) for partial volume effects.

reference concentration, since the ratio of signal intensities is unitless. Therefore, for instance, in the external reference approach, if the external reference concentration is measured in millimolar (i.e. millimoles solute per liter solution) then the in vivo concentration will be returned in the same units (equivalent to millimoles per liter brain volume). Traditionally, metabolite concentrations in tissue determined by conventional biochemical techniques are more often expressed in units such as millimoles per kg tissue wet or dry weight. To convert millimolar to millimoles per kg wet weight, it is necessary to divide by the tissue density (1.05 kg/liter for normal brain). To convert to millimoles per dry weight, it is necessary to know the wet/dry weight ratio. Note that these approaches assume that the entire volume of sample is occupied by solid brain tissue; in practice, this may well not be the case, since the large voxel sizes used for in vivo MRS often contain appreciable cerebrospinal fluid (CSF) contamination.

CSF normally contains much lower levels of metabolites than brain; therefore CSF contamination (without appropriate correction methods) will lead to underestimation of brain metabolite concentrations. Fortunately, several methods now exist for estimating voxel CSF content and applying appropriate correction factors. One method makes use of measuring the voxel water signal as a function of multiple different echo times.[24] Since CSF has a much longer T_2 than brain water, bi-exponential fitting of the echo signal versus time can estimate the relative fractions of brain and CSF water. This information can then be used to estimate true tissue metabolite concentrations. Alternatively, MR imaging-based segmentation methods can also be used to estimate brain and CSF volumes within the MRS voxel. One particularly simple approach is to use long echo time (for instance, $TE = 500$ msec) fast-spin echo (FSE) MRI, which essentially only contains signal from the long T_2 CSF, the brain signal having decayed to the noise level at this TE.[25] More sophisticated approaches use multiple FSE scans with different contrast in combination with appropriate multi-spectral post-processing methods, in order to estimate not only CSF content, but also fractional gray and white matter content.[26] An example of brain and CSF segmentation using rapid FSE imaging is shown in Figure 3.5.

Common artifacts (and other problems) in MRS

MRS is very sensitive to the technique used, and is commonly degraded by artifacts if either not performed correctly, or in conditions unfavorable for MRS. Some common artifacts and other problems are discussed below.

Inadequate field homogeneity

Probably the single most important factor that determines the quality of a spectrum is the homogeneity of the main magnetic field, B_0, over the volume of interest (voxel). Poor field homogeneity causes resonances to overlap, lowers the signal-to-noise ratio, and

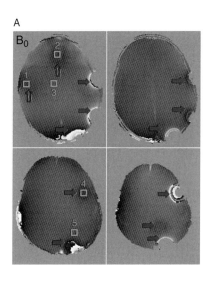
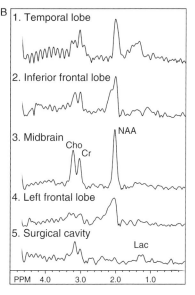
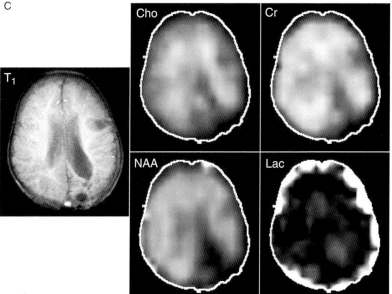

Figure 3.6. Effects of metal artifacts and air–tissue interfaces on field homogeneity in the brain, in a 2-year-old female with prior resection for primitive neuroectodermal tumor (PNET). (A) Maps of B_0 field show inhomogeneities above the sphenoid sinus and auditory canals (open arrows), as well as major local field inhomogeneities caused by surgical staples applied during craniotomy (closed arrows). (B) Multi-voxel spectra (MRSI, TR/TE 2300/280 msec) have poor quality in these regions compared to others (e.g. mid-brain). The left parietal lesion has low levels of all metabolites, and a slight elevation of lactate, consistent with resection cavity and no evidence of recurrent tumor. Metabolic images and localizer T_1-weighted images are shown in (C).

has other effects such as worsening water suppression. Ideally, the width of singlet peaks such as NAA or Cr (without linebroadening) should be on the order of 0.05 ppm (e.g. less than ~5 Hz at 1.5 T, or ~10 Hz at 3 T [note to convert ppm to Hz, use Hz = ppm × 42.6 × T]). Linewidths broader than this may lead to serious degradation of spectral quality, leading to uninterpretable data.

Poor linewidths can be the result of many factors, including shimming failure, the presence of paramagnetic materials (e.g. dental hardware (braces), shunts, post-surgical staples, hemorrhage, or calcifications), or simply due to tissue–air magnetic susceptibility differences. Known regions of bad field inhomogeneity in normal subjects include the mesial anterior temporal and inferior frontal lobes, and superior to the auditory canal, because of their proximity to airspaces within the head (Figure 3.6). Field homogeneity is also altered by the presence of iron in certain brain structures, mainly the deep gray matter nuclei, such as the caudate nucleus, globus pallidus and putamen, and substantia nigra – all these structures typically give broader linewidths on MRS than other brain regions (see Chapter 4).

As noted above, the linewidth is expected to increase (approximately linearly) with increasing field strength. The reason for this is believed to be due to increasing *microscopic* field inhomogeneity, most likely arising from the "BOLD" effect (paramagnetic deoxyhemoglobin) within the microvascular tissue. The observed linewidth can be expressed as the sum of three terms:

$$\Delta v_{1/2}(Hz) = \frac{1}{\pi T_2} + \sum_v \gamma \Delta B_0(v_{micro}) + \sum_v \gamma \Delta B_0(v_{macro}) \quad (3.12)$$

where $1/\pi T_2$ is considered the "natural" (minimum) linewidth that would occur in the absence of any field homogeneity, the v_{micro} term is the microscopic tissue inhomogeneity, and v_{macro} is the "global" magnet field homogeneity over the volume of the voxel (v) – this term should be near zero if successful shimming is applied.

In a clinical MRS exam, inadequate field homogeneity can occur quite frequently for the multiple reasons described above. Sometimes, a shim algorithm failure can be overcome by simply repeating the procedure, or by moving the region-of-interest slightly and re-shimming, or by using high-order (as opposed to linear) shims. In other situations, e.g. when lesions are in very unfavorable locations or when paramagnetic substances are present, MRS may simply not be feasible.

Inadequate water suppression

Since water in biological tissues such as the brain is present in very large quantities (e.g. a proton concentration of 110 M in CSF, ranging from approximately 71.6 to 86.6 M in white and gray matter, respectively) and is a much larger signal than the small metabolite signals (e.g. less than 10 mM), water suppression is important for avoiding dynamic range problems and other artifacts. Common water suppression methods are described in Chapter 2. Poor water suppression itself is not a problem unless the "wings" of the water peak (or water in regions of very poor field homogeneity) overlap with peaks of interest in the spectrum.

Poor water suppression may be the result of bad field homogeneity, or the result of a failed optimization of water suppression pulse flip angles. Repeating the calibration procedures, or slightly changing the voxel position, may result in better water suppression. Spurious water signals from outside the volume-of-interest (resulting from imperfect spatial localization) may cause problems, particularly if in regions of poor field homogeneity – these can often be suppressed using outer-volume suppression (OVS) saturation bands.

Inadequate lipid suppression

As with water, lipid signals (at least in the scalp) are 2–3 orders of magnitude larger than metabolites. Failure to suppress lipids leads to inability to detect lactate, and may also interfere with the NAA signal to 2.02 ppm. Lipid suppression methods are described in Chapter 2. Lipid contamination is often a problem if head motion is present (particularly for MRSI), resulting in lipid signals smeared across the brain. Lipid suppression may also be improved using carefully placed OVS bands.

Effects of head motion

Just as in MRI, head motion may have a disastrous effect on MRS, with loss of resolution and SNR, lipid and water suppression problems, and erroneous voxel location. It is a good idea to repeat localizer images at the end of the experiment if there is suspicion that head motion has occurred during MRS acquisition.

What's the vertical scale?

One of the problems with MRS is that spectra are, for the most part, displayed without any Y-axis (vertical) scale. Many software packages (particularly those commercially supplied) offer little control over the vertical scale, and usually automatically adjust it so that the largest peak fills the viewable area. This is good for visualizing the spectrum, but may be misleading when comparing different voxels within the same patient, particularly when all metabolites may be simultaneously increased or decreased. Ideally, spectra from different regions of interest in the same patient should always be displayed with the same vertical scaling factor to avoid this problem. A simple way to check for this is to make sure the noise levels look similar in different spectra; if they are not, most likely the vertical scaling factor has changed.

Incorrect voxel (or MRSI slice) placement

An inherent limitation of nearly all MRS(I) methods to date is that they have limited brain coverage, limited to a few voxels or slices. Thus the clinical information obtained is critically dependent on the correct choice of voxel location a priori. This is also an issue in longitudinal studies where spectra may be followed over time; evidently, scans will only be readily interpreted if they are recorded from the same location and with the same technical parameters. In practice, this requires careful technologist supervision and tight control over MRS protocols, which unfortunately does not always happen in real life. In the future, automated voxel and/or slice placement routines (based on real-time registration of localizer images to standard anatomical templates) may become increasingly used for studies of this type.

Incorrect choice of protocol

The choice of protocol for a particular study (e.g. short or long echo time, single- or multi-voxel) may also be problematic. Some recommended protocols, and the criteria for choosing them, are given in Chapter 2. Evidently, short echo time spectra should be used when compounds such as mI or Glx are required, while multi-voxel methods are important for pathologies such as brain tumors, which are often heterogeneous. Technologists should be carefully supervised to ensure that the protocol selected is appropriate for the clinical issue to be addressed.

Lesion too small, unfavorable location

It should be realized that not all lesions or brain regions are amenable to study by MRS. A lesion needs to be probably at least 1.0–1.5 cm in diameter to be evaluated by MRS, at least using conventional methodology at 1.5 T. In addition, several regions of the brain (see above) are unfavorable for MRS because of magnetic susceptibility effects from nearby air spaces, or other artifacts. In these instances, MRS cannot be used.

Out-of-voxel magnetization

Ideally, the slice selective pulses used in localization sequences such as STEAM or PRESS sequence would have perfect, "box-car" slice profiles, and all magnetization from outside the voxel would be dephased by very strong crusher gradients. In practice, slice profiles are not perfect and there will always be a small amount of residual magnetization originating from "outside the voxel". This can be a significant problem if these regions contain strong signals (e.g. unsuppressed water, or lipids). There are several solutions to this problem; one is to use OVS pulses (which in fact should be standard for most single voxel or MRSI protocols), another is to ensure proper crushing is used in conjunction with an optimum ordering of slice select gradients,[27] and another is to use the so-called "spectroscopic imaging acquisition mode" ("SIAM").[28]

The presence of out-of-voxel magnetization should be considered when evaluating the presence of artifacts in localized MR spectra. An artifact very commonly seen near the water peak is a broad line (i.e. water in a region of poor field homogeneity, not suppressed by the water suppression pulses because it is too far from resonance) with rapidly changing phase. This arises from a spurious echo from two (usually first and third) of the three slice selective pulses; it refocuses later than the principle echo, and hence has a different phase. An example of this type of artifact is shown in Figure 3.7.

Chemical shift displacement effects

When a slice-selective pulse is applied, the position of the slice depends on the frequency offset (from resonance, $\delta - \delta_{metab}$) and the gradient strength (dB_0/dz). The thickness of the slice will depend on the ratio of the RF pulse bandwidth (BW) to the gradient strength:

$$Position(mm) = \frac{(\delta - \delta_{metab})}{\gamma dB_0/dz} \quad (3.13)$$

$$Thickness(mm) = \frac{BW}{\gamma dB_0/dz} \quad (3.14)$$

Since δ_{metab} will be different for each metabolite, this means that the voxel position of each metabolite will be slightly different, depending on the chemical shift and the strength of the gradient used (the higher the bandwidth, and stronger the gradient, the smaller the chemical shift dispersion). Since chemical shift dispersion (measured in Hz) increases linearly with increasing magnetic field strength, this is more of a problem at higher fields than lower. Figure 3.8 shows chemical shift dispersion effects for a 1 kHz bandwidth slice selective pulse at 3 T.

There are several solution to this problem; one is to use the highest possible bandwidth RF pulses (including frequency swept pulses).[29] However, these pulses tend to require high RF power levels, and may exceed specific absorption rate (SAR) requirements for scanning in humans, or encounter hardware limitations. Another solution is to use high bandwidth saturation pulses (which typically have higher bandwidths and lower SAR than either excitation or refocusing pulses) to sharpen the edges of the STEAM or PRESS excitation box (Figure 3.9).[30,31,32]

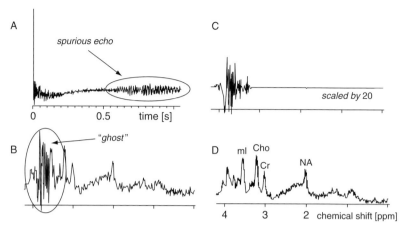

Figure 3.7. Effects of spurious echoes on single-voxel localized MRS. Insufficient amplitude of gradient crusher pulses in combination with local B_0 inhomogeneities can lead to the refocusing of unwanted echoes (e.g. 2 pulse echo in a PRESS sequence). (A) The FID from a PRESS acquisition (TE 20 ms, TR 3 s). The encircled part of the FID originates from an unwanted echo. (B) The typical appearance of spurious echoes, often called ghosts, in the spectrum. (C) The particular phase evolution proved the spurious signal to arise from a two-pulse echo of the initial 90° and last 180° pulse. In the current case, elimination of the ghosting artifact can easily be accomplished by zeroing the latter half of the FID. The resulting spectrum is plotted in (D). Reproduced with permission from [37].

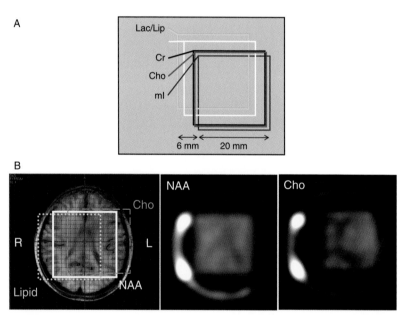

Figure 3.8. Chemical shift displacement effects. (A) Example of calculated chemical shift displacements at 3 T (1 ppm = 128 Hz) from *myo*-inositol (ml) to lactate/lipids for a 20 mm voxel and a 1 kHz RF pulse bandwidth. There is a shift of 6 mm between the ml voxel and the lac/lipid voxel. (B) Experimental demonstration of chemical shift displacements for lipid, NAA and Cho in a 2D-PRESS-MRSI experiment. Note that the lipid voxel is shifted to the right, while the Cho voxel is shifted to the left. A higher bandwidth pulse is used in the anterior-posterior direction, so the displacement is smaller in this direction. Note also that the very intense lipid signals are seen in the NAA and Cho images, particularly on the right hand side (figure courtesy of Dr James Murdoch).

Chemical shift dispersion effects can also perturb modulation patterns which may be required for assigning molecules such as lactate, or editing for GABA. For instance, the lactate doublet at 1.3 ppm is expected to be inverted at an echo time of 140 msec; however, this only occurs if both the 1.3 ppm resonance and the coupled 4.1 ppm methine both experience 180° pulses. This will not be the case in all regions of the localized voxel unless very high bandwidth slice selective pulses are used. Depending on the relative displacement of the 1.3 and 4.1 ppm voxels, the lactate signal may be either positive, negative, or even zero (Figure 3.10). One solution to this problem is to use high-bandwidth refocusing pulses; if these are not available, another approach is to use inner-volume suppression (IVS) to remove spatial regions of the localized volume where undesired modulation patterns are located (Figure 3.10).

Chapter 3: Spectral analysis methods, quantitation, and common artifacts

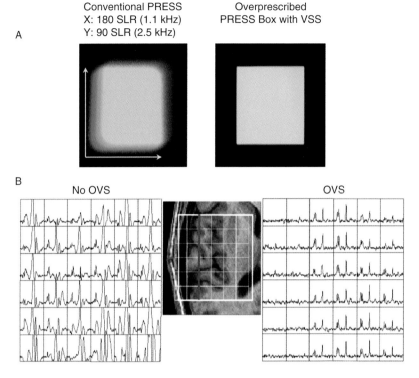

Figure 3.9. (A) A phantom PRESS-MRSI experiment (with frequency offsets of +100, 0 and −100 Hz) with and without very high bandwidth slice selective outer volume suppression (OVS) pulses ("VSS" pulses). (B) In vivo 2D-PRESS-MRSI data with and without pulses used for lipid suppression. Without OVS, lipid contamination renders the brain spectra uninterpretable (figure courtesy of Dr Ralph Hurd).

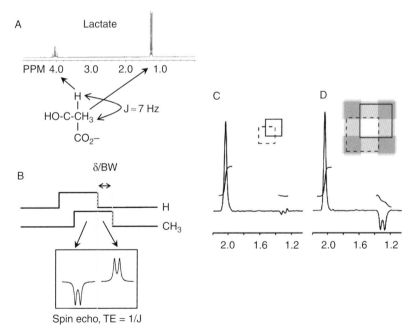

Figure 3.10. Effects of chemical shift displacement on the modulation of the lactate signal. (A) Spectrum and structure of the lactate molecule. There is a 7 Hz coupling between the methyl (1.3 ppm) and methine (4.1 ppm) resonances. (B) Part (1−δ/BW) of the methyl doublet is inverted, while part is positive. (C) NAA and lactate phantom: under typical conditions (3 T, PRESS, TE 144, 900 Hz BW slice selective pulses) extensive signal cancellation occurs (δ/BW ≈ 0.4). (D) Using inner-volume saturation (four 4.3 kHz, 2 msec, saturation pulses) ≈ 98% signal detection efficiency of lactate is achieved.

Can MRS be performed after administration of Gd contrast?

This is a commonly asked question. While there have only been a few studies on this subject, the general consensus appears to be that Gd-chelates have little or no detectable effect on the MRS signal,[33,34,35] and therefore it is quite possible to perform MRS post-Gd.

In normal brain, Gd stays within the vasculature, so its only effect on metabolites (which are believed to

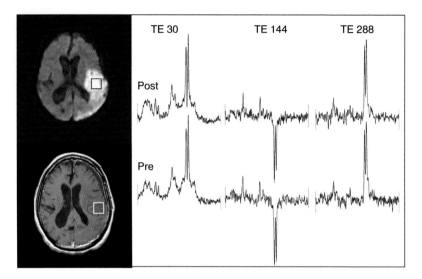

Figure 3.11. Spectra obtained before and after the intravenous administration of contrast medium of an enhancing lesion (sub-acute infarct) at TE 30, 144, and 288. Spectra are dominated by Lac, as well other signals from Glu, Cr, and Cho. Within the SNR of these scans, post-contrast spectra are not significantly different relative to pre-contrast (figure courtesy of Dr Ralph Hurd and Dr Orest Boyko).

be of intracellular origin) is to cause a slight line broadening at typical, steady-state concentrations. Even in enhancing brain lesions, Gd is most likely in the extravascular, extracellular space, and does not appear to cause significant changes in metabolite T_1 or T_2 relaxation, since most metabolites are within the intracellular space. Figure 3.11 shows an example of MRS performed before and after Gd administration in an enhancing brain lesion. For pre- and post-Gd comparisons, it may be necessary to readjust the shim values post-Gd to account for any susceptibility differences.

Conclusions

In summary, in considering experimental design for both research studies as well as clinical MRS exams, there should be an equal (or perhaps even greater) degree of thought applied to the data analysis methods to be used as to the acquisition methods. There are a variety of approaches to both spectral fitting and quantitation; each has their own sets of advantages and disadvantages. The methods ultimately chosen will depend on the question to be asked, the experimental conditions used to record the data, and the analysis software available to the investigator.

References

[1] Ordidge RJ, Cresshull ID. The correction of transient B0 field shifts following the application of pulsed gradients by phase correction in the time domain. *J Magn Reson Imaging* 1986; **69**: 151–5.

[2] de Graaf AA, van Dijk JE, Bovee WM. QUALITY: quantification improvement by converting lineshapes to the Lorentzian type. *Magn Reson Med* 1990; **13**: 343–57.

[3] Marion D, Ikura M, Bax A. Improved solvent suppression in one- and two-dimensional NMR spectra by convolution of time-domain data. *J Magn Reson* 1989; **84**: 425–30.

[4] Lowe MJ, Sorenson JA. Spatially filtering functional magnetic resonance imaging data. *Magn Reson Med* 1997; **37**: 723–9.

[5] Derby K, Hawryszko C, Tropp J. Baseline deconvolution, phase correction, and signal quantification in Fourier localized spectroscopic imaging. *Magn Reson Med* 1989; **12**: 235–40.

[6] Soher BJ, Young K, Govindaraju V, Maudsley AA. Automated spectral analysis III: application to in vivo proton MR spectroscopy and spectroscopic imaging. *Magn Reson Med* 1998; **40**: 822–31.

[7] Soher BJ, van Zijl PC, Duyn JH, Barker PB. Quantitative proton MR spectroscopic imaging of the human brain. *Magn Reson Med* 1996; **35**: 356–63.

[8] Mierisova S, Ala-Korpela M. MR spectroscopy quantitation: a review of frequency domain methods. *NMR Biomed* 2001; **14**: 247–59.

[9] Marshall I, Bruce SD, Higinbotham J, MacLullich A, Wardlaw JM, Ferguson KJ, et al. Choice of spectroscopic lineshape model affects metabolite peak areas and area ratios. *Magn Reson Med* 2000; **44**: 646–9.

[10] Soher BJ, Maudsley AA. Evaluation of variable line-shape models and prior information in automated 1H

spectroscopic imaging analysis. *Magn Reson Med* 2004; **52**: 1246–54.

[11] Provencher SW. Estimation of metabolite concentrations from localized in vivo proton NMR spectra. *Magn Reson Med* 1993; **30**: 672–9.

[12] de Beer R, van den Boogaart A, van Ormondt D, Pijnappel WW, den Hollander JA, Marien AJ, *et al.* Application of time-domain fitting in the quantification of in vivo 1H spectroscopic imaging data sets. *NMR Biomed* 1992; **5**: 171–8.

[13] Barker PB, Soher BJ, Blackband SJ, Chatham JC, Mathews VP, Bryan RN. Quantitation of proton NMR spectra of the human brain using tissue water as an internal concentration reference. *NMR Biomed* 1993; **6**: 89–94.

[14] Freeman R. *A Handbook of Nuclear Magnetic Resonance*. Harlow, England: Longman Scientific and Technical, 1987.

[15] Hetherington HP, Mason GF, Pan JW, Ponder SL, Vaughan JT, Twieg DB, *et al.* Evaluation of cerebral gray and white matter metabolite differences by spectroscopic imaging at 4.1 T. *Magn Reson Med* 1994; **32**: 565–71.

[16] Stockler S, Holzbach U, Hanefeld F, Marquardt I, Helms G, Requart M, *et al.* Creatine deficiency in the brain: a new, treatable inborn error of metabolism. *Pediatr Res* 1994; **36**: 409–13.

[17] Thulborn KR, Ackerman JJH. Absolute molar concentrations by NMR in inhomogeneous B_1. A scheme for analysis of in vivo metabolites. *J Magn Reson* 1983; **55**: 357–71.

[18] Christiansen P, Henriksen O, Stubgaard M, Gideon P, Larsson HB. In vivo quantification of brain metabolites by 1H-MRS using water as an internal standard. *Magn Reson Imaging* 1993; **11**: 107–18.

[19] Alger JR, Symko SC, Bizzi A, Posse S, DesPres DJ, Armstrong MR. Absolute quantitation of short TE brain 1H-MR spectra and spectroscopic imaging data. *J Comput Ass Topogr* 1993; **17**: 191–9.

[20] Ordidge RJ, Cresshull ID. The correction of transient B0 field shifts following the application of pulsed gradients by phase correction in the time domain. *J Magn Reson* 1986; **69**: 151–5.

[21] Webb PG, Sailasuta N, Kohler SJ, Raidy T, Moats RA, Hurd RE. Automated single-voxel proton MRS: technical development and multisite verification. *Magn Reson Med* 1994; **31**: 365–73.

[22] Soher BJ, Hurd RE, Sailasuta N, Barker PB. Quantitation of automated single-voxel proton MRS using cerebral water as an internal reference. *Magn Reson Med* 1996; **36**: 335–9.

[23] Mathews VP, Barker PB, Blackband SJ, Chatham JC, Bryan RN. Cerebral metabolites in patients with acute and subacute strokes: concentrations determined by quantitative proton MR spectroscopy. *Am J Radiol* 1995; **165**: 633–8.

[24] Ernst T, Kreis R, Ross B. Absolute quantitation of water and metabolites in the human brain. I. Compartments and water. *J Magn Reson B* 1993; **102**: 1–8.

[25] Horská A, Calhoun VD, Bradshaw DH, Barker PB. Rapid method for correction of CSF partial volume in quantitative proton MR spectroscopic imaging. *Magn Reson Med* 2002; **48**: 555–8.

[26] Horska A, Jacobs MA, Calhoun V, Arslanoglu A, Barker PB. A fast method for image segmentation: application to quantitative proton MRSI at 3 Tesla. *ISMRM, 11th Scientific Meeting and Exhibition*; Toronto, Canada; 2003.

[27] Ernst T, Chang L. Elimination of artifacts in short echo time H MR spectroscopy of the frontal lobe. *Magn Reson Med* 1996; **36**: 462–8.

[28] Hurd R, Sailasuta N. Elimination of artifacts in short echo proton spectroscopy. *5th Annual Meeting of the International Society of Magnetic Resonance in Medicine*; Vancouver, BC, Canada; 1997. p. 1453.

[29] Murdoch JM. Still Iterating … and iterating … to solve pulse design problems. *International Society of Magnetic Resonance in Medicine (ISMRM) 10th Annual Meeting*; Honolulu, Hawai'i; 2002. p. 923.

[30] Edden RA, Barker PB. Spatial effects in the detection of gamma-aminobutyric acid: improved sensitivity at high fields using inner volume saturation. *Magn Reson Med* 2007; **58**: 1276–82.

[31] Edden RA, Schar M, Hillis AE, Barker PB. Optimized detection of lactate at high fields using inner volume saturation. *Magn Reson Med* 2006; **56**: 912–7.

[32] Tran TK, Vigneron DB, Sailasuta N, Tropp J, Le Roux P, Kurhanewicz J, *et al.* Very selective suppression pulses for clinical MRSI studies of brain and prostate cancer. *Magn Reson Med* 2000; **43**: 23–33.

[33] Lin AP, Ross BD. Short-echo time proton MR spectroscopy in the presence of gadolinium. *J Comput Assist Tomogr* 2001; **25**: 705–12.

[34] Sijens PE, Oudkerk M, van Dijk P, Levendag PC, Vecht CJ. 1H MR spectroscopy monitoring of changes in choline peak area and line shape after Gd-contrast administration. *Magn Reson Imaging* 1998; **16**: 1273–80.

[35] Sijens PE, van den Bent MJ, Nowak PJ, van Dijk P, Oudkerk M. 1H chemical shift imaging reveals loss of brain tumor choline signal after administration of Gd-contrast. *Magn Reson Med* 1997; **37**: 222–5.

[36] Windham JP, Abd-Allah MA, Reimann DA, Froelich JW, Haggar AM. Eigenimage filtering in MR imaging. *J Comput Assist Tomogr* 1988; **12**: 1–9.

[37] Kreis R. Issues of spectral quality in clinical 1H-magnetic resonance spectroscopy and a gallery of artifacts. *NMR Biomed* 2004; **17**: 361–81.

Chapter 4

Normal regional variations: brain development and aging

Key points

- Substantial regional variations in proton brain spectra exist; differences between gray and white matter, anterior–posterior gradients, and differences between the supra- and infra-tentorial brain are common.

- Spectra change rapidly over the first few years of life; at birth, NAA is low, and choline and *myo*-inositol are high. By about 4 years of age, spectra from most regions have a more "adult-like" appearance.

- In normal development, only subtle age-related changes are found between the ages of 4 and 20 years.

- In normal aging, only subtle age-related changes are found. A recent meta-analysis indicated the most common findings are mildly increased choline and creatine in frontal brain regions of elderly subjects (> 68 years), and stable or slightly decreasing (parietal regions only) NAA.

Introduction

Interpretation of spectra from patients with neuropathology requires a knowledge of the normal regional and age-related spectral variations seen in the healthy brain. This is a difficult issue, since spectra are quite dependent on the technique used to record them (particularly choice of echo time, and field strength), and also show quite large regional and age-related (at least in young children) dependencies. However, while there still remain some gaps in the literature (e.g. detailed, regional studies in very young children), for the most part regional and age-related changes in brain spectra are now well-characterized. This chapter reviews what is known about regional metabolite variations, as well as metabolic changes associated with brain development, and aging.

Anatomical variations in brain spectra: young adults

The figures included in this chapter include three different acquisition protocols in normal, young adults. Figures 4.1–4.4 show the results of a 4-slice MRSI study performed at long echo time (TE 280 msec) at 1.5 T in one adult individual. Figure 4.5 shows the average results from a 1.5 T whole brain EPSI study (14 subjects, age range 27–48 (average 36)), recorded at TE 70 msec. Figures 4.6 and 4.7 are from a 3 T single-voxel study performed at short echo time (TE 35 msec) in 35 normal subjects (age range 20–41 years (average 31)).

At the level of the lateral ventricles and above (Figures 4.2–4.4), brain spectra show fairly characteristic patterns for gray and white matter,[1,2,3,4] although there are some anterior–posterior differences,[5,6] in particular with higher Cho in frontal brain regions. Depending on the quantification technique used (and if partial volume correction for CSF is applied or not), most studies have found that the Cho is higher in white matter than cortical gray matter, while Cr levels are lower in white matter than gray matter (Figure 4.5). NAA levels (if measured without CSF correction) are typically quite similar between gray and white matter, but since cortical gray matter voxels typically have more CSF contamination than white matter voxels, after CSF correction, gray matter NAA concentrations are usually higher than white matter. At the level of the third ventricle (Figure 4.1) and below, significant anatomical variations exist in brain spectra. High levels of Cho are found in the insular cortex, thalamus, and hypothalamus.[7] Occipital Cho in the region of the visual cortex is generally low. The pons has high levels of NAA and Cho, and low levels of Cr, perhaps due to its high density of fiber bundles.[8] Cerebellar levels of Cr and Cho are significantly higher than supratentorial values,[3,8] and temporal lobe has been reported to have lower NAA values.[9] Significant anterior–posterior differences have also been reported in normal hippocampal metabolite

Chapter 4: Normal regional variations: brain development and aging

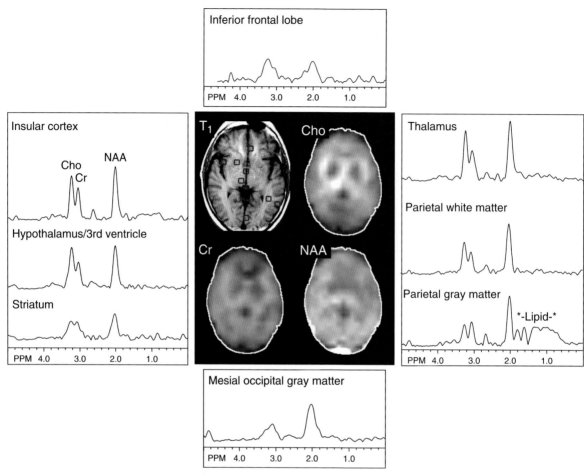

Figure 4.1. Normal volunteer, 49 years old, slice 1 of a multi-slice MRSI data set (TR 2300/TE 272 msec, nominal voxel size 0.8 cm^3) using spin-echo acquisition with OVS lipid suppression. Metabolic images show appreciable regional variations, in particular the Cho image which shows high levels in the insular cortex, thalamus, and hypothalamus. The striatum, because of its high iron content, shows decreased signal intensity (for all metabolites) resulting from decreased T_2 and T_2^*. Poor field inhomogeneity due to susceptibility effects is also apparent in the inferior frontal lobe and occipital lobe. Parietal gray matter shows a lower Cho signal than other regions; the voxel chosen here close to the scalp also shows mild lipid contamination.

concentrations, with lower NAA and higher Cho in the anterior regions of the hippocampus.[10,11] It appears that metabolites are highly symmetric between the left and right hemispheres in normal subjects, and that there are either no (or minimal) gender differences.[12,13]

The metabolic changes described above (for Cho, Cr and NAA) are beautifully depicted in the representation of whole-brain EPSI data from 14 subjects in Figure 4.5. The axial view clearly shows higher Cr and NAA (CSF corrected) in cortical gray matter, while also apparent on the axial view is the high Cho signal in the mesial frontal gray matter. The sagittal and coronal views show the high levels of Cr and Cho in the cerebellum, as well as the thalamus, hypothalamus, and basal ganglia.

The regional distribution of mI and Glx (only observable in short TE spectra) has not received as much attention as Cho, Cr, and NAA. One recent study was performed at 3 T using TE 35 msec single-voxel MRS found higher levels of Glx in gray matter than white matter (as would be expected, since the major constituent of Glx in normal brain is glutamate),[14] and with the highest level in the cerebellar vermis (Figures 4.6 and 4.7). Regional variations in mI were less clear, although there were trends for mI to be higher in gray matter than white matter, perhaps surprisingly since there have been some studies suggesting that mI is found predominantly in glial cells.

While brain metabolite concentrations will vary to some degree on the quantitation method used to

Chapter 4: Normal regional variations: brain development and aging

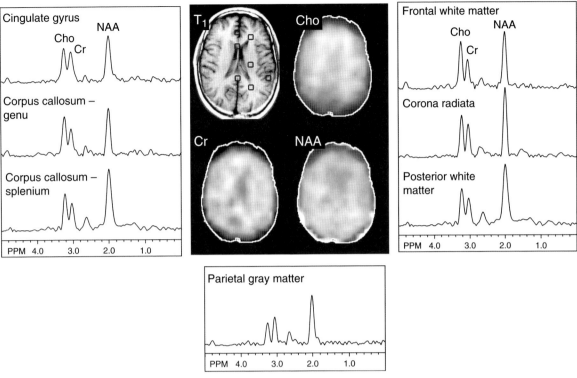

Figure 4.2. Same scan as Figure 4.1, slice 2 at the level of the lateral ventricles. Except for in the mesial frontal gray matter (cingulate gyrus), the Cho signal is larger than the Cr signal in white matter regions compared to gray. Cho is highest in the frontal lobe, decreasing posteriorly. NAA is relatively uniformly distributed.

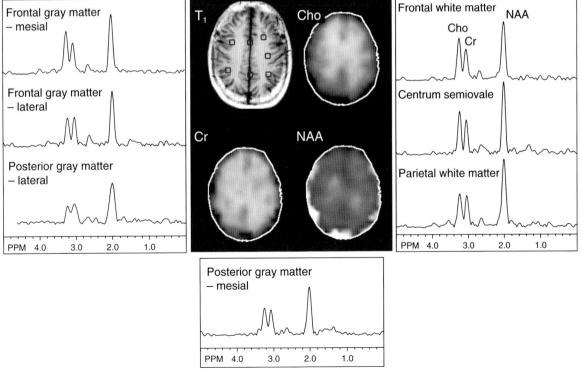

Figure 4.3. Same scan as Figure 4.1, slice 3 at the level of the centrum semiovale. Similar spectral patterns (gray/white matter differences) are seen as in slice 2.

Chapter 4: Normal regional variations: brain development and aging

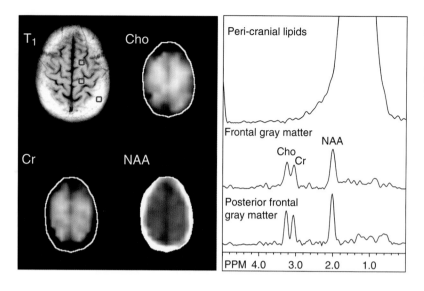

Figure 4.4. Same scan as Figure 4.1, slice 4 at the vertex. Large lipid signal is present in the scalp, even after the use of OVS for lipid suppression. Spectra from cortical gray matter show near equal Cho and Cr signals in this example, with approximately 2-to-1 ratio of NAA/Cr.

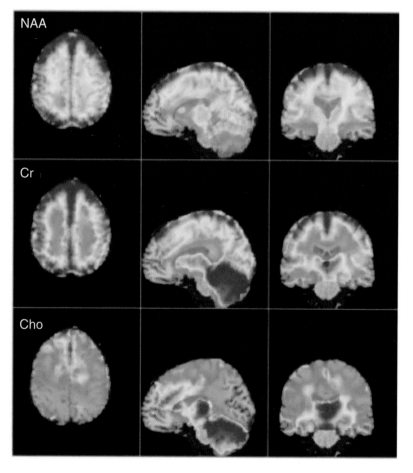

Figure 4.5. Average, CSF-corrected metabolic images of Cho, Cr, and NAA presented in axial, sagittal, and coronal views from a 1.5 T whole-brain EPSI study (14 subjects, age range 27–48 years (average 36)), recorded at TE 70 msec, reproduced with permission from [26]. Highest NAA levels are found in cortical gray matter (after CSF correction), with lower levels in white matter, and anterior temporal lobe, and cerebellum. Cr is highest in gray matter, cerebellum, and basal ganglia. Cho shows high levels in anterior mesial gray matter, basal ganglia, and cerebellum. Some brain regions (brain stem, anterior frontal lobe) are not included since spectra did not meet minimal acceptable quality in these regions.

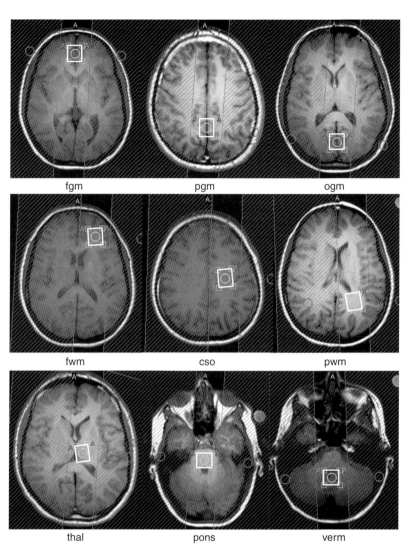

Figure 4.6. Voxel locations used for the spectra depicted in Figure 4.7.

estimate them (see Chapter 3), Table 4.1 may be of some value in determining the typical range of regional metabolite concentrations found in normal, young adult brain. It is cautioned that these values may vary somewhat from scanner to scanner, so each user is encouraged to collect their own control subjects using specific scanners and protocols for direct comparison to values in patients.

Age-related variations in brain spectra: early brain development in children

Several papers have been published on the changes that occur in proton spectra in the developing brain, and most of the results are in good agreement.[15,16,17,18] At birth, NAA is low, while Cho and mI are high, and over the first few years of life there is a gradual normalization towards adult values (Figure 4.8).[17] Similar patterns are seen for both gray and white matter, although regional developmental changes have yet to be studied in detail. The major metabolic changes clearly occur within the first year of life, with slower changes occurring thereafter, with full adult values not being reached until about 20 years of age.[19] Some regions may also develop more slowly than others, such as, for instance, frontal lobe white matter.[20] One study found a maximum NAA/Cho ratio in gray matter at about 10–12 years of age, after which it began decreasing slowly. This

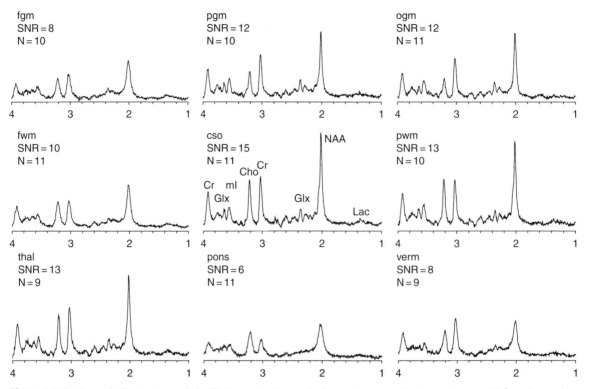

Figure 4.7. Single voxel, 3 T spectra recorded at TE 35 msec in 35 normal subjects (age range 20–41 years (average 31)) from the voxel locations indicated in Figure 4.6. Number of subjects per region and average NAA signal-to-noise ratios (SNR) are indicated.

is interesting since this is also the age at which blood flow and glucose supply to the brain is maximal, and may be related to dendrite development (up to age 10–12) followed by the onset of synaptic pruning thereafter.

Age-related variations in brain spectra: the elderly brain

In contrast to studies of developing brain, studies of normal aging by MRS are somewhat less concordant. Some groups find lower NAA with increasing age,[21,22] which may reflect neuronal loss, while others find no change in NAA with age.[4,23] In one study, NAA was only reduced in subjects who also had cerebral atrophy as identified by MRI.[24] Some groups have also found increased levels of Cr or Cho in older subjects, perhaps reflecting increased glial cell density.[4,23] One of the earliest studies to report this finding was a quantitative, multi-slice MRSI protocol which examined correlations between metabolite concentrations and age (range 8–74 years). Significant positive correlations were found between Cho concentrations and age in both the genu of the corpus callosum and the putamen ($P < 0.02$, Figure 4.9). Some regions showed trends for decreasing NAA (for instance, posterior white matter), but these did not reach statistical significance. The discrepancies between different studies could be due to technical factors in data collection and analysis, but probably also reflect the wide physiological variations of normal human aging, and hence depend on the study population. While more studies are required to definitively establish the spectroscopic characteristics of normal aging, it is apparent that the metabolic changes associated with normal aging are much more subtle than those associated with early brain development.

A recent meta-analysis of MRS studies of aging identified 18 studies (out of a total of 231 potentially relevant studies) that quantified metabolite concentrations as a function of age.[25] These 18 studies included data from 703 healthy subjects, who were split into younger (age range 4–56 years) and older groups (68–89 years). Consistent with the above discussion, it was found that between group differences

Chapter 4: Normal regional variations: brain development and aging

Table 4.1. Regional metabolite concentrations (mM + SD), with comparison to other authors' results. From [6], 2 T SV STEAM study, ages 18–39 years, no correction for T_1, T_2, or CSF partial volume ($\bar{N}=22$ for fwm, $N=20$ for CSO, $N=27$ for pwm, $N=12$ for fgm, $N=19$ for pgm, $N=14$ for ogm, $N=16$ for thal, and $N=8$ for verm). From [19], 2 T SV STEAM study, ages 18–39 years, no correction for T_1, T_2, or CSF partial volume ($N=45$ for pgm, $N=63$ for pwm). From [3], 2 T SV STEAM study, ages 21–32 years, no correction for T_1, same T_2 correction applied to all locations, CSF partial volume correction for gray matter only ($N=41$ for cso, $N=30$ for pgm, $N=7$ for thal, $N=5$ for pons; *they report Glu only, not Glx). From [27], 4 T SV STEAM study, ages 26±1 years, no correction for T_1 or T_2, CSF partial volume correction applied ($N=11$ for pgm and cso). From [28], 1.5 T SV PRESS study, ages 25–78 years, no correction for T_1, location-specific T_2 correction applied, no correction for CSF partial volume ($N=7$ for pons, $N=5$ for verm). For the 2 T STEAM studies, lack of T_1 correction results in an underestimate on the order of < 3%, while lack of T_2 correction results in an underestimate of about 12%.

Location	Source	Field	mI	tCho	tCr	Glx	tNAA
frontal WM	[14]	3 T	3.74 ± 0.65	2.03 ± 0.39	7.21 ± 1.06	8.39 ± 2.02	11.28 ± 1.14
frontal WM	[6]	2 T	3.8 ± 0.9	1.78 ± 0.41	5.7 ± 0.5	8.8 ± 3.1	9.6 ± 1.1
CSO	[14]	3 T	2.89 ± 0.41	1.65 ± 0.25	6.69 ± 0.37	6.77 ± 1.90	12.13 ± 0.78
CSO	[6]	2 T	3.1 ± 0.6	1.68 ± 0.27	5.7 ± 0.6	8.2 ± 2.2	10.6 ± 0.8
CSO	[3]	2 T	4.7 ± 1.0	1.8 ± 0.3	6.1 ± 0.8	*8.1 ± 1.5	11.2 ± 1.2
CSO	[27]	4 T	4.24 ± 0.49	1.83 ± 0.14	6.63 ± 0.31	7.12 ± 0.72	9.47 ± 0.42
par-occ. WM	[14]	3 T	3.3 ± 0.6	1.6 ± 0.24	6.14 ± 0.92	6.48 ± 1.58	10.97 ± 1.19
par-occ. WM	[6]	2 T	4.1 ± 0.8	1.64 ± 0.21	5.5 ± 0.8	8.2 ± 2.0	10.4 ± 0.9
par-occ. WM	[19]	2 T	5.2 ± 0.7	1.6 ± 0.2	4.9 ± 0.6	7.6 ± 1.7	8.2 ± 0.7
frontal GM	[14]	3 T	4.4 ± 0.92	1.78 ± 0.59	8.35 ± 1.22	11.77 ± 1.92	11.8 ± 1.42
frontal GM	[6]	2 T	4.3 ± 0.9	1.38 ± 0.17	6.4 ± 0.7	12.9 ± 1.7	8.4 ± 1.0
parietal GM	[14]	3 T	4.3 ± 0.79	1.35 ± 0.16	8.95 ± 1.13	12.2 ± 2.66	11.86 ± 0.92
parietal GM	[6]	2 T	4.3 ± 0.7	1.10 ± 0.14	6.5 ± 0.6	12.0 ± 1.8	8.7 ± 0.8
parietal GM	[19]	2 T	4.4 ± 0.6	1.2 ± 0.1	6.5 ± 0.5	12.9 ± 1.7	9.1 ± 0.7
parietal GM	[3]	2 T	6.2 ± 1.1	1.4 ± 0.3	8.2 ± 1.4	*12.5 ± 3.0	11.7 ± 2.2
parietal GM	[27]	4 T	5.20 ± 0.44	1.70 ± 0.14	7.87 ± 0.44	11.21 ± 0.89	10.57 ± 0.68
occipital GM	[14]	3 T	4.77 ± 0.64	1.02 ± 0.09	9.31 ± 0.86	10.86 ± 1.81	13.23 ± 1.13
occipital GM	[6]	2 T	4.1 ± 0.6	0.88 ± 0.10	6.9 ± 0.7	12.5 ± 1.6	10.6 ± 0.8
thalamus	[14]	3 T	3.53 ± 0.52	1.89 ± 0.21	9.22 ± 1.16	10.33 ± 1.40	13.56 ± 0.71
thalamus	[3]	2 T	4.9 ± 0.6	1.8 ± 0.2	7.5 ± 0.6		11.8 ± 0.7
thalamus	[6]	2 T	3.6 ± 0.9	1.81 ± 0.21	6.8 ± 0.7	9.8 ± 2.8	10.5 ± 0.7
pons	[14]	3 T	4.8 ± 1.45	2.61 ± 0.44	6.72 ± 1.47	9.86 ± 3.52	12.91 ± 1.99
pons	[3]	2 T	5.4 ± 0.6	2.9 ± 0.7	6.0 ± 0.3		12.5 ± 0.8
pons	[28]	1.5 T		3.1 ± 0.8	6.7 ± 2.0		12.6 ± 2.5
inferior vermis	[14]	3 T	4.22 ± 0.91	2.1 ± 0.37	11.95 ± 1.15	12.89 ± 2.99	11.08 ± 1.02
superior vermis	[6]	2 T	5.6 ± 1.1	2.19 ± 0.30	9.0 ± 0.9	12.9 ± 2.6	8.7 ± 0.9
superior vermis	[28]	1.5 T		3.0 ± 1.0	12.8 ± 2.9		9.8 ± 2.7

Abbreviations: CSF, cerebrospinal fluid; CSO, centrum semiovale; fgm, frontal gray matter; Glu, glutamate; Glx, glutamate and glutamine; GM, gray matter; mI, myo-inositol; ogm, occipital gray matter; par-occ, WM parieto-occipital white matter; pgm, parietal gray matter; pwm, parietal white matter; SV, single voxel; tCho, total choline; tCr, total creatine; thal, thalamus; tNAA, total N-acetyl aspartate; verm, cerebellar vermis; WM, white matter.

were subtle, but indicated increases in Cho and Cr in frontal regions with age, and decreasing NAA only in the parietal region. The same study also reported that Glx and mI have been less studied, and generally report no change with aging.[25]

Conclusions

In summary, because of significant technique-, regional-, or age-related changes, it is advisable that spectroscopy studies for clinical or research purposes

Chapter 4: Normal regional variations: brain development and aging

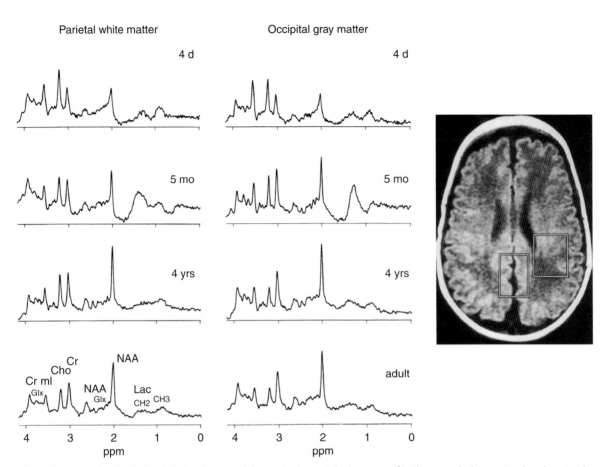

Figure 4.8. Age-related variations in MRS – the normal developing brain. At birth, spectra of both gray and white matter show low signals from NAA and elevated levels of Cho and mI. As the brain develops, NAA increases and Ch and mI decrease so that by about 4 years of age (in these locations) the spectra are essentially indistinguishable from those in young adults. Reproduced with permission from [17]. Spectra recorded at 1.5 T.

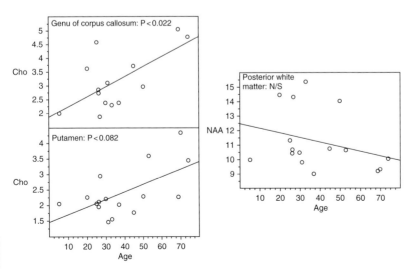

Figure 4.9. Age-related variations in MRS – the aging brain. Significant increases of Cho with age in genu of corpus callosum and putamen, and a trend for decreased NAA in posterior white matter. Adapted with permission from [4].

should always use age- and anatomically matched spectra from control subjects for comparison. These spectra should be recorded with identical techniques and on the same scanner as those performed in patients. In addition, spectroscopy scans of focal brain lesions are often much easier to interpret if spectra from presumed normal brain in the contralateral hemisphere are available for comparison. Finally, the interpretation of spectra from very young children (term and preterm neonates, and children less than 1–2 years of age) are particularly challenging because of the rapid changes in brain metabolism that occur in these age ranges.

References

[1] Hetherington HP, Mason GF, Pan JW, Ponder SL, Vaughan JT, Twieg DB, et al. Evaluation of cerebral gray and white matter metabolite differences by spectroscopic imaging at 4.1 T. *Magn Reson Med* 1994; **32**: 565–71.

[2] Kreis R, Ernst T, Ross BD. Absolute quantitation of water and metabolites in the human brain. II. Metabolite concentrations. *J Magn Reson Ser B* 1993; **102**: 9–19.

[3] Michaelis T, Merboldt KD, Bruhn H, Hanicke W, Frahm J. Absolute concentrations of metabolites in the adult human brain in vivo: quantification of localized proton MR spectra. *Radiology* 1993; **187**: 219–27.

[4] Soher BJ, van Zijl PC, Duyn JH, Barker PB. Quantitative proton MR spectroscopic imaging of the human brain. *Magn Reson Med* 1996; **35**: 356–63.

[5] Degaonkar MN, Pomper MG, Barker PB. Quantitative proton magnetic resonance spectroscopic imaging: regional variations in the corpus callosum and cortical gray matter. *J Magn Reson Imaging* 2005; **22**: 175–9.

[6] Pouwels PJ, Frahm J. Regional metabolite concentrations in human brain as determined by quantitative localized proton MRS. *Magn Reson Med* 1998; **39**: 53–60.

[7] Barker PB, Szopinski K, Horska A. Metabolic heterogeneity at the level of the anterior and posterior commissures. *Magn Reson Med* 2000; **43**: 348–54.

[8] Jacobs MA, Horska A, van Zijl PC, Barker PB. Quantitative proton MR spectroscopic imaging of normal human cerebellum and brain stem. *Magn Reson Med* 2001; **46**: 699–705.

[9] Breiter SN, Arroyo S, Mathews VP, Lesser RP, Bryan RN, Barker PB. Proton MR spectroscopy in patients with seizure disorders. *Am J Neuroradiol* 1994; **15**: 373–84.

[10] Vermathen P, Laxer KD, Matson GB, Weiner MW. Hippocampal structures: anteroposterior *N*-acetylaspartate differences in patients with epilepsy and control subjects as shown with proton MR spectroscopic imaging. *Radiology* 2000; **214**: 403–10.

[11] Arslanoglu A, Bonekamp D, Barker PB, Horska A. Quantitative proton MR spectroscopic imaging of the mesial temporal lobe. *J Magn Reson Imaging* 2004; **20**: 772–8.

[12] Charles HC, Lazeyras F, Krishnan KRR, Boyko OB, Patterson LJ, Doraiswamy PM, et al. Proton spectroscopy of human brain: effects of age and sex. *Prog Neuro-Psychopharmacol Biol Psychiat* 1994; **18**: 995.

[13] Nagae-Poetscher LM, Bonekamp D, Barker PB, Brant LJ, Kaufmann WE, Horska A. Asymmetry and gender effect in functionally lateralized cortical regions: a proton MRS imaging study. *J Magn Reson Imaging* 2004; **19**: 27–33.

[14] Baker EH, Basso G, Barker PB, Smith MA, Bonekamp D, Horska A. Regional apparent metabolite concentrations in young adult brain measured by (1)H MR spectroscopy at 3 Tesla. *J Magn Reson Imaging* 2008; **27**: 489–99.

[15] Huppi PS, Posse S, Lazeyras F, Burri R, Bossi E, Herschkowitz N. Magnetic resonance in preterm and term newborns: 1H-spectroscopy in developing brain. *Pediatric Res* 1991; **30**: 574–8.

[16] Kimura H, Fujii Y, Itoh S, Matsuda T, Iwasaki T, Maeda M, et al. Metabolic alterations in the neonate and infant brain during development: evaluation with proton MR spectroscopy. *Radiology* 1995; **194**: 483–9.

[17] Kreis R, Ernst T, Ross BD. Development of the human brain: in vivo quantification of metabolite and water content with proton magnetic resonance spectroscopy. *Magn Reson Med* 1993; **30**: 424–37.

[18] van der Knaap MS, van der Grond J, van Rijen PC, Faber JAJ, Valk J, Willemse K. Age-dependent changes in localized proton and phosphorus MR spectrscopy of the brain. *Radiology* 1990; **176**: 509–15.

[19] Pouwels PJ, Brockmann K, Kruse B, Wilken B, Wick M, Hanefeld F, et al. Regional age dependence of human brain metabolites from infancy to adulthood as detected by quantitative localized proton MRS. *Pediatr Res* 1999; **46**: 474–85.

[20] Horska A, Kaufmann WE, Brant LJ, Naidu S, Harris JC, Barker PB. In vivo quantitative proton MRSI study of brain development from childhood to adolescence. *J Magn Reson Imaging* 2002; **15**: 137–43.

[21] Christiansen P, Toft P, Larsson HBW, Stubgaard M, Henriksen O. The concentration of *N*-acetyl aspartate,

creatine+phosphocreatine, and choline in different parts of the brain in adulthood and senium. *Magn Reson Imaging* 1993; **11**: 799.

[22] Lim KO, Spielman DM. Estimating NAA in cortical gray matter with applications for measuring changes due to aging. *Magn Reson Med* 1997; **37**: 372–7.

[23] Chang L, Ernst T, Poland RE, Jenden DJ. In vivo proton magnetic resonance spectroscopy of the normal aging human brain. *Life Sci* 1996; **58**: 2049.

[24] Lundbom N, Barnett A, Bonavita S, Patronas N, Rajapakse J, Tedeschi, Di Chiro G. MR image segmentation and tissue metabolite contrast in 1H spectroscopic imaging of normal and aging brain. *Magn Reson Med* 1999; **41**: 841–5.

[25] Haga KK, Khor YP, Farrall A, Wardlaw JM. A systematic review of brain metabolite changes, measured with (1)H magnetic resonance spectroscopy, in healthy aging. *Neurobiol Aging* 2009; **30**: 353–63.

[26] Maudsley AA, Darkazanli A, Alger JR, Hall LO, Schuff N, Studholme C, *et al.* Comprehensive processing, display and analysis for in vivo MR spectroscopic imaging. *NMR Biomed* 2006; **19**: 492–503.

[27] Kaiser LG, Schuff N, Cashdollar N, Weiner MW. Age-related glutamate and glutamine concentration changes in normal human brain: 1H MR spectroscopy study at 4 T. *Neurobiol Aging* 2005; **26**: 665–72.

[28] Mascalchi M, Brugnoli R, Guerrini L, Belli G, Nistri M, Politi LS, *et al.* Single-voxel long TE 1H-MR spectroscopy of the normal brainstem and cerebellum. *J Magn Reson Imaging* 2002; **16**: 532–7.

Chapter 5
MRS in brain tumors

Key points

- Imaging of brain tumors has evolved into a multimodal tool providing improved diagnostic and prognostic accuracy, fundamental in disease monitoring and assessing response to therapy.
- Proton MR spectroscopic imaging (^{1}H-MRSI) combines the spatial localization capabilities of MR imaging with the biochemical information of ^{1}H-MR spectroscopy, and provides a valuable clinical tool for brain tumors by depicting metabolic changes reflective of cellular density, anaplasia, and mitotic index.
- Choline is elevated in all tumor types due to altered membrane metabolism, and shows correlation with cellular density and indices of cell proliferation. N-acetyl-aspartate (NAA) decreases with tumor infiltration and substitution of normal neural and glial cells. The Cho/NAA ratio is, therefore, a useful parameter particularly in most adult and pediatric primary brain tumors, with a higher ratio correlating with higher cell density and generally associated with a poor prognosis.
- While ^{1}H-MRS can show different metabolic patterns in different tumor types, it is not used as a primary diagnostic tool.
- Increasing Cho/NAA and Cho/Cr ratios in serial exams of a primary astrocytoma are suggestive of transformation to a higher grade. By following metabolic changes, ^{1}H-MRS can be useful in monitoring disease progression or response to therapy.
- ^{1}H-MRSI allows the evaluation of spatial heterogeneity and the macroscopic boundary of a mass, and may provide guidance for targeted biopsy, surgery, or therapy.
- ^{1}H-MRS studies can also be particularly useful in distinguishing neoplastic from non-neoplastic lesions, and differentiating recurrent tumor from predominantly delayed radiation necrosis.

Introduction

The role of imaging in patients with brain tumors has changed dramatically in the last decade. From a relatively simple descriptor of morphology and anatomy, imaging of brain tumors has evolved into a multimodal tool capable of measuring metabolic, hemodynamic, and microscopic structural alterations. Clinical applications of MR spectroscopy, MR perfusion, and MR diffusion in brain tumors have increased exponentially as indicated by the number of scientific publications on these topics. Much has been learned about the pearls and pitfalls of these imaging techniques. Many studies have focused on diagnostic accuracy, and on grading of neoplasms, primary central nervous system (CNS) gliomas in particular, using histopathology as a reference standard. More recent studies have addressed accuracy of imaging in measuring prognosis and patient outcome. Despite a lack of large and multicenter studies supporting evidence that the new imaging methods have an impact on outcome measurements, management of patients with a brain tumor without advanced imaging methods has become unthinkable.

New concepts about brain tumor initiation, transformation, and diffusion

For a long time, the classification of brain tumors has been based on histogenesis. The success of the classification proposed by Bailey and Cushing in 1926[1] has been associated with the hypothesis that the supposed cell of origin and the degree of maturation of tumor cells are determinant factors for prognosis.

Histopathologic criteria, immunohistochemistry and, more recently, molecular genetics have improved characterization of the supposed cell of origin and have further refined the prognostic criteria. The debate about the origin of brain tumors is again of great interest after the introduction of neural stem cells and their implications for new therapeutic strategies.

The natural history of brain tumors can be divided into four main stages: initiation, progression, transformation, and diffusion. The *initiation* of glial tumor is not yet understood, and is the focus of intense research. When a macroscopic lesion is first identified in humans, it is already organized as a tumor. Experimental studies in rats have shown that induced tumors arise from primitive neuroepithelial (neural stem) cells of the ventricular zone (VZ), from its derivative subventricular zone (SVZ), or from a renewal pool of cells in the remnants of SVZ or subependymal layer, hippocampus, cerebellum, and first cortical layer. Neural stem cells are characterized by their self-renewal capability and multipotency. Both neurons and glial cells derive from neural stem cells of the VZ and SVZ, and they differentiate along their respective pathways under extrinsic and intrinsic stimuli. Growth factor signaling also controls their passage from one stage to the next. Availability of markers for glia or neurons affirms that neurogenesis and gliogenesis continue in the adult mammalian brain; demonstration of the existence of neural stem cells also in the adult has changed our notion of oncogenesis dramatically. Glial tumors may arise from stem cells either in the VZ or SVZ. New hypotheses on glioma origin postulate that astrocytes and radial glia (astroglial lineage) might act as multipotent stem cells both in embryo and in adulthood.[2] The vulnerability of stem cells to undergo neoplastic transformation depends on the interaction of several factors, including the number of replicating cells, the duration a cell population remaining in cycle, and the state of cellular differentiation. Another important concept is that the more genetic alterations are needed for a tumor to develop the more advanced the progenitor cells' stage of differentiation. Then the histology of a tumor may be more a reflection of the environment and time of initiation than of the cell of origin, and the former factors would determine whether a tumor ultimately becomes an astrocytoma or an oligodendroglioma.[3] Early genetic events differ between astrocytic and oligodendroglial neoplasms (Figure 5.1), but all tumors have an initially invasive phenotype, which complicates therapy.[4] Tumor cells are heterogeneous and variably express differentiated antigens typical of the cell of origin, but only a minority of them are self-renewing, multipotent, clonogenic, and continuously replenishing mature cells.[5] The discovery that only a proportion of tumor cells are clonogenic when xenografted has introduced another important new concept: the existence of cancerous stem cells.[3]

The growth and transformation of tumors can be summarized as follows. Tumors are supposed to initiate as monoclonal expansion and, due to genetic instability, they will develop genetic heterogeneity that is associated with increasing mutation rate and proliferation capacity. Once the genome becomes unstable, as the cell divides, genetic material that codes for growth promotion (i.e. oncogenes of which protein products serve to accelerate cell growth) will accumulate, whereas genetic material that codes for growth control (i.e. suppressor genes of which protein products serve as brakes on cell growth) will be lost. These events result in phenotypic heterogeneity. New clones will arise and compete with older clones. They may adapt better to the tissue microenvironment and show greater proliferation potential after having lost differentiating capacity. Anaplasia is a feature of the neoplastic tissue that has lost such capacity typical of a given cytogenetic stage, and regresses to a less differentiated stage. Various molecular genetic alterations have been linked to the following pathologic events occurring with tumor progression: proliferation and apoptosis in diffuse astrocytoma; deregulation of cell cycle in anaplastic astrocytoma; necrosis, angiogenesis, and clonal selection in glioblastoma (GBM).[6] Cell migration and invasion, angiogenesis, necrosis, and apoptosis are neoplastic events that have been consistently associated with poor prognosis in gliomas. New therapeutic strategies can be explored when the molecular pathways regulating these events and their phenotypic imaging characteristics become better understood.

Several molecular biology techniques are used to identify genetic aberrations in tumors. At present, the most commonly applied technique is genotype analysis with fluorescent microsatellite markers that evaluate loss of heterozygosity (LOH). Preliminary results suggest that in the near future, molecular genetic tumor analysis together with clinical, quantitative imaging and histopathologic phenotypes will allow a more accurate prediction of survival and will be of great importance in selecting and developing the appropriate therapy. At present, the use of quantitative imaging and molecular genetics for the classification

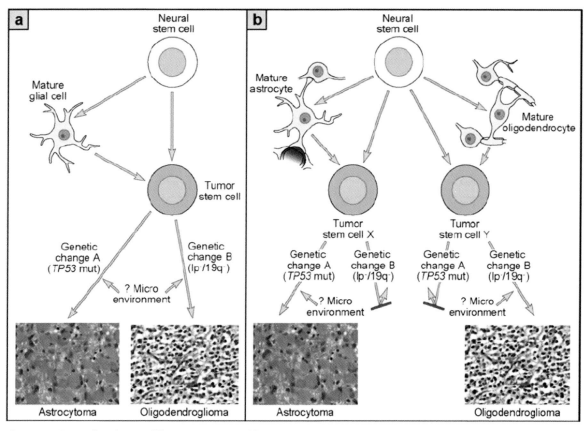

Figure 5.1. Stem cells and tumor differentiation: two models.
(a) A single type of tumor stem cell can arise either from a neural stem cell or from a mature glial cell. A genetic mutation in the TP53 pathway will generate the final phenotype of an astrocytoma. A genetic change causing chromosome 1p and 19q loss of heterozygosity will generate an oligodendroglioma.
(b) In this model, one tumor stem cell (X) is permissive for neoplastic transformation only in the setting of TP53 mutation while other genetic changes (i.e. 1p and 19q loss) are lethal, hence the restricted lineage to an astrocytoma. Another stem cell (Y) undergoes tumorigenesis only in the setting of 1p and 19q loss, yielding an oligodendroglioma.
(Modified from [4]).

of the tumor type, subtype, or grade remains a challenge because large validation studies are still lacking. It is also unknown if the correlation found between molecular genetics data and survival can be prospectively applied to each individual case at the time of diagnosis.[7]

Growth depends on the balance between cell proliferation and cell loss, and is regulated by cell cycle time, growth fraction, tumor doubling time, necrosis, and apoptosis. Necrosis is a sudden event in which many cells are killed at the same time because of hypoxia, energy depletion, and inflammatory response. Apoptosis denotes a programmed cell death comprised of three separate complex regulatory pathways,[8] and is considered the major cause of cell loss in gliomas and other tumors. However, apoptosis is more often associated with a high proliferation rate. The apoptotic index increases in the spectrum from low-grade diffuse (infiltrative) astrocytoma to GBM; in oligodendroglioma, the index is higher than in astrocytoma, and it also increases with anaplasia. Moreover, a high apoptotic index is found in PNETs, lymphomas, and metastases. Whether the finding of increased number of cells undergoing apoptosis might indicate tumor regression and better prognosis is still controversial.[9] Failure of apoptosis may be responsible for tumor development and may be linked to breakage of the pathway regulated by tumor suppressor p53. In contrast, induction of apoptosis in glioma cells could be instrumental to therapies.

Diffusion and infiltration into the adjacent brain tissue are two other important properties of gliomas and make their treatment much more difficult. Motility of glioma cells and invasion are facilitated by extracellular matrix and adhesion molecules. Cell motility is associated with poor prognosis. Velocity of solid tumor expansion is linear with time and varies from 4 mm/year in low-grade glioma (LGG) to 3 mm/month in high-grade glioma (HGG).[10] There is a cell density gradient decreasing from the center of the mass towards its periphery at the boundary with normal tissue. How far neoplastic cells can be found from the macroscopic edge of the tumor is another very important issue. The 2 cm distance detected by computerized tomography has been considered the safety margin according to the classic report by Burger in 1988.[11] It is about time that new studies with more updated imaging modalities define with greater accuracy the extent of a tumor. Cell motility and invasion capability have great implications for planning of surgery and post-surgical adjunct therapy.

Cell invasion is not necessarily a consequence of cell proliferation. Examples of infiltrative but not highly proliferative gliomas are common and have been described in detail by Darlymple *et al.*[12] In solid tumors, a cellular density gradient is much more frequently found between the center and the cortex than toward the white matter. Similar gradient properties can be found in the frequency of mitoses and nuclei stained for proliferation markers, MIB_1. When the tumor border is clear-cut the gradient is also steep. In infiltrated cortex, the MIB_1 labeling index may be very low, because there is a dissociation between migratory and proliferation capacities.[13] Recognition of asymmetry in tumor infiltration and proliferation also has implications for therapy planning.

Histopathological classification of brain tumors

The classification of tumors of the nervous system released by the World Health Organization (WHO) in 2007 is the most comprehensive to date and has received a large consensus among neuropathologists.[14] It divides brain tumors into seven categories: tumors arising from neuroepithelial tissue, from peripheral nerves, from the meninges, lymphomas and hematopoietic neoplasms, germ cell tumors, tumors of the sellar region, and metastatic tumors.

Among tumors of neuroepithelial tissue, the gliomas are by far the most common and best studied. Gliomas include tumors arising from neural stem cells in the VZ or SVZ or from neoplastic transformation of precursor or mature glial cells. The group of neuroepithelial tumors also includes ependymomas, neuronal and mixed neuronal–glial tumors (such as ganglioglioma), and embryonal tumors (such as medulloblastoma). Gliomas are classified by their histologic features, according to the presumptive cell of origin, differentiation, and grade of malignancy. At the current state of knowledge, cytogenesis is more a theoretical concept than a definitive basis for tumor classification.

Astrocytomas are believed to originate from astrocytes, which are stellate branched cells permeating the interstitium and interacting with blood vessels and neurons. Astrocytes have indeed a very important stromal role: the extracellular matrix reacts through cell adhesion molecules and other factors with the intracellular space, modulating cell migration, differentiation, proliferation, and apoptosis. In view of the abundance of astrocytes and their multiple functions, astrocytomas make up a heterogeneous group of tumor subtypes, with different biological behavior. According to the WHO classification, glial tumors are graded on the basis of the most malignant area identified on histopathologic specimens. Fibrillary astrocytoma (WHO grade II) is characterized by increased cellularity with a monomorphic population of cells infiltrating the neuropil. Anaplastic astrocytoma (WHO grade III) is characterized by nuclear polymorphism and mitoses. The occurence of angiogenesis and necrosis are features of GBM (grade IV). GBM can arise de novo (primary) or transform from a pre-existing LGG (secondary).

Anatomic location and patient's age at initial presentation are also very important factors for diagnosis and prognosis. According to a recent study by Duffau and Capelle, LGGs more frequently than GBMs are found in the cortex of secondary functional areas, especially within the supplementary motor area (SMA) and insula.[15] This preferential location may be associated with a high risk of adverse postoperative sequelae, and it may be one reason that resection of LGGs remains controversial.

Oligodendrogliomas are moderately cellular and composed of monotonous, round, and homogeneous nuclei with a clear cytoplasm. They have a dense network of branching capillaries. It is not uncommon for oligodendrogliomas to bleed, and they may present as an intracranial hemorrhage. Additional histologic features include calcifications and mucoid/cystic degeneration. Estimates of incidence of oligodendroglioma vary enormously in different series since diagnosis depends on use of permissive or restrictive criteria. If restricted histologic criteria are used, diagnostic signs are the honeycomb appearance of the cells and the high density and chicken-wire distribution of small vessels. When more permissive criteria are used,[16,17] the incidence of oligodendrogliomas obviously increases while that of diffuse astrocytomas decreases. Burger has masterly illustrated this diagnostic conflict.[18] Today more than ever, the correct diagnosis of oligodendroglioma is important because effective chemotherapy has become available. Allelic LOH in the 1p and 19q chromosomes have been associated with longer survival and a favorable response to chemotherapy with procarbazine, lomustine, and vincristine (PVC).[19] In contrast, LOH on 17p and TP53 mutations characteristic of astrocytic tumors are rare in oligodendrogliomas and practically mutually exclusive with LOH 1p and 19q. The differential diagnosis also has important prognostic implications, since the number of mitoses and nuclei positive for proliferation markers found in the two tumors may receive a different weight by the neuropathologists. In diffuse astrocytomas, mitoses are absent or very low in number and a high mitotic index indicates anaplasia. In grade II oligodendrogliomas, the number of mitoses allowed, on the other hand, is definitely higher. Therefore the same mitotic index may suggest anaplasia or not depending on the diagnosis of astrocytoma or oligodendroglioma.[20,21] The apoptotic index is also much higher in oligoastrocytoma than in astrocytomas.[22] As we will see later, this asynchronous biologic behavior between the two tumor types is a recurrent confounding factor in neuroimaging as well: the choline (Cho) signal measured with ^{1}H-MRS and the cerebral blood volume (rCBV) measured with perfusion MR are generally higher in grade II oligodendrogliomas than in astrocytomas.

Mixed oligoastrocytoma (MOA) were first recognized as an entity by Cooper in 1935.[23] The diagnosis of MOA requires recognition of two different glial components, both of which must be unequivocally neoplastic. They may be divided into biphasic (compact) and intermingled (diffuse) variants. In the former, distinct areas of both cell types are juxtaposed, while in the latter variant the two components are intimately admixed. Estimates of their incidence vary with the diagnostic criteria used and must be interpreted with caution. Their incidence may vary from 1.8% in North American, 9.2% in Norwegian, and 10–19% in German series. On conventional MR imaging, MOA demonstrate no special features that would allow a reliable distinction from oligodendrogliomas. About 30–50% of MOA are characterized by LOH on 1p and 19q. About 30% carry mutations of the TP53 gene and/or LOH on 17p that are frequently found in astrocytomas. The presence of LOH 10q has also been associated with shorter overall survival in MOA.[24] Apparently these genetic alterations are consistent throughout every individual MOA and suggest that they are monoclonal neoplasms originating from a single precursor cell rather than tumors that have developed concurrently.[25]

GBM is the most malignant astrocytic tumor, composed of poorly differentiated neoplastic cells. It is the most frequent brain tumor, accounting for 12–15% of all intracranial tumors. GBM may manifest at any age, but about 80% of patients are between 45 and 80 years old. It may develop from diffuse astrocytomas, anaplastic astrocytomas, but more frequently they occur de novo after a short clinical history. Primary GBM accounts for the vast majority of cases in older people, while secondary GBM typically develops in younger patients (less than 45 years). The time to progression varies considerably, ranging from less than 1 year to more than 10 years, with a mean interval of 4–5 years. There is increasing evidence that these two subtypes represent distinct disease entities, which evolve through different genetic pathways, and are likely to respond differently to therapy.[26] GBM is a very heterogeneous disease and it is often multifocal. Multifocal GBM is suggestive of a more invasive and migratory tumor phenotype, a feature more common to stem cell-derived cancer. According to a recent report GBM originating from the SVZ and extending into the cortex has a higher rate of developing multifocal disease, while GBM growing in the cortex has a higher rate of local recurrence.[27] Tumor location may suggest the presumed origin of the tumor from SVZ neural stem cells in the former phenotype and from transformation of mature glial cells in the latter. Stem cell-derived GBM may require treatment that attends to both the primary lesion and the SVZ.

WHO grading and patient survival

Histological grading is not only a practical way of predicting the biological behavior of a tumor in an individual patient, but also an effort to define, in conjunction with other parameters, a homogeneous group of patients. Significant indicators of anaplasia in gliomas include nuclear atypia, mitotic activity, cellularity, vascular proliferation, and necrosis.[28] These histopathological features are condensed in a three-tiered scheme.[14] A simple and reproducible grading scheme is of paramount importance to planning therapy as well as for the interpretation of response to multiple therapeutic regimens. The validity and reproducibility of any grading system depends on the homogeneity of the lesions within each class. Grading is only one component of a combination of criteria used to predict a response to therapy and outcome. Other criteria include the patient's age, neurological performance status, tumor location, extent of surgical resection, proliferation indices, genetic alterations, and one radiological feature: contrast enhancement.[14] Whether ^{1}H-MRSI and other imaging parameters will be included in this list will depend on their unique contribution to characterize neoplasms in a homogeneous group. The two biggest assets of in vivo MR imaging are the possibility to follow non-invasively the biological behavior of individual tumors and guide therapy.

For each tumor type, the combination of the above criteria contribute to an overall approximation of prognosis. The median survival for grade II diffuse astrocytoma is around 5 years; however, the range of survival is broad and unpredictable.[29] Despite the initial low proliferative index, most of these patients die from progression to GBM. Patients with grade III anaplastic astrocytoma survive for 2–3 years; the majority of patients with GBM, in particular the elderly, have a median survival of less than 1 year. The outcome of 676 GBM patients over a 7-year period at a single institution has been reported recently: survival probabilities were 57% at 1 year, 16% at 2 years, and 7% at 3 years.[30]

Studies of patients with grade II oligodendrogliomas reported a median survival of about 10 years. A recent series of 106 patients yielded a median survival of 16 years,[31] probably due to earlier diagnosis following the advent of MRI. Malignant progression is not uncommon, although it is considered less frequent compared to diffuse astrocytomas. A median survival time of 6.3 years and 5- and 10-year survival rates of 58% and 32%, respectively, have been reported in a study of 60 patients with grade II MOA.

Conventional MR imaging

MR, with its excellent soft tissue contrast, is the modality of choice for the characterization of brain lesions. It is the most sensitive imaging modality in detecting a mass, in defining the location and apparent size, and in interrogating the integrity of the blood–brain barrier with an exogenous contrast agent. In most medical centers neurosurgeons and neuro-oncologists make their management and therapy planning decisions based on conventional MR images.

In 1990, it was reported that neuroradiologists succesfully predicted tumor grade on a three-tiered grading system about 80% of the time using the gestalt approach. In 36 gliomas, the study found that positive MR predictors were mass effect ($P = 0.0000$) and cyst formation or necrosis ($P = 0.0512$) and MR accuracy rate approached that of neuropathologic diagnosis, which is subjected to sampling errors.[32] Perilesional edema, signal heterogeneity, hemorrhage, border definition, and crossing of the midline were among other MR features which were not found to be significant positive predictors. Two major limitations of that study were that the number of patients was very small and that only one subtype of brain tumors (i.e. astrocytomas) was included. In 1995, another study compared diagnostic accuracy of experienced radiologists with two objective and reproducible methods: neural network models and multiple linear regression.[33] Diagnostic accuracy of grading glioma was 61% for the neural network model, 59% for multiple regression, and 57% for radiologist's experience. All three methods were better than random (50%), but the study showed that the relationship between MRI features and tumor grade is not strong enough to allow sufficiently accurate determination of grade. Other studies have reported an MR imaging accuracy of diagnosing supratentorial neoplasms to vary between 30% and 90%.[34]

At the current state of the art, conventional MR imaging features cannot be used to determine underlying tissue histopathology. The limitations of conventional MRI are well known: it does not allow accurate diagnosis of tumor type; it cannot distinguish infiltrating tumor from vasogenic edema since both appear bright on T_2-weighted images; it cannot distinguish infiltrating tumor from functioning brain tissue at tumor bordeline; it cannot measure cell density, cell anaplasia, mitotic index, or angiogenesis. As a consequence, it cannot accurately predict individual patient time-to-progression and survival.

Semiotics of proton MR spectroscopy in brain tumors

Proton MR spectroscopic imaging (^{1}H-MRSI) combines the spatial localization capabilities of MR imaging with the biochemical information of ^{1}H-MR spectroscopy. A multitude of spectroscopic studies has been published in the literature in the last 20 years. [35,36,37,38,39,40,41,42] ^{1}H-MRSI in brain tumors is feasible, reasonable, and a valuable clinical tool. These studies have consistently shown that the Cho signal is elevated in all tumor types because of altered membrane metabolism.[43] The Cho (3.24 ppm) signal increases with cellular density and, according to some authors, also correlates with indices of cell proliferation (Ki-67).[41,45,46] The signal intensity of N-acetyl-aspartate (NAA, 2.02 ppm), a marker of neurons and their processes, decreases with tumor infiltration and substitution of normal neural and glial cells. Changes in the creatine (Cr, 3.02 ppm) signal may vary with tumor type: there is often a mild increase in LG astrocytomas, followed by progressive depletion with increasing anaplasia. Unfortunately, at the spatial and spectral resolutions available with clinical studies, no metabolite is tumor type-specific.

Elevation of the Cho signal has been recognized as an important surrogate marker of tumor progression [38] and response to therapy.[47] Precursors and catabolites of phospholipid metabolism are altered in tumors. The total choline peak detected by clinical ^{1}H-MRS is the sum of the signals of free Cho and Cho-containing compounds: phosphocholine (PC), and glycerol 3-phosphocholine (GPC), with perhaps small contributions also from phosphoethanolamine (PE), and glycerol 3-phosphoethanolamine (GPE). Elevation of PC and PE, intracellular phosphomonoesters (PME), suggests enhanced cell membrane synthesis during cellular growth, while elevation of GPC and GPE, phosphodiesters (PDE), is due to abnormal rates of membrane synthesis, catabolism, and metabolic turnover. Oncogene expression and malignant transformation have common endpoints in choline phospholipid metabolism. They often determine a shift in PC and GPC that results in increasing PC/GPC ratio and total choline signal. Experimental data suggest that increased PC/GPC is primarily related to malignant degeneration; conversely, reduced PC/GPC is related to growth arrest.[48] Ex vivo studies have also shown that PC elevation is likely related to malignancy, since it is found at a twofold greater level in HGG compared to LGG and normal tissue. [49] Therefore, MRS appears to be more sensitive to up-regulation of the anabolic pathway than acceleration of the catabolic pathways.

In vivo ^{1}H-MRSI studies have shown that Cho elevation is highest in the center of a solid, non-necrotic glioma with a Cho/NAA gradient declining toward the periphery of the tumor. The slope of this gradient is frequently shallow toward the cortex, while it is frequently steeper toward the white matter. One study in 18 glioma patients demonstrated a significant linear correlation between normalized choline (nCho) signal and cell density, not with proliferative index.[50] The same study found a significant inverse linear correlation between cell density and the apparent diffusion coefficient (ADC) measured by diffusion MR. The Cho signal is consistently low in areas of necrosis.

In a recent prospective study specifically devoted to the characterization of diffuse WHO grade II gliomas, a significant correlation between Ki-67 labeling index and single-voxel MRS data was found.[46] It was shown that increasing Cho/NAA and Cho/Cr ratios correlated with three classes of Ki-67 indices. The presence of lactate was found only in LGG with Ki-67 > 4%; the presence of lipids was found only in LGG with Ki-67 > 8%. Although the Ki-67 labeling index is not formally considered as a prognostic marker in the WHO classification, an index over 4% is predictive of a more aggressive clinical behavior. This study provides evidence that ^{1}H-MRS may be a reliable tool for predicting the proliferative activity of WHO grade II gliomas. Observation of increasing Cho/NAA and Cho/Cr ratios in consecutive ^{1}H-MRSI exams is highly suggestive of LGG transformation (Figure 5.2). Tedeschi et al. have shown that serial ^{1}H-MRSI examinations can detect transformation of LGG and distinguish between progressive and stable gliomas.[51] In this longitudinal study of 27 patients, the authors demonstrated an interval increase in Cho signal of greater than 45% in patients with progressive disease, while interval Cho increase was 20% or less in stable patients (Figure 5.3).

The Cr signal is the sum of Cr and phosphocreatine (PCr). These two molecules play an important role in short-term energy storage, with the Cr kinase reaction generating ATP. Cr is synthesized in the liver and used in the brain (slightly higher in gray than in white matter), in the kidneys, heart, and skeletal muscles. It is absent in all other tissues. In brain tumors Cr signal may change according to tumor type and glioma grade. Cr is virtually absent in lymphomas (Figure 5.4)[47] and metastases, very

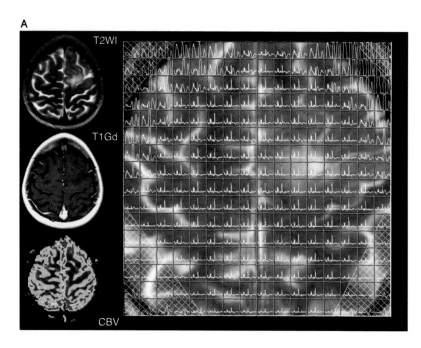

Figure 5.2A. Metabolic and imaging changes during tumor progression.

Axial T_2-weighted MR imaging shows a small infiltrating mass in the left superior frontal gyrus of a 20-year-old woman presenting with a focal seizure of the right arm. There is no evidence of enhancement after gadolinium injection on T_1-weighted MR imaging. MR perfusion showed low cerebral blood volume (rCBV) in the same region.

MRSI (PRESS: TR/TE = 1200/135 ms; 24 × 24 matrix; FOV = 200 × 200 × 15 mm^3) was acquired at the level of the mass. The grid with multiple spectra gives an overview of the metabolic changes within the infiltrating tumor and adjacent parenchyma, in comparison with the normal contralateral hemisphere.

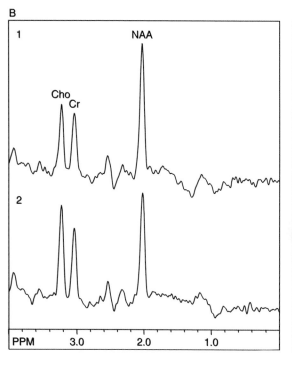
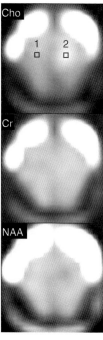

Figure 5.2B. Cho, Cr, and NAA maps with two selected spectra from the same data set in Figure 5.2A are illustrated. Note mild Cho increase and moderate NAA decrease in the tumor (2) compared with the contralateral normal spectrum (1). The findings are compatible with a low-grade infiltrating astrocytoma.

low in meningiomas and oligodendrogliomas. In LG astrocytic tumors[38,52] and in gliomatosis cerebri [53] Cr signal may be elevated. In the author's experience, Cr may be as elevated as the Cho signal in grade II glioma infiltrating the thalamus and other gray matter structures (Figure 5.5). However, Cr signal may drop substantially as signs of anaplasia increase. The Cho/Cr ratio is often much greater

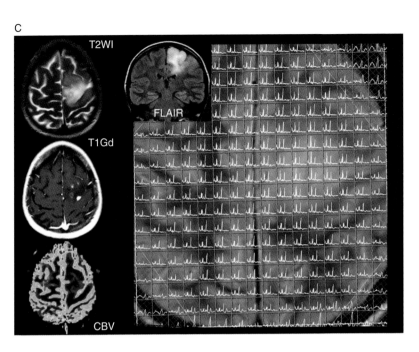

Figure 5.2C. At follow-up seven months later, T_2-weighted and FLAIR images show that the mass has grown and patches of enhancement have appeared on the post-contrast T_1-weighted MR image. On perfusion MR note that CBV is now elevated compared with the contralateral white matter.

The follow-up MRSI grid confirms extensive growth of the tumor with several additional abnormal spectra compared with the earlier study.

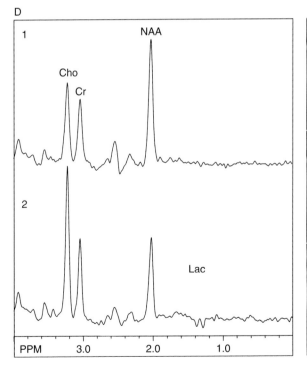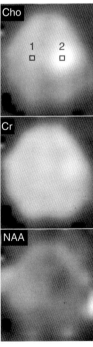

Figure 5.2D. Cho, Cr, and NAA maps with two selected spectra from the same data set as in Figure 5.2C are illustrated. Note that Cho has significantly increased and NAA decreased in the tumor (2) compared with the contralateral normal spectrum (1). Note also appearance of a small lactate peak in the spectrum from the tumor, suggesting increased glycolytic tumor metabolism.

The neoplasm was surgically removed and a grade III anaplastic MOA was diagnosed.

than unity in primary de novo and secondary GBM (Figure 5.6).

Mobile lipids (1.40 and 0.9 ppm) are a characteristic feature of GBM, metastases, lymphomas, and abscesses.[54,55,56] Abnormal signals resonating at 1.4 and 0.9 ppm have been associated with lipid droplets in areas of extracellular necrosis.[57] Primary cerebral lymphoma in immunocompetent patients mimics the infiltrative behavior of glial neoplasms. The diagnosis of lymphoma is difficult with conventional MRI, and it has important diagnostic and therapeutic implications: surgery should be avoided, chemo- and radiotherapy are the treatments of choice. A spectrum with high Cho, absent Cr and NAA and elevated lipid is the hallmark of cerebral lymphomas;[47,58] however, it is not the rule in all patients with CNS lymphoma (Figure 5.4). Large amount of lipids can also be found in areas that have been treated with radiotherapy and have evolved into delayed radiation necrosis (Figure 5.7).

The hypothesis that accumulation of lactate (1.34 ppm) may correlate with higher tumor grade was investigated in the early days of MR spectroscopy. As originally described by Warburg, neoplastic cells may develop bioenergetic aberrations, such as elevated anaerobic glycolysis.[59] This behavior is mainly characteristic of HGG that have lost aerobic cell respiration capability. Anaerobic glycolysis is less efficient and leads to increased production of lactate (Figure 5.8). ^{1}H-MRS studies have shown that lactate accumulation occurs in about one third of gliomas.[36] It is found more frequently in HGG, but its presence is not a reliable indicator of tumor grade.[38] Lactate can be found occasionally in grade II gliomas. Lactate accumulation may occur in areas of tumor with low rCBV and slow flow (low mean transit time [rMTT]) (Figure 5.9). The detection of lactate accumulation in an LGG may be a sign of anaerobic glycolysis occuring in proliferating cells, or due to relative hypoxia.[46] This lactate sign may precede increased vascular density and angiogenesis that is detected by MR perfusion on CBV maps. Thus lactate accumulation in LGG may be a transient phenomenon that will eventually disappear if induction of angiogenesis facilitates lactate clearance through the venous drainage. Increased lactate within necrotic pseudocysts and in areas of the tumor where venoular outflow is obstructed is also a relatively frequent finding. Lactate may also be detected in the necrotic areas of GBM and metastases.

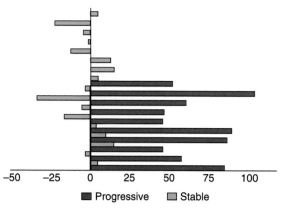

Figure 5.3. Choline changes in progressive versus stable brain tumor patients.

The bar graph shows percentage changes of Cho signal between two consecutive studies in patients with progressive (red) and stable (green) disease. Note that all patients with Cho signal changes greater than 20% had progressive disease.
(Modified with permission from [51].)

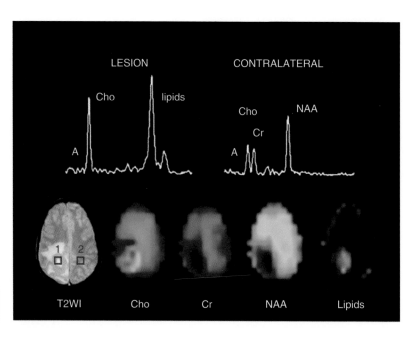

Figure 5.4. Absent Cr signal in CNS lymphoma.

MRSI (PRESS: TR/TE = 2000/144 ms; 32 × 32 matrix; FOV = 240 × 240 × 15 mm^3) was acquired at the level of the mass in the right parietal lobe. Cho, Cr, NAA, and lipids maps with two selected spectra are illustrated. The spectrum from the mass (1) shows very high elevation of Cho and mobile lipids signal, associated with absent Cr and NAA signals. The metabolite maps nicely illustrate the abnormal changes. (Modified with permission from [47].)

Chapter 5: MRS in brain tumors

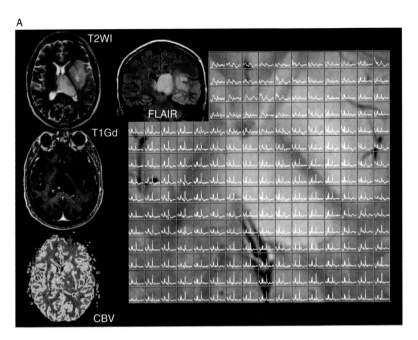

Figure 5.5A. Elevated Cr signal in grade II astrocytoma infiltrating the thalamus.

Axial T_2-weighted and coronal FLAIR MR images show a mass infiltrating the left thalamus, lenticular nucleus, and insular cortex in a 62-year-old man. There is no evidence of enhancement after gadolinium injection on T_1-weighted MR imaging. MR perfusion showed low cerebral blood volume (rCBV).

MRSI (PRESS: TR/TE = 1200/135 ms; 24 × 24 matrix; FOV = 200 × 200 × 15 mm³) was acquired at the level of the basal ganglia. The grid with multiple spectra provides an easy overview of the metabolic changes in the infiltrating tumor and in the contralateral healthy hemisphere. Note elevated Cho and Cr signals in the left thalamus, lenticular nucleus, and insula. In most spectra Cr signal is more elevated than Cho.

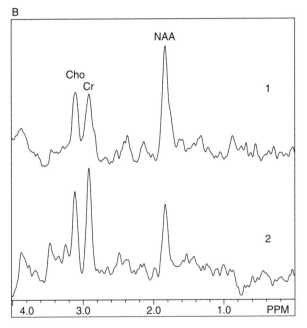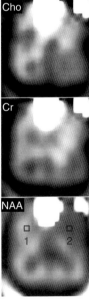

Figure 5.5B. Cho, Cr, and NAA maps with two selected spectra are illustrated. Note elevated Cr and Cho signals associated with moderate NAA signal loss in the tumor infiltrating the left putamen (2) compared with a normal spectrum in the right putamen (1).

The NAA signal is a surrogate marker of neurons and its processes in gray matter and of axons in white matter. Tumor infiltration is associated with displacement and destruction of normal tissue; thus, NAA is markedly reduced in the core of most solid brain tumors. In the core of meningioma, metastasis and lymphoma absence of the NAA peak on the spectrum is a frequent finding since these tumor types do not contain NAA. In gliomas, which grow by infiltration, the detection of different amounts of NAA signal in the spectra may indicate residual neurons or axons (Figure 5.10). Whether this

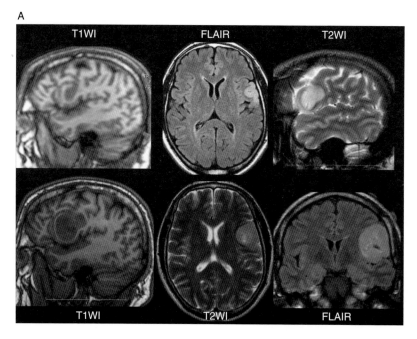

Figure 5.6A. Fast growing GBM with low Cr.

In the top row, sagittal T_1-weighted, axial FLAIR, and sagittal T_2-weighted MR images show a small mass in the left ventrolateral prefrontal cortex (VLPFC). There was no evidence of enhancement after gadolinium injection on T_1-weighted MR imaging (not shown). At this time, a stereotaxic biopsy yielded a diagnosis of grade II astrocytoma. Three months later (bottom row) the mass is still well-defined, without peri-lesional edema, but it has enlarged significantly.

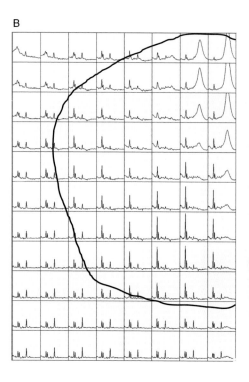
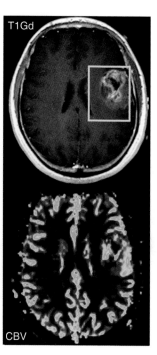

Figure 5.6B. MRSI (PRESS: TR/TE = 1200/135 ms; 24 × 24 matrix; FOV = 200 × 200 × 15 mm^3) was acquired at the level of the lateral ventricles. The grid with multiple spectra gives an overview of the metabolic changes within the mass in the left VLPFC and in the adjacent normal tissue. The curved line outlines the tumor boundary on the spectral grid. Note the very high Cho peak associated with low Cr and NAA signals. Very high Cho/NAA values can be found in the solid components of the mass, especially when the neoplastic cells are still well perfused and oxygenated, as demonstrated by elevated rCBV on MR perfusion. The T_1-weighted MR image after gadolinium injection shows heterogeneous enhancement of the mass. At second surgery the histopathological diagnosis of GBM was made.

residual neural tissue is viable cannot be determined with MRS. Other imaging methods could be used to address this question. Another explanation for residual NAA signals within gliomas is partial volume average with adjacent normal tissue due to the limited spatial resolution of MRS.

The role of *myo*-inositol (mI) in LGG remains controversial. In a very recent study it was shown

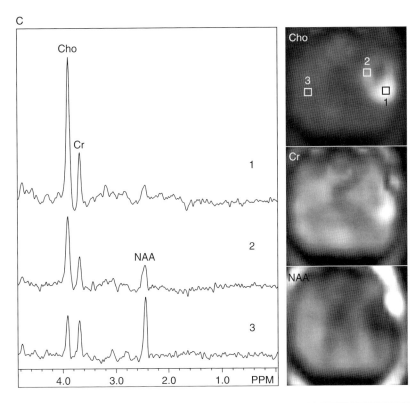

Figure 5.6C. Cho, Cr, and NAA maps with three selected spectra are illustrated. Metabolite ratios from the spectrum with maximum abnormalities are measured. In this patient, peak integration showed Cho/NAA = 17.10; Cho/Cr = 3.32 (normal values are Cho/NAA = 0.60; Cho/Cr = 1.00); nCho = 3.54; nCr = 1.02; nNAA = 0.13.

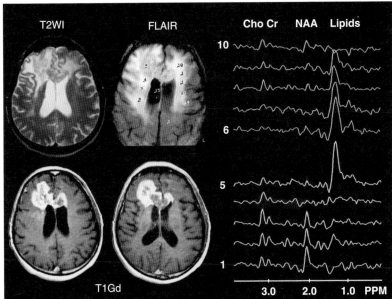

Figure 5.7. Elevated lipids signal in delayed radiation necrosis.
Axial post-contrast T_1-weighted MR images show growth of a heterogeneously enhancing mass with butterfly appearance in the genu of the corpus callosum with moderate peri-lesional T_2 hyperintensity on axial T_2-weighted image in a 47-year-old man who had been treated with surgery followed by radiotherapy for anaplastic gemistocytic astrocytoma. On conventional MR imaging the differential diagnosis between recurrent glioma and delayed radiation necrosis (DRN) was raised.
MRSI (PRESS: TR/TE = 1500/135 ms; 32 × 32 matrix; FOV = 200 × 200 × 15 mm^3) was acquired at the level of the enhancing lesion. Multiple selected spectra are illustrated. Spectra from the lesion in the corpus callosum (5–9) shows very high lipid signal (1.4 ppm) associated with depletion of all major metabolites. Spectra from the T_2-hyperintensity surrounding the lesion (2–4, 10–13) show mild NAA and Cr signal drop with no evidence of increased Cho signal. Follow-up imaging and clinical exams confirmed the diagnosis of DRN.

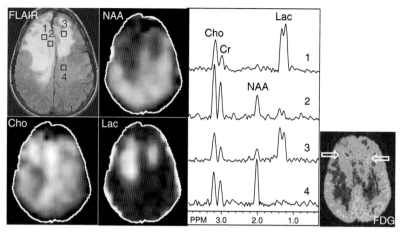

Figure 5.8. Recurrent GBM with high Cho and high lactate.
Axial FLAIR shows a recurrent lesion in the right frontal lobe extending to the contralateral frontal lobe in a 55-year-old woman who had surgery for GBM 8 months earlier.
MRSI (PRESS: TR/TE = 1500/135 ms; 32 × 32 matrix; FOV = 240 × 240 × 15 mm^3) was acquired at the level of the centrum semiovale. NAA, Cho, and Lac maps with four selected spectra are illustrated. Note very high choline and lactate in spectra from the T_2-hyperintense lesion (1–3) with marked NAA signal loss. A normal spectrum (4) is shown for comparison. The diagnosis of recurrent GBM was confirmed by ^{18}FDG-PET scan which showed two hot spots of FDG uptake in both frontal lobes (arrows).

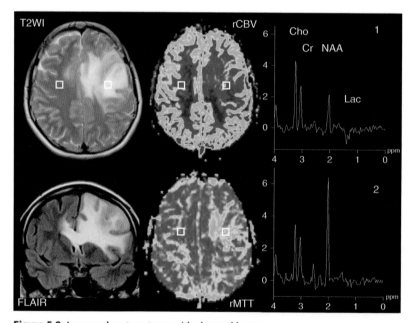

Figure 5.9. Low-grade astrocytoma with elevated lactate.
Axial T_2-weighted and coronal FLAIR MR images show a large mass infiltrating the left frontal lobe and the genu of the corpus callosum in a 26-year-old woman.
DSC-MR perfusion shows homogeneous low rCBV values and an area of delayed mean transit time on the rMTT map.
MRSI (PRESS: TR/TE = 1200/135 ms; 24 × 24 matrix; FOV = 200 × 200 × 15 mm^3) was acquired at the level of the centrum semiovale. The selected spectrum from the left frontal mass in (1) shows moderate elevation of Cho, moderate NAA signal loss, and unchanged Cr as compared with the contralateral spectrum (2). Moderate accumulation of lactate within the mass is likely due to reduced clearance from venous congestion secondary to the mass effect and edema. It is important to keep in mind that most gliomas produce lactate; however, lactate detection on MRS may also be attributed to delayed clearance by the venous outflow.

Chapter 5: MRS in brain tumors

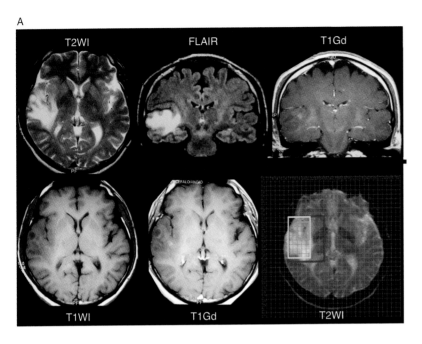

Figure 5.10A. Metabolic and histopathologic heterogeneity in low-grade astrocytoma.

Axial T_2-weighted and coronal FLAIR MR images show a mass infiltrating the right temporal lobe with a small focus of enhancement on T_1-weighted MR image after gadolinium injection in a 47-year-old man who had a partial complex seizure with sensation of déjà vu at the end of a marathon. On the T_2-weighted MR localizer different VOI indicates groups of spectra illustrated in Figure 5.10B.

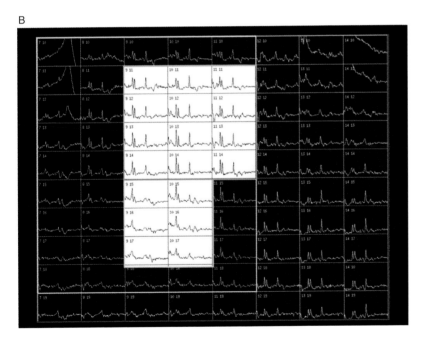

Figure 5.10B. MRSI (PRESS: TR/TE = 1200/135 ms; 24 × 24 matrix; FOV = 200 × 200 × 15 mm^3) was acquired at the level of the temporal lobes. The spectral grid gives an overview of the heterogeneous metabolic changes within the infiltrating tumor. There is moderate Cho elevation, associated with marked Cr and NAA signal loss in the six spectra from the VOI outlined in blue. In the anterior temporal lobe there is only mild Cho and Cr elevation associated with mild NAA signal loss in the VOI outline in green. Normal spectra are seen in the voxels adjacent to the tumor, immediately outside of the T_2-hyperintensity. The majority of metabolic changes are localized within the area of T_2-signal abnormality.

that low-grade oligodendroglioma may have higher mI/Cr ratio than astrocytoma.[60] However, previous studies have shown abnormally elevated mI in LG astrocytomas as well. Elevation of mI and Cr has been reported also in gliomatosis cerebri.[53] In the author's experience, elevation of mI and Cr in LGG is a relatively common finding; however, it appears to be more reflective of the pattern of tumor infiltration in the gray matter rather than of the tumor type.

Alanine resonates at 1.47 ppm and is occasionally found in the spectrum of meningioma and abscesses. Alanine is a J-coupled, doublet peak (similar to lactate)

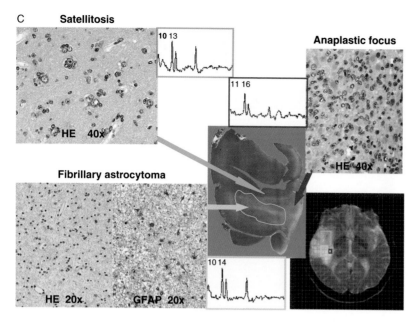

Figure 5.10C. At surgery, histopathologic specimens were coregistered with the presurgical MRSI exam. The location of three histopathologic specimens is indicated in the axial hematoxylin–eosin (HE) stained slice of the resected mass. Moderate cellular density (HE staining) with reactive gliosis on GFAP staining were found in the specimen from the VOI outlined in blue. A focal area with anaplasia was found in the specimen from the VOI outlined in red. Only reactive glial cells surrounding neurons (satellitosis/gliosis) were found in the specimen from the VOI outlined in green in the anterior temporal lobe, where the spectrum shows only mild Cho and Cr elevation associated with mild NAA signal loss. This case nicely illustrates the potential of MRSI to detect tumor heterogeneity.

and it will be inverted at TE values between ~135 and 144 ms (1/J), which can help with assignment. It may overlap with lactate and form an apparent "triplet" peak that will be inverted at TE = 135–144 and positive at TE = 270–288 ms. In a recent study, alanine was found in 10 of 31 meningiomas (32%).[60] Alanine and lactate were overlapped in 8 of the 10 cases. Nine of the meningiomas were benign, one was atypical. However, the reported occurrence of alanine varied greatly according to different studies, from 0/6 to 21/23.[61] One study showed that this variance may arise from two technical aspects: failure to recognize the overlapping peak of alanine and lactate, and the small voxel size.[60] Alanine is produced during transamination between alanine and α-ketoglutarate, instead of glycolysis. Alanine is less commonly detected in malignant meningiomas probably because glycolysis plays a greater role for the large energy supply in these tumors.

There are several indications for an MRS study: differentiation of neoplastic versus non-neoplastic lesions, definition of brain tumor type, grading of glioma, assessing response to therapy, and differentiation of recurrent tumor versus delayed radiation necrosis. In the evaluation of brain tumors, multivoxel (2D and 3D ^{1}H-MRSI[62]) is superior to single-voxel methods for at least three reasons. ^{1}H-MRSI allows the evaluation of spatial heterogeneity and the identification of the macroscopic boundary of a mass, and it is less prone to partial volume artifacts. Spectroscopic imaging is valuable in characterizing areas of T_2-weighted signal hyperintensity: it may delineate areas with high cellular density; it may distinguish areas with prevalent vasogenic edema from areas with neoplastic invasion or necrosis. The choice of the echo time (TE) is also important. Historically, the majority of ^{1}H-MRSI studies have been performed with intermediate (136 ms) or long (272 ms) TE. The longer TE usually provides simpler, cleaner spectra that are easy to interpret, with flatter baselines and less lipid contamination from the scalp. Short TE (30 ms) is recommended when evaluation of metabolites with shorter T_2 values such as myo-inositol, glutamate/glutamine, and lipids are needed.

Interpretation of ^{1}H-MRSI studies will also be more robust if spectra are acquired with high spatial resolution. Many of the spectra illustrated in this chapter were recorded with a nominal spatial resolution of at least $8 \times 8 \times 15\,\text{mm}^3$ (TR/TE = 1200/136; 24×24 matrix, FOV = 200 mm) with a scan time of approximately 14 min.

Five frequently asked questions

Is it a tumor?

This is one of the most frequently asked questions (FAQ) to radiologists. Spectroscopy is valuable in cases when conventional MR imaging and the clinical

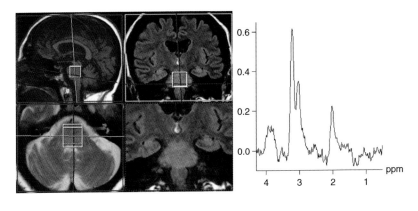

Figure 5.11. Elevated Cho in Erdheim–Chester disease.

Single-voxel MRS (PRESS: TR/TE = 1500/135 ms; VOI = 20 × 20 × 20 mm^3) was acquired from the pons. Coronal FLAIR MR image shows subtle signal hyperintensity in a 60-year-old man with progressive gait ataxia and mild pyramidal signs. The spectrum shows elevated Cho associated with moderate NAA signal loss and mild lactate accumulation. Diagnosis of Erdheim–Chester disease, a rare non-Langherans cell histiocytosis, was confirmed by histopathology. In this systemic disease, histiocytes may infiltrate bones and soft tissues (including retroperitoneum, kidney, lung, skin, orbits, and brain). It is important to recognize the nature of the systemic disease and avoid biopsy of the pontine lesion.

history are ambiguous. When the differential diagnosis includes stroke, focal cortical dysplasia or herpes encephalitis and neoplasm, the finding of an elevated Cho peak makes the diagnosis of tumor more likely. There are caveats as well, and the radiologist must be well aware that some non-neoplastic focal lesions may show elevated Cho. For example, an acute giant demyelinating plaque could mimic an HGG on both MRI and ^{1}H-MRS. Acute demyelinating lesions may show elevated Cho and decreased NAA signal.[63] Successful classification of resonance profiles in 66 gliomas and 5 acute demyelinating lesions using a leave-one-out linear discriminant analysis has been reported.[64] Cho elevation in the ^{1}H-MR spectrum has been also been reported (Figure 5.11) in Erdheim–Chester disease (ECD), a rare non-Langherans histiocytosis that presents in adults. ECD is considered an inflammatory process that is believed to originate from mononuclear phagocytes which proliferate and infiltrate multiple organs, including the brain. Infiltrative nodules or masses in the pons, cerebellum, and hypothalamic-pituitary axis may simulate an infiltrating enhancing tumor on post-contrast MRI.[65]

Is it a GBM, metastasis, or an abscess?

The second FAQ is the differential diagnosis of a ring enhancing mass. The best strategy is to use a multivoxel PRESS sequence with intermediate TE to look for elevation of Cho in the enhancing rim and in the peri-lesional T_2 hyperintensity. If Cho is elevated in both areas, a likely diagnosis of GBM may be suggested.[66] Elevation of Cho in the enhancing rim but not in the surrounding tissue would suggest the diagnosis of metastasis. In spectra derived from the necrotic/cystic core of the mass, accumulation of lipids or lactate without elevated Cho is not a specific finding; thus, the acquisition of an additional single voxel spectrum with short TE would be helpful to detect the presence of other minor peaks beyond lactate: succinate (2.42 ppm), acetate (1.9 ppm), or amino acids such as leucine (3.6 ppm), alanine (1.5 ppm), and valine (0.9 ppm).[67,68] The detection of few or all of these peptides and amino acids confirms the diagnosis of a pyogenic abscess (see also Chapter 7).

What's the grade of this glioma?

Whether ^{1}H-MRSI is useful for grading of gliomas or not remains a controversial issue. There is a body of evidence in the literature that both Cho/NAA and Cho/Cr increase with cellular density and mitotic index. However, in the individual cases it may be difficult to assign a grade to a mass on the basis of ^{1}H-MRS alone (Figure 5.12). It is therefore useful to review changes in metabolic profile occurring during the malignant transformation from diffuse to anaplastic astrocytoma: the NAA signal falls to the baseline while the Cho signal increases with higher cell density and proliferation. The Cho/NAA ratio is likely the most sensitive index for tumor cell density and proliferation. This ratio can be used as a marker of tumor infiltration. Cho/NAA reaches the highest values in anaplastic astrocytomas and GBM. Elevated Cho/NAA values can be found in the solid

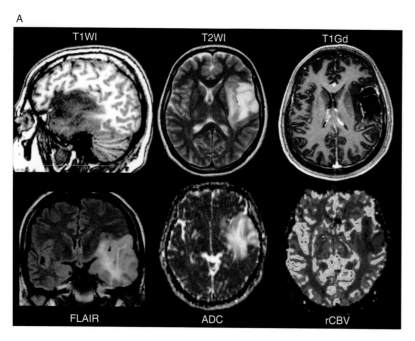

Figure 5.12A. Elevated Cho/NAA in mixed oligoastrocytoma.
In the individual case it may be difficult to assign the grade to a mass on the basis of 1H-MRS alone. Sagittal T_1-weighted, axial T_2-weighted and coronal FLAIR MR images show a large mass infiltrating the right temporal lobe in a 41-year-old woman. There is no evidence of enhancement on T_1-weighted MR image after gadolinium injection. Diffusion imaging shows higher ADC in the lobar white matter compared to the infiltrated temporal gyri. MR perfusion shows a homogeneous mass with low cerebral blood volume (rCBV).

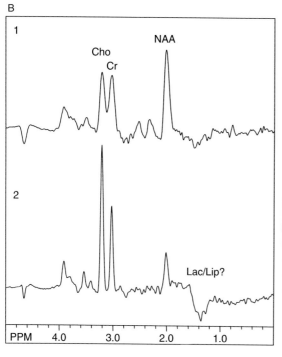
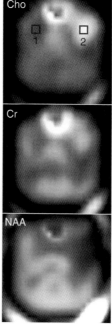

Figure 5.12B. MRSI (PRESS: TR/TE = 1200/135 ms; 24 × 24 matrix; FOV = 200 × 200 × 15 mm³) was acquired at the level of the temporal lobes. Cho, Cr, and NAA maps with two selected spectra are illustrated. In the left temporal mass (2) note elevated Cho and Cr in association with moderate NAA signal loss; the normal spectrum in the contralateral temporal lobe (1) is shown for comparison. In this patient, peak integration showed Cho/NAA = 3.20; Cho/Cr = 1.84 (normal values are Cho/NAA = 0.60; Cho/Cr = 1.00); nCho = 1.58; nCr = 0.63; nNAA = 0.28.

The inverted signal around 1.44 ppm may be an artifact due to lipid signal leaking from the skull base, an undesired but common finding obtained on multivoxel MRSI at the level of the temporal lobes if shimming is not optimal.

components of the mass when the tissue is still well perfused and oxygenated (Figure 5.6). Once components of the mass become hypoxic or their apoptotic index increases, a significant drop of the Cho signal occurs in those areas, sometime associated with accumulation of lipids. The Cr signal also changes during this malignant transformation. Cr signal is usually normal or slightly elevated in differentiated and oxygenated astrocytomas. Elevation of Cr is more commonly seen in astrocytoma infiltrating the cortex compared to those growing in the white matter. In astrocytomas infiltrating the gray matter in the cortex or in the basal ganglia and thalami, both Cho and Cr may be elevated with the Cho/Cr approaching unity or even slightly below (Figure 5.5). Then Cr may drop significantly when new clones with greater proliferation and less differentiating capacities will arise and prevail. Despite several ^{1}H-MRSI studies which have reported high diagnostic accuracy in glioma grading, [38,40,62,69,70,71] the possibility that ^{1}H-MRS may soon replace tissue diagnosis and grading is remote. Only one of the above studies, however, has compared accuracy of ^{1}H-MRSI with contrast-enhanced MR imaging,[69] which is considered the reference standard. Multivoxel studies have shown consistently that there is a trend with HGG having higher Cho/NAA and Cho/Cr compared with LGG. This is especially true if the voxel with the maximum Cho/NAA ratio is used for the analysis. However, 95% confidence intervals are wide and substantial overlapping of results between WHO grade II and III and between grades III and IV makes assignment of grade difficult on the individual patient (Figure 5.13). This is confirmed by ranking all gliomas by the Cho/NAA or Cho/Cr ratio, respectively (Figure 5.14). There is a continuum from slow-growing to fast-growing gliomas, with the LGG more frequently represented on the left side of the plot, the WHO III in the middle and the HGG on the right side. The introduction of cut-offs for Cho/NAA and Cho/Cr has been unsuccessful because of too much overlap between different grades. In addition, different medical centers typically use different acquisition protocols and analysis techniques, and threshold values (if they exist) may be difficult to compare between sites. A further complicating factor is metabolic heterogeneity within lesions – for tumor grading purposes, it is unclear exactly where to measure the spectrum from. Logically, it would appear that the use of MRSI to search for the most abnormal voxel (e.g. highest Cho) within the lesion should find the part of the lesion with the highest density of tumor cells. In many cases, the center of the lesion may be necrotic and therefore not a good location to measure the spectrum from.

In conclusion, metabolite abnormalities in gliomas are distributed along a spectrum from grade II through grade IV. However, higher Cho/NAA and Cho/Cr ratios generally suggest a faster growing neoplasm and higher grade neoplasm.

Can MRS predict patient survival?

Early in this chapter, it has been remarked that the long-standing success of the cytogenetic classification proposed by Bailey and Cushing in 1926[1] is due to a strong correlation between histopathological criteria for grading and prognosis. Whether MRS parameters can predict survival rate has only been the focus of limited studies – one recent study found that high

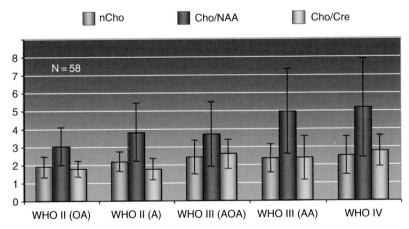

Figure 5.13. Metabolite ratios in pure and mixed astrocytomas.
nCho, Cho/NAA and Cho/Cr were measured in the voxel with maximal spectral abnormality in 58 presurgical patients studied at the Fondazione IRCCS Istituto Neurologico Besta. Multivoxel studies have shown consistently that there is a trend, with high-grade gliomas showing higher Cho/NAA and Cho/Cr compared with low-grade gliomas. This is especially true if the voxel with the maximum Cho/NAA ratio is used for the analysis. Unfortunately, 95% confidence intervals are wide and overlapping of results between different grades makes grade assignment difficult in the individual patient. Note that Cho/NAA is higher in pure astrocytomas than in mixed oligodendroastrocytoma of the same grade. Abbreviations: oligoastrocytoma (OA), astrocytoma (A), anaplastic oligoastrocytoma (AOA), anaplastic astrocytoma (AA).

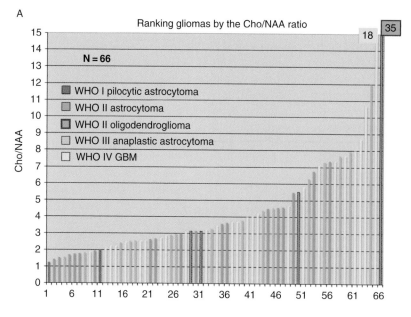

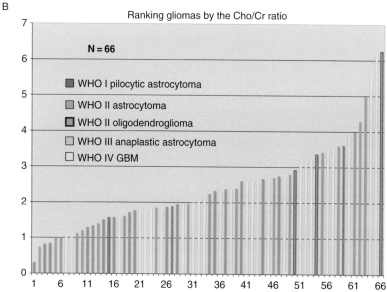

Figure 5.14A–B. Ranking gliomas by Cho/NAA and Cho/Cr.

Presurgical MRSI (PRESS: TR/TE = 1200/135 ms; 24 × 24 matrix; FOV = 200 × 200 × 15 mm^3) were acquired at the level of the mass. Cho/NAA and Cho/Cr were measured in the voxel with maximal spectral abnormality in 66 presurgical patients with glioma studied at the Fondazione IRCCS Istituto Neurologico Besta.

The bar graph shows that there is a spectral distribution of glioma. The LGG are more frequently represented on the left side of the plot (lower Cho/NAA and Cho/Cr), the anaplastic astrocytoma in the middle, and the HGG on the right side. Note that also LG oligodendrogliomas are spread out along the X-axis: in both charts the tumor with the highest ratio (Cho/NAA = 35.64; Cho/Cr = 6.25 is a LG oligodendroglioma (see also Figure 5.15). In this cohort of patients the sensitivity was high (93%), but specificity was very low (35%), with a threshold of Cho/NAA = 2.0. The sensitivity and specificity were 81% and 48%, respectively, with a threshold of Cho/Cr =1.8.

creatine levels in grade II gliomas were a predictor of malignant transformation and decreased survival time.[72] Other studies in pediatric brain tumors have also found high Cho levels (e.g. normalized Cho, or ratio of Cho/NAA) to be prognostic of poor outcome and decreased survival time.[73,74]

Is it an oligodendroglioma?

The fifth and last FAQ is a fascinating question, because it is also the fundamental question patients often ask at presentation: is this an indolent tumor that will eventually respond to therapy? Diagnosis of oligodendroglioma with histopathology is challenging since even the most experienced neuropathologists have not yet agreed on diagnostic criteria. An oligodendroglioma in Paris may be called a diffuse astrocytoma in Baltimore or an oligoastrocytoma in Milan. Oligodendroglioma is a neoplasm with higher cellular density, vascular density, mitotic and apoptotic indices than a diffuse astrocytoma. Notwithstanding, the prognosis is better for oligodendrogliomas than

Chapter 5: MRS in brain tumors

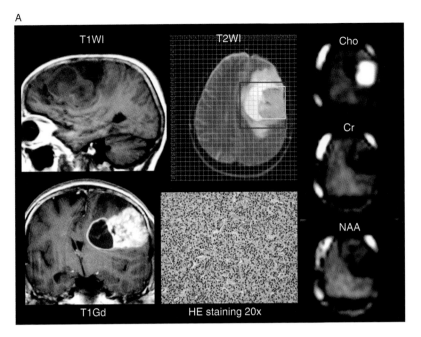

Figure 5.15A. Sky-rocketing choline in an oligodendroglioma.

Sagittal and coronal T_1-weighted and axial T_2-weighted MR images show a large enhancing mass in the left frontal lobe in a 64-year-old woman.

MRSI (PRESS: TR/TE = 1500/144 ms; 32 × 32 matrix; FOV = 200 × 200 × 15 mm^3) was acquired at the level of the mass. Cho, Cr, and NAA maps are illustrated: in the mass, note the very high Cho signal with complete depletion of Cr and NAA. Surprisingly at histopathology the diagnosis of low-grade oligodendroglioma was made. The very high Cho signal is likely related to the very high cellular density that was demonstrated on the HE-stained histologic slide. The rate of mitosis and anaplasia was very low in this oligodendroglioma.

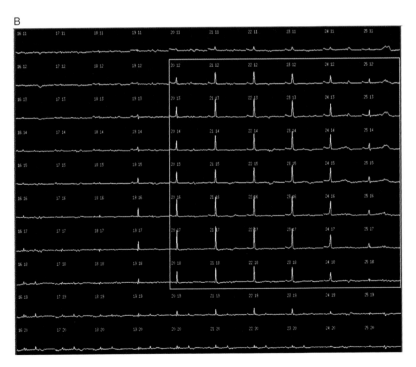

Figure 5.15B. The spectral grid gives an overview of the dramatic elevation of the Cho signal in the tumor (spectra outlined in yellow) in comparison with the adjacent normal tissue.

their astrocytoma "cousins." The oligodendroglioma is ambigous also on MR spectroscopy: the Cho peak may be very highly elevated (Figure 5.15), Cr may be absent, and lactate present – all features of a more malignant neoplasm, yet the prognosis is generally more favorable compared to an astrocytoma. Oligodendrogliomas are "mischievous" also on MR perfusion: rCBV may be very elevated because of high capillary density despite low level of angiogenesis. On post-contrast MRI, oligodendroglioma may sometimes be enhanced as well, although this is not typically seen in low-grade oligodendrogliomas. LOH on chromosome arms 1p and 19q is an early genetic event in the histogenesis of oligodendroglioma and it is considered a genetic hallmark: this feature is present in up to 90% of oligodendrogliomas and 50% of oligoastrocytomas.[24,25]

Diagnostic accuracy of MRS

In general, multivoxel ^{1}H-MRSI studies[40,42] using linear discriminant analysis have shown higher diagnostic accuracy than single-voxel ^{1}H-MRS studies.[39] In one large MRSI study,[40] Preul et al. showed that it correctly classified 104 of 105 patients with 4 tumor histotypes (LG astrocytoma, HGG, meningioma, and metastasis). Tate et al., with single-voxel ^{1}H-MRS and linear discriminant analysis, correctly classified 133 of 144 patients in 3 rather broad tumor types (LG astrocytoma, HG tumors [AA, GBM and metastasis] and meningioma).[75] Herminghaus et al., with single-voxel ^{1}H-MRS and linear discriminant analysis in 94 consecutive gliomas, reported a success rate of 86% in grading glioma and a 95% success rate in differentiating LGG from HGG.[76] Recently, in a retrospective ^{1}H-MRSI study on 69 patients, Hourani et al. have shown a high rate of success (84%) for correctly classifying brain tumors from other non-neoplastic brain lesions (i.e. stroke, demyelination, stable undiagnosed lesions).[77] When MR perfusion was added to the analysis, diagnostic accuracy was unchanged with a sensitivity of 72% and specificity of 92%. ^{1}H-MRSI was also shown to be more accurate than MR perfusion in a subset of patients in the same study. However, it is fair to note that this study showed that both ^{1}H-MRSI and MR perfusion techniques may misclassify patients with LGG as non-neoplastic lesions, and vice versa.

The accuracy of ^{1}H-MRSI, integrated with other MR imaging methods, in diagnosing intraxial focal brain masses with a strategically designed algorithm was determined in a study with 111 patients.[78] Diagnosis for each patient was made after collecting imaging data with contrast-enhanced MRI, diffusion MR, perfusion MR, and ^{1}H-MRSI in this order. Accuracy, sensitivity, and specificity of the strategy, respectively, were 90%, 97%, and 67% for discrimination of neoplastic from non-neoplastic processes, 90%, 88%, and 100% for discrimination of HG from LG neoplasms, and 85%, 84%, and 87% for discrimination of HG neoplasms and lymphoma from LG neoplasms and non-neoplastic diseases. The algorithm was strategically built with eight nodes which required imaging input.

MRS imaging-guided therapy planning and monitoring

When the diagnosis of glioma is suspected, surgery is usually the treatment of choice. When the mass is located in the dominant hemisphere and in particular near or within eloquent areas, the risk of postoperative sequelae is higher. However, in the majority of patients, surgery is performed to obtain a pathological specimen so as to establish a definitive diagnosis, to offer a prognosis and to evaluate possible additional or alternative treatment.

^{1}H-MRS has been proposed as a valuable presurgical planning tool to identify the most aggressive components within the tumor volume.[79] Integration of segmented Cho/NAA maps fused with three-dimensional T_1-weighted MR images in a neuronavigational system (BrainLab, Germany) has been accomplished.[80] These hybrid images have been used for frameless stereotaxy and MR spectroscopy-guided biopsy sampling. A correlation study of ^{1}H-MRSI with histopathology in 76 biopsy specimens has found a negative linear correlation ($r = -0.905$, $P < 0.001$) of NAA concentration and a positive exponential correlation for Cho ($r = 0.769$, $P < 0.001$) and Cho/NAA ($r = 0.885$, $P < 0.001$) with increasing tumor infiltration (indicated by tumor cell nuclei/whole cell nuclei on histopathology).[80]

In patients diagnosed with WHO II glioma who had a total tumor resection, there is no indication for additional therapy. Chemotherapy with procarbazine, CCNU and vincristine (PCV) may be indicated in patients older than 40 years, with a large residual tumor volume after surgery, or when a diagnosis of oligodendroglioma, MOA, or gemistocytic astrocytoma is made. As already mentioned, these tumors will respond well to chemotherapy, especially if molecular genetics shows LOH on chromosome 1p and 19q. Chemotherapy with PVC is mandatory in anaplastic oligodendroglioma and MOA (WHO III). Additional treatment with radiotherapy and chemotherapy with temozolomide (TMZ) is administered in patients diagnosed with anaplastic astrocytoma (WHO III) and GBM (WHO IV). The benefit of a second surgery at recurrence is uncertain, and new clinical trials are needed to assess its effectiveness.[30] Upon tumor recurrence, chemotherapy with TMZ or PVC may improve survival and it is a reasonable option.

Spectroscopic imaging is valuable to target volumes for radiotherapy and to evaluate response to therapy. Incorporation of ^{1}H-MRSI into the treatment planning process may have the potential to improve control while minimizing side effects and complications.[81] ^{1}H-MRSI may be useful in monitoring therapeutic response in patients with brain tumors, as it was demonstrated in a longitudinal study in a woman with non-Hodgkin lymphoma who had a complete and persistent favorable response to radiotherapy (Figure 5.17).[47]

In following patients post-radiation therapy, the differential diagnosis between recurrent tumor and delayed radiation necrosis (DRN) is one common dilemma. ^{1}H-MRSI has been shown to be useful to improve diagnostic acumen, while conventional MR imaging often cannot differentiate the two entities. [82,83,84] Detection of an abnormally elevated Cho signal suggests the diagnosis of recurrent tumor (Figure 5.16). Alternatively, if an elevated Cho peak is not found within areas of T_2-signal abnormality or contrast enhancement, the diagnosis of DRN (or predominantly radiation necrosis) would be suggested (Figure 5.7). In one of the first applications of ^{1}H-MRSI as a presurgical planning tool, Cho elevation was accurate in depicting areas of recurrent tumor within lesions that may be confused with DRN on conventional MR imaging.[85] The detection of lipid signals is not a discriminant sign, despite lipid being usually found in a larger amount in areas of DRN than in areas of recurrent tumor. Lactate may be found in recurrent tumors, but is not a discriminant factor, either.

Special considerations about pediatric brain tumors

Brain tumors are the second most common group of neoplasms in childhood, following leukemia. Among the primary neuroepithelial brain neoplasms, the percentage of LG neoplasms is much higher in children than in adults. In contrast, metastatic brain tumors are rare in children, while they represent 30% of neoplasms in adults. Age at onset and location are two very important diagnostic and prognostic factors. Low-grade astrocytomas (pilocytic or diffuse astrocytoma, pleomorphic xanthoastrocytoma, and subependymal giant cell astrocytoma – usually in association with tuberous sclerosis) are the most frequent (about 35–40%), followed by primitive neuroectodermal tumors (PNET) or medulloblastoma (about 20%) and ependymoma (about 10–15%). Craniopharyngioma, HGG, ganglioglioma, and germ cell tumors are less common primary brain tumors in children. Pilocytic astrocytoma is the most common tumor in children and may occur in cerebellum, hypothalamus, and optic nerves. Diffuse astrocytomas are relatively frequently located in the brain stem and, in particular, in the pons, where they are usually poorly defined and cause diffuse enlargement. PNET is the second most frequent and it

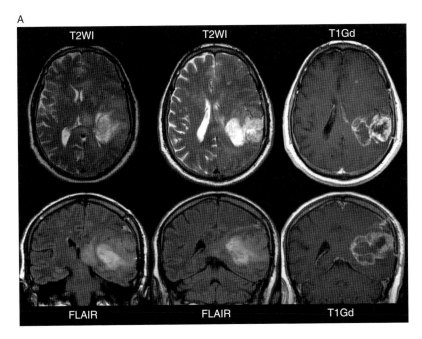

Figure 5.16A. Metabolic heterogeneity in a recurrent GBM.

A 56-year-old man who had chemotherapy after partial surgical resection of a GBM. The surgeon wants to rule out the possibility of an abscess.

Axial T_2-weighted and coronal FLAIR MR images show a large mass infiltrating the right temporal lobe and the inferior parietal lobule. Note a thick rim of enhancement around the surgical cavity on T_1-weighted MR image after gadolinium injection. A deeper seeded component of the mass with a rim of enhancement reaches the wall of the ipsilateral ventricle. More subtle T_2-signal abnormalities are seen in the adjacent temporal and parietal gyri in the ipsilateral thalamus, in the corpus callosum, and left frontal lobe.

Chapter 5: MRS in brain tumors

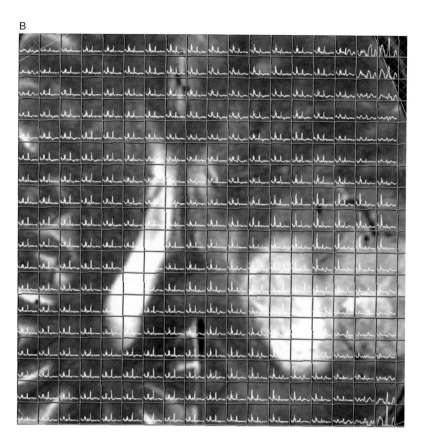

Figure 5.16B. MRSI (PRESS: TR/TE = 1200/135 ms; 24 × 24 matrix; FOV = 200 × 200 × 15 mm^3) was acquired at the level of the lateral ventricles. The spectral grid gives an overview of the dramatic heterogeneous metabolic changes. In the surgical cavity the spectrum is flat with marked depletion of the main metabolites. In the deeper seeded enhancing component, Cho is only moderately elevated, while the dominant peak is lipid, a sign of necrosis. Note the marked elevation of Cho in multiple spectra of the surrounding tissue despite the subtle hyperintensity on T_2-weighted MR images. Note elevated Cho signal also in the left frontal lobe where a small nodule of enhancement is seen on post-Gd T_1-weighted MR images. Normal spectra are seen in the voxels adjacent to the tumor, immediately outside of the T_2-hyperintensity. Also in this case metabolic changes are localized within the area of T_2-signal abnormality.

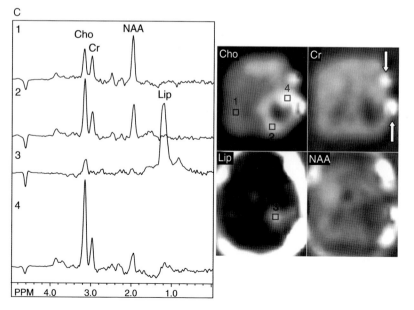

Figure 5.16C. Cho, Cr, Lip, and NAA maps with four selected spectra are illustrated. A normal spectrum from the contralateral temporal lobe (1) is shown for comparison. Note the very high Cho/NAA in two non-enhancing areas of the mass (2, 4), a likely sign of high cellular density and poliferation without necrosis. In the deeper seeded enhancing component of the mass (3) there is a strong signal from mobile lipids, a sign of necrosis, in association with depletion of Cr and NAA; the Cho signal is also relatively weak. A susceptibility artifact at the craniotomy site is hyperintense on the Cho and Cr maps (arrows). Bright signal around the brain on the Lip and NAA maps arises from residual lipid signals from the scalps.

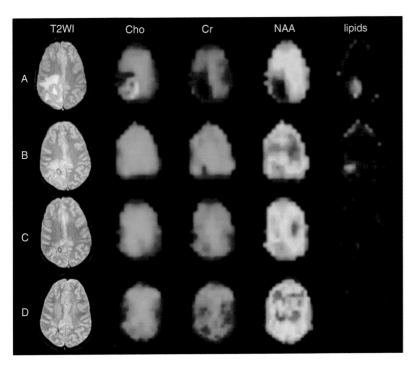

Figure 5.17. Response to therapy in lymphoma.

1H-MRSI maps of Cho, Cr, NAA, and lipids from a serial exams in a patient with non-Hodgkin lymphoma (A) before, (B) 17 days and (C) 28 days after initiation of radiation therapy, and (D) at 33-month follow-up. The pretreatment study shows a right parietal white matter mass containing elevated Cho and lipid, and absent Cr and NAA. There is a progressive reduction of Cho and lipid signals from the lesion during treatment, accompanied by restoration of normal brain metabolites Cr and NAA. At 33 months after diagnosis, Cho and lipid signals within the mass returned to normal levels. Reproduced with permission from Bizzi et al. [47].

occurs in the cerebellar vermis during the first decade of life. Supratentorial locations are rare. These tumors are composed of densely packed cells with hyperchromatic nuclei and scant cytoplasm. Focal areas of hemorrhage and necrosis are frequently found. The presence of leptomeningeal metastases is often associated with PNET, therefore staging with MRI of the spine must be performed before surgery. Ependymoma is relatively more common in childhood than in adults: 65% occur in the posterior fossa, 25% supratentorially, and 10% in the spinal cord. Histologically, they are very well circumscribed and separated from the brain. Ependymoma may show features of anaplasia with high mitotic rate, cellular pleomorphism, and necrosis in about 25% of cases.

In 1995, Wang et al. proposed the use of single-voxel ^{1}H-MRS at long TE to differentiate 30 pediatric brain tumors occurring in the posterior fossa.[86] It was shown that the Cho/NAA ratio was higher in PNET, while Cho/Cr was significantly higher in ependymomas. More recently, it has been shown that an abnormally elevated taurine signal (3.4 ppm) can be detected with single voxel ^{1}H-MRS at short TE in PNET, but not in other posterior fossa tumors.[87] In this study on 29 children, large standard deviations in Cho, Cr, and *myo*-inositol were found in PNET and other types of tumors: Cho/NAA was significantly higher in PNET ($P < 0.001$). In a study on 14 children with hemispheric brain neoplasms, elevated Cho/NAA and Cho/Cr were associated with shorter survival and poor prognosis.[88]

In pediatric patients, as in adults, the Cho/NAA ratio is higher in tumors with higher cell density, associated with a poor prognosis, and this is the most valuable parameter. Typical short TE spectra from various pediatric brain tumors are shown in Figure 5.18. As in adults, definitive diagnosis of tumor type and grade cannot usually be made based on the spectrum alone because of too much overlap between groups; however, MRS may provide useful, complementary information to assist other imaging modalities in making a differential diagnosis.

Conclusions

MR spectroscopy and MRSI are useful techniques that provide unique metabolic information for characterization of brain tumors in vivo. ^{1}H-MRS is a valuable tool in differentiating active and recurrent tumor from non-neoplastic lesions, and from therapy-related tissue injury. ^{1}H-MRS also provides an estimate of cell tumor density and proliferation, and may detect transformation of tumor to a more malignant subtype.

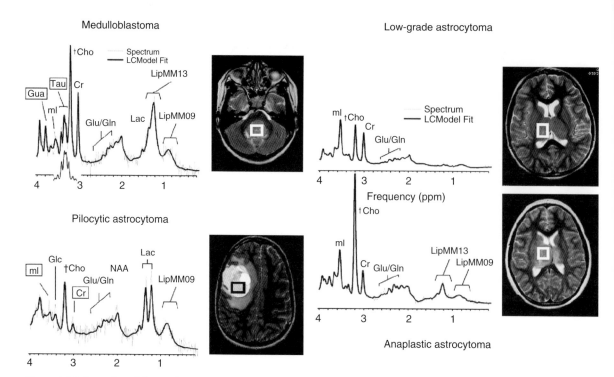

Figure 5.18. Examples of short echo time single voxel spectra from different pediatric brain tumors. The medulloblastoma shows low levels of NAA, as well as elevated levels of Cho, lactate, and lipids, and peaks assigned to taurine (Tau) and guanadinoacetate (Gua). Pilocytic astrocytomas typically have low levels of Cr, as well as elevated lactate in this example. As in adults, high-grade astrocytomas show increased Cho compared to low grade, while NAA is absent in both examples. Reproduced with permission from Panigrahy et al., Am J Neurol Radiol 2006; **27**: 560–72.

Several technical, logistical, and financial issues somewhat inhibit the widespread application of this methodology, including the lack of widely accepted, standardized acquisition and analysis protocols that could be used in multicenter studies. Despite these challenges, ^{1}H-MRS will continue to be an important tool for physicians working to improve brain tumor diagnosis, prognosis, and therapy.

References

[1] Bailey P, Cushing H. *A Classification of the Tumors of the Glioma Group on a Histogenetic Basis with a Correlation Study of Prognosis.* Philadelphia: Lippincott, 1926.

[2] Doetsch F. The glia identity of neural stem cells. *Nature Neurosci* 2003; **6**: 1127–34.

[3] Recht L, Jang T, Litofsky NS. Neural stem cells and neurooncology: Quo vadis? *J Cell Biochem* 2003; **88**: 11–9.

[4] Louis DN. Molecular pathology of malignant gliomas. *AnnuRev Pathol Mech Dis* 2006; **1**: 97–117.

[5] Reya T, Morrison RJ, Clarke MF, Weissman IL. Stem cells, cancer, and cancer stem cells. *Nature* 2001; **414**: 105–11.

[6] Louis DN. A molecular genetic model of astrocytoma histopathology. *Brain Pathol* 1997; **7**: 755–64.

[7] Schiffer D. *Brain Tumor Pathology: Current Diagnostic Hotspots and Pitfalls.* Dordrecht, The Netherlands: Springer, 2006.

[8] Hengartner MO. Biochemistry of apoptosis. *Nature* 2002; **2000**: 770–6.

[9] Steinmach JP, Weller M. Mechanisms of apoptosis in central nervous system tumors: application to theory. *Curr Neurol Neurosci Rep* 2002; **2**: 246–53.

[10] Swanson CR, Bridge C, Murray JD, Alvord AC. Virtual and real brain tumors: using mathematical modeling to quantify glioma growth and invasion. *J Neurol Sci* 2003; **216**: 1–10.

[11] Burger PC, Heinz ER, Shibata T, Kleihues PC. Topographic anatomy and CT correlations in the untreated glioblastoma multiforme. *J Neurosurg* 1988; **68**: 698–704.

[12] Darlymple JS, Parisi JE, Roche PC, Ziesmer SC, Scheithauer BW, Kelly PJ. Changes in proliferating cell nuclear antigen expression in glioblastoma multiforme cells along a stereotactic biopsy trajectory. *Neurosurgery* 1994; **35**: 1036–45.

[13] Schiffer D, Cavalla P, Dutto A, Borsotti L. Cell proliferation and invasion in malignant gliomas. *Anticancer Res* 1997; **17**: 61–70.

[14] Louis DN, Ohgaki H, Wiestler OD, Cavenee WK, Burger PC, Jouvet A, *et al.* The 2007 WHO classification of tumours of the central nervous system. *Acta Neuropathol (Berl)* 2007; **114**: 97–109.

[15] Duffau H, Capelle L. Preferential brain locations of low-grade gliomas. *Cancer* 2004; **100**: 2622–6.

[16] Daumas-Duport C, Varlet P, Tucker ML, Beuvon F, Cervera P, Chodkiewicz J-P. Oligodendroglioma. Part I: Pattern of growth, histological diagnosis, clinical and imaging correlations: a study of 153 cases. *J Neuro-Oncology* 1997; **34**: 37–59.

[17] Daumas-Duport C, Tucker ML, Kolles H, Cervera P, Beuvon F, Varlet P, *et al.* Oligodenrogliomas. Part II: A new grading system based on morphological and imaging criteria. *J. Neuro-Oncology* 1997; **95**: 493–504.

[18] Burger PC. What's an oligodendroglioma? *Brain Pathol* 2002; **12**: 257–9.

[19] Cairncross JG, Ueki K, Zlatescu MC, Lisle DK, Finkelstein DM, Hammond RR, *et al.* Specific genetic predictors of chemotherapeutic response and survival in patients with anaplastic oligodendrogliomas. *J Natl Cancer Inst* 1998; **90**: 1473–9.

[20] Schiffer D, Dutto A, Cavalla P, Bosone I, Chiò A, Villani R, *et al.* Prognostic factors in oligodendroglioma. *Can J Neurol Sci* 1997; **24**: 313–9.

[21] Louis DN, Ohgaki H, Wiestler OD, Cavenee WK, Burger PC, Jouvet A, *et al.* The 2007 WHO classification of tumours of the central nervous system. *Acta Neuropathol* 2007; **114**: 97–109.

[22] Schiffer D, Dutto A, Cavalla P, Chiò A, Migheli A, Piva R. Role of apoptosis in the prognosis of oligodendrogliomas. *Neurochem Int* 1997; **2**: 245–50.

[23] Cooper ERA. The relation of oligodendrocytes and astrocytes in cerebral tumors. *J Pathol Bacteriol* 1935; **41**: 259–66.

[24] Bissola L, Eoli M, Pollo B, Merciai BM, Silvani A, Salsano E, *et al.* Association of chromosome 10 losses and negative prognosis in oligoastrocytomas. *Ann Neurol* 2002; **52**: 842–5.

[25] Kraus JA, Jkoopman J, Kaske P, Maintz D, Brandner S, Schramm J, *et al.* Shared allelic losses on chromosomes 1p and 19q suggest a common origin of oligodendroglioma and oligoastrocytoma. *J Neuropathol Exp Neurol* 1995; **54**: 91–4.

[26] Kleihues P, Ohgaki H. Primary and secondary glioblastoma: from concept to clinical diagnosis. *Neuro-Oncology* 1999; **1**: 44–51.

[27] Lim DA, Mayo MC, Chen M-H, Keles E, Berger MS. Relationship of glioblastoma multiforme to neural stem cell regions predicts invasive and multifocal tumor phenotype. *Neuro-Oncology* 2007; **9**: 424–9.

[28] Daumas-Duport C, Scheithauer B, O'Fallon J, Kelly P. Grading of astrocytomas: a simple and reproducible method. *Cancer* 1988; **62**: 2152–65.

[29] Bauman G, Lote K, Larson D, Stalpers L, Leighton C, Fisher B, *et al.* Pretreatment factors predict overall survival for patients with low-grade glioma: a recursive partitioning analysis. *Int J Radiat Oncol Biol Phys* 1999; **45**: 923–9.

[30] Filippini G, Falcone C, Boiardi A, Broggi G, Bruzzone MG, Caldiroli D, *et al.* Prognostic factors for survival in 676 consecutive patients with newly diagnosed primary glioblastoma. *Neuro-Oncology*, 2007 Nov 9 [Epub ahead of print].

[31] Olson JD, Riedel E, Deangelis LM. Long-term outcome of low-grade oligodendroglioma and mixed glioma. *Neurology* 2000; **54**: 1442–8.

[32] Dean BL, Drayer BP, Bird CR, Flom RA, Hodak JA, Coons SW, *et al.* Gliomas: classification with MR imaging. *Radiology* 1990; **174**: 411–5.

[33] Christy PS, Tervonen O, Scheithauer BW, Forbes GS. Use of a neural network and a multiple regression model to predict histologic grade of astrocytoma from MRI appearances. *Neuroradiology* 1995; **37**: 89–93.

[34] Mihara F, Numaguchi Y, Rothman M, Sato S, Fiandaca MS. MR imaging of adult supratentorial astrocytomas: an attempt of semiautomatic grading. *Radiat Med* 1995; **13**: 5–9.

[35] Bruhn H, Frahm J, Gyngell ML, Merboldt KD, Hanicke W, Sauter R, *et al.* Noninvasive differentiation of tumors with use of localized H-1 MR spectroscopy in vivo: initial experience in patients with cerebral tumors. *Radiology* 1989; **172**: 541–8.

[36] Alger JR, Frank JA, Bizzi A, Fulham MJ, Desouza BX, Duhaney MO, *et al.* Metabolism of human gliomas: assessment with H-1 MR spectroscopy and F-18 fluorodeoxyglucose PET. *Radiology* 1990; **177**: 633–41.

[37] Demaerel P, Johannik K, Van Hecke P, Van Ongeval C, Verellen S, Marchal G, *et al.* Localized 1H NMR spectroscopy in fifty cases of newly diagnosed intracranial tumors. *J Comput Assist Tomogr* 1991; **15**: 67–76.

[38] Fulham MJ, Bizzi A, Dietz MJ, Shih HH, Raman R, Sobering GS, et al. Mapping of brain tumor metabolites with proton MR spectroscopic imaging: clinical relevance. *Radiology* 1992; **185**: 675–86.

[39] Negendank W. Studies of human tumors by MRS: a review. *NMR Biomed* 1992; **5**: 303–24.

[40] Preul MC, Caramanos Z, Collins DL, Villemure JG, Leblanc R, Olivier A, et al. Accurate, noninvasive diagnosis of human brain tumors by using proton magnetic resonance spectroscopy. *Nat Med* 1996; **2**: 323–5.

[41] Tamiya T, Kinoshita K, Ono Y, Matsumoto K, Furuta T, Ohmoto, T. Proton magnetic resonance spectroscopy reflects cellular proliferative activity in astrocytomas. *Neuroradiology* 2000; **42**: 333–8.

[42] De Edelenyi FS, Rubin C, Esteve F, Grand S, Decorps M, Lefournier V, et al. A new approach for analyzing proton magnetic resonance spectroscopic images of brain tumors: nosologic images. *Nat Med* 2000; **6**: 1287–9.

[43] Ackerstaff E, Glunde K, Bhujwalla ZM. Choline phospholipid metabolism: a target in cancer cells? *J Cell Biochem* 2003; **90**: 525–33.

[44] Podo F. Tumour phospholipid metabolism. *NMR Biomed* 1999; **12**: 413–39.

[45] Shimizu H, Kumabe T, Shirane R, Yoshimoto T. Correlation between choline level measured by proton MR spectroscopy and Ki-67 labeling index in gliomas. *Am J Neuroradiol* 2000; **21**: 659–65.

[46] Guillevin R, Menuel C, Duffau H, Kujas M, Capelle L, Aubert A, et al. Proton magnetic resonance spectroscopy predicts proliferative activity in diffuse low-grade gliomas. *J Neuro-oncol* 2007 Dec 28 [Epub ahead of print].

[47] Bizzi A, Movsas B, Tedeschi G, Phillips CL, Okunieff P, Alger JR, et al. Response of non-Hodgkin lymphoma to radiation therapy: early and long-term assessment with H-1 MR spectroscopic imaging. *Radiology* 1995; **194**: 271–6.

[48] Bhakoo KK, Williams SR, Florian CL, Land H, Noble MD. Immortalization and transformation are associated with specific alterations in choline metabolism. *Cancer Res* 1996; **56**: 4630–5.

[49] Usenius JP, Vainio P, Hernesniemi J, Kauppinen RA. Choline-containing compounds in human astrocytomas studied by 1H NMR spectroscopy in vivo and in vitro. *J Neurochem* 1994; **63**: 1538–43.

[50] Gupta RK, Cloughesy TF, Sinha U, Garakian J, Lazareff J, Rubino G, et al. Relationships between choline magnetic resonance spectroscopy, apparent diffusion coefficient and quantitative histopathology in human glioma. *J Neurooncol* 2000; **50**: 215–26.

[51] Tedeschi G, Lundbom N, Raman R, Bonavita S, Duyn JH, Alger JR, et al. Increased choline signal coinciding with malignant degeneration of cerebral gliomas: a serial proton magnetic resonance spectroscopy imaging study. *J Neurosurg* 1997; **87**: 516–24.

[52] Stadlbauer A, Gruber S, Nimsky C, Fahlbusch R, Hammen T, Buslei R, et al. Preoperative grading of gliomas by using metabolite quantification with high-spatial-resolution proton MR spectroscopic imaging. *Radiology* 2006; **238**: 958–69.

[53] Galanaud D, Chinot O, Nicoli F, Confort-Gouny S, Le Fur Y, Barrie-Attarian M, et al. Use of proton magnetic resonance spectroscopy of the brain to differentiate gliomatosis cerebri from low-grade glioma. *J Neurosurg* 2003; **98**: 269–76.

[54] Sijens PE, Levendag PC, Vecht CJ, Van Dijk P, Oudkerk M. 1H MR spectroscopy detection of lipids and lactate in metastatic brain tumors. *NMR Biomed* 1996; **9**: 65–71.

[55] Poptani H, Gupta RK, Roy R, Pandey R, Jain VK, Chhabra DK. Characterization of intracranial mass lesions with in vivo proton MR spectroscopy. *Am J Neuroradiol* 1995; **16**: 1593–603.

[56] Kuesel AC, Sutherland GR, Halliday W, Smith IC. 1H MRS of high grade astrocytomas: mobile lipid accumulation in necrotic tissue. *NMR Biomed* 1994; **7**: 149–55.

[57] Zoula S, Herigault G, Ziegler A, Farion R, Decorps M, Remy C. Correlation between the occurrence of 1H-MRS lipid signal, necrosis and lipid droplets during C6 rat glioma development. *NMR Biomed* 2003; **16**: 199–212.

[58] Harting I, Hartmann M, Jost G, Sommer C, Ahmadi R, Heiland S, et al. Differentiating primary central nervous system lymphoma from glioma in humans using localised proton magnetic resonance spectroscopy. *Neurosci Lett* 2003; **342**: 163–6.

[59] Warburg O. On the origin of cancer cells. *Science* 1956; **123**: 309–14.

[60] Yue Q, Isobe T, Shibata Y, Anno I, Kawamura H, Yamamoto Y, et al. New observations concerning the interpretation of magnetic resonance spectroscopy of meningioma. *Eur Radiol* 2008; **12**: 2901–11.

[61] Demir MK, Iplikcioglu AC, Dincer A, Arslan M, Sav A. Single voxel proton MR spectroscopy findings of typical and atypical intracranial meningiomas. *Eur J Radiol* 2006; **60**: 48–55.

[62] Li X, Lu Y, Pirzkall A, Mcknight T, Nelson SJ. Analysis of the spatial characteristics of metabolic abnormalities in newly diagnosed glioma patients. *J Magn Reson Imaging* 2002; **16**: 229–37.

[63] De Stefano N, Matthews P, Antel JP, Preul MC, Francis G, Arnold DL. Chemical pathology of acute demyelinating lesions and its correlation with disability. *Ann Neurol* 1995; **38**: 901–09.

[64] De Stefano N, Caramanos Z, Preul MC, Francis G, Antel JP, Arnold DL. In vivo differentiation of astrocytic brain tumors and isolated demyelinating lesions of the type seen in multiple sclerosis using 1H magnetic resonance spectroscopic imaging. *Ann Neurol* 1998; **44**: 273–8.

[65] Salsano E, Savoiardo M, Nappini S, Maderna E, Pollo B, Chinaglia D, et al. Late-onset sporadic ataxia, pontine lesion, and retroperitoneal fibrosis: a case of Erdheim–Chester disease. *Neurol Sci* 2008; **29**: 263–7.

[66] Law M, Cha S, Knopp EA, Johnson G, Arnett J, Litt AW. High-grade gliomas and solitary metastases: differentiation by using perfusion and proton spectroscopic MR imaging. *Radiology* 2002; **222**: 715–21.

[67] Grand S, Passaro G, Ziegler A, Estève F, Boujet C, Hoffmann D, et al. Necrotic tumor versus brain abscess: importance of amino acids detected at 1H MR spectroscopy – initial results. *Radiology* 1999; **213**: 785–93.

[68] Kapsalaki EZ, Gotsis ED, Fountas KN. The role of proton magnetic resonance spectroscopy in the diagnosis and categorization of cerebral abscesses. *Neurosurg Focus* 2008; **24**: E3.

[69] Law M, Yang S, Wang H, Babb JS, Johnson G, Cha S, et al. Glioma grading: sensitivity, specificity, and predictive values of perfusion MR imaging and proton MR spectroscopic imaging compared with conventional MR imaging. *Am J Neuroradiol* 2003; **24**: 1989–98.

[70] Astrakas LG, Zurakowski D, Tzika AA, Zarifi MK, Anthony DC, De Girolami U, et al. Noninvasive magnetic resonance spectroscopic imaging biomarkers to predict the clinical grade of pediatric brain tumors. *Clin Cancer Res* 2004; **10**: 8220–8.

[71] Lukas L, Devos A, Suykens JA, Vanhamme L, Howe FA, Majós C, et al. Brain tumor classification based on long echo proton MRS signals. *Artif Intell Med* 2004; **31**: 73–89.

[72] Hattingen E, Raab P, Franz K, Lanfermann H, Setzer M, Gerlach R, et al. Prognostic value of choline and creatine in WHO grade II gliomas. *Neuroradiology* 2008; **50**: 759–67.

[73] Marcus KJ, Astrakas LG, Zurakowski D, Zarifi MK, Mintzopoulos D, Poussaint TY, et al. Predicting survival of children with CNS tumors using proton magnetic resonance spectroscopic imaging biomarkers. *Int J Oncol* 2007; **30**: 651–7.

[74] Warren KE, Frank JA, Black JL, Hill RS, Duyn JH, Aikin AA, et al. Proton magnetic resonance spectroscopic imaging in children with recurrent primary brain tumors. *J Clin Oncol* 2000; **18**: 1020–6.

[75] Tate AR, Majos C, Moreno A, Howe FA, Griffiths JR, Arus C. Automated classification of short echo time in in vivo 1H brain tumor spectra: a multicenter study. *Magn Reson Med* 2003; **49**: 29–36.

[76] Herminghaus S, Dierks T, Pilatus U, Moller-Hartmann W, Wittsack J, Marquardt G, et al. Determination of histopathological tumor grade in neuroepithelial brain tumors by using spectral pattern analysis of in vivo spectroscopic data. *J Neurosurg* 2003; **98**: 74–81.

[77] Hourani R, Brant LJ, Rizk T, Weingart JD, Barker PB, Horska A. Can proton MR spectroscopic and perfusion imaging differentiate between neoplastic and nonneoplastic brain leisons in adults? *Am J Neuroradiol* 2008; **29**: 366–72.

[78] Al-Okaili RN, Krejza J, Woo JH, Wolf RL, O'Rourke DM, Judy KD, et al. Intraaxial brain masses: MR imaging-based diagnostic strategy – initial experience. *Radiology* 2007; **243**: 539–50.

[79] Dowling C, Bollen AW, Noworolski SM, McDermott MW, Barbaro NM, Day MR, et al. Preoperative proton MR spectroscopic imaging of brain tumors: correlation with histopathologic analysis of resection specimens. *Am J Neuroradiol* 2001; **22**: 604–12.

[80] Stadlbauer A, Moser E, Gruber S, Nimsky C, Fahlbusch R, Ganslandt O. Integration of biochemical images of a tumor into frameless stereotaxy achieved using a magnetic resonance imaging/magnetic resonance spectroscopy hybrid data set. *J Neurosurg* 2004; **101**: 287–94.

[81] Pirzkall A, McKnight TR, Graves EE, Carol MP, Sneed PK, Wara WW, et al. MR-spectroscopy guided target delineation for high-grade gliomas. *Int J Radiat Oncol Biol Phys* 2001; **50**: 915–28.

[82] Taylor JS, Langston JW, Reddick WE, Kingsley PB, Ogg RJ, Pui MH, et al. Clinical value of proton magnetic resonance spectroscopy for differentiating recurrent or residual brain tumor from delayed cerebral necrosis. *Int J Radiat Oncol Biol Phys* 1996; **36**: 1251–61.

[83] Schlemmer HP, Bachert P, Henze M, Buslei R, Herfarth KK, Debus J, et al. Differentiation of radiation necrosis from tumor progression using proton magnetic resonance spectroscopy. *Neuroradiology* 2002; **44**: 216–22.

[84] Rock JP, Hearshen D, Scarpace L, Croteau D, Gutierrez J, Fisher JL, et al. Correlations between magnetic resonance spectroscopy and image-guided histopathology, with special attention to radiation necrosis. *Neurosurgery* 2002; **51**: 912–9; discussion 919–20.

[85] Preul MC, Leblanc R, Caramanos Z, Kasrai R, Narayanan S, Arnold DL. Magnetic resonance spectroscopy guided brain tumor resection: differentiation between recurrent glioma and radiation change in two diagnostically difficult cases. *Can J Neurol Sci* 1998; **25**: 13–22.

[86] Wang Z, Sutton LN, Cnaan A, Haselgrove JC, Rorke LB, Zhao H, *et al.* Proton MR spectroscopy of pediatric cerebellar tumors. *Am J Neuroradiol* 1995; **16**: 1821–33.

[87] Kovanlikaya A, Panigrahy A, Krieger MD, Gonzalez-Gomez I, Ghugre N, McComb JG, *et al.* Untreated pediatric primitive neuroectodermal tumor in vivo: quantitation of taurine with MR spectroscopy. *Radiology* 2005; **236**: 1020–5.

[88] Girard N, Wang ZJ, Erbetta A, Sutton LN, Phillips PC, Rorke LB., *et al.* Prognostic value of proton MR spectroscopy of cerebral hemisphere tumors in children. *Neuroradiology* 1998; **40**: 121–5.

Chapter 6
MRS in stroke and hypoxic–ischemic encephalopathy

Key points
- MRS is highly sensitive to metabolic changes associated with hypoxic or ischemic injury to the brain.
- Lactate is elevated during acute hypoxia or ischemia, and may also increase during "secondary" energy failure after reperfusion.
- NAA decreases with prolonged hypoxia or ischemia.
- In ^{31}P MRS, high-energy phosphates decrease, inorganic phosphate increases, and pH decreases during acute hypoxia and ischemia.
- Both ^{1}H and ^{31}P MRS offer prognostic information in HIE, which may be complementary to, and sometimes easier to interpret than, conventional or diffusion-weighted MRI in the neonatal brain.
- DWI may be better for evaluating small HIE lesions than MRS.

Introduction: MR spectroscopy in stroke

A stroke is the rapidly developing loss of brain function due to vascular failure to supply adequate blood nutrients to the brain. Stroke can be due to ischemia (lack of blood supply) caused by thrombosis or embolism, or due to a hemorrhage (vasculature rupture). Acute stroke is a medical emergency, and imaging plays an important role in confirming the clinical diagnosis of stroke, categorizing it as either ischemic or hemorrhagic, and identifying the underlying pathophysiology. Increasingly, imaging is also being used to guide therapeutic interventions and monitor their success. Traditionally, X-ray computed tomography has been the imaging modality of choice, primarily because of its speed and widespread availability. MRI has been used increasingly because of its excellent soft tissue contrast, high sensitivity and multimodal capabilities (e.g. diffusion, perfusion, MR angiography, etc.); however, it is still not particularly widely used worldwide, because of limited access in the acute setting in many hospitals.

The first reports of the application of proton magnetic resonance spectroscopy (^{1}H-MRS) to human stroke appeared in the late 1980s.[1,2] Although there have been reports of ^{31}P,[3] ^{23}Na,[4] and ^{13}C[5] spectroscopy in human stroke, the majority of studies to date have utilized the proton nucleus, both because of its high sensitivity and the fact that it can be readily combined with conventional MRI. While proton MRS studies performed in the early 1990s showed promise for its diagnostic value in acute stroke, MRS has had relatively little clinical impact since then. Several reasons may explain this lack of clinical use. A major obstacle is the technical difficulty of performing MRS in a timely fashion in the acute stroke population. In addition, other MRI techniques (such as diffusion and perfusion MRI) can be performed more easily and provide much of the information needed to make treatment decisions. Nevertheless, it is important to be aware of the spectroscopic correlates of acute and chronic infarction, and on occasion MRS may be helpful, particularly when trying to distinguish ischemic from non-ischemic lesions.

Most proton MRS studies of human stroke have focused on the signals from N-acetyl aspartate (NAA) and lactate, as potential surrogate markers of neuronal integrity and ischemia, respectively. However, there are also often changes in the other metabolite signals, particularly in the chronic stages of brain infarction. The significance of changes in brain metabolites in the context of ischemia and infarction is briefly discussed below (for a more detailed review of these compounds and their significance to other pathologies, please see Chapter 1).

NAA

The evidence for and against NAA (2.02 ppm) as a surrogate neuronal marker is discussed in Chapter 1.

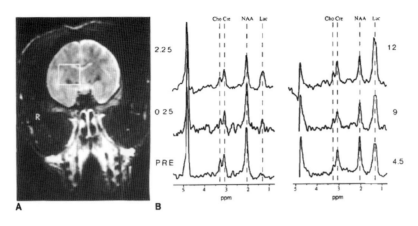

Figure 6.1. Single-voxel proton spectra (TR/TE 2000/280 msec) recorded as a function of time in the basal ganglia of the baboon after permanent occlusion of the right middle cerebral artery. Lactate is seen to increase and NAA decrease steadily over a time period of 12 h. Choline and creatine signals are relatively unchanged. Reproduced with permission from [79].

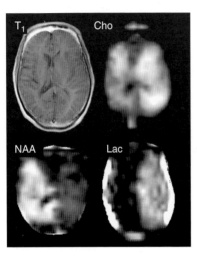
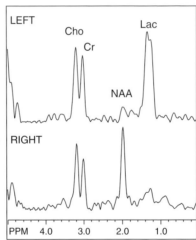

Figure 6.2. Proton MRSI in a patient with embolic, tandem left ICA and MCA occlusion 24 h after stroke onset. T_1-weighted MRI shows swelling of left hemisphere, while MRSI shows strongly elevated lactate and an absence of NAA throughout the hemisphere. Such a large and severe metabolic abnormality has a very poor prognosis; the patient died 3 days later.

Various studies have looked at the time-course of NAA changes following the induction of ischemia. [6,7,8,9] In animal models of focal ischemia (middle cerebral artery occlusion), it has been shown consistently that NAA declines quite slowly over a time scale of hours after the induction of ischemia. Typically, NAA may be reduced to 50% of its pre-ischemic value after approximately 10–12 h (Figure 6.1),[7] although this time course is probably highly sensitive to the degree of cerebral blood flow reduction. For instance, NAA reductions were found to be greater in the core of the ischemic region compared to the periphery in the study by Higuchi et al. [10] Several papers have also described an initial rapid decrease in NAA of about 10% within the first few minutes, [9,10,11] followed by a slower decrease in NAA with a time constant of hours. The origin of this is uncertain, but it has been suggested that this may be due to the presence of more than one pool of NAA, or perhaps changes in other molecules (e.g. glutamate, glutamine, GABA, etc.) which overlap with the spectral resonance of NAA. It might also be due to the dilutional effect of cytotoxic edema.

Since it is generally thought that irreversible brain damage in most of these models occurs over the first 1–3 h of ischemia, the reduction in NAA may occur more slowly than the loss of tissue viability. In infarcted brain (e.g. weeks to months after stroke onset), either in animal models or in humans, NAA is generally either very low or completely absent, consistent with the complete loss of neuronal tissue from these regions.[1,12,13] Figure 6.2 shows an example of a hemispheric loss of NAA in a patient with a tandem occlusion of the left internal and middle cerebral arteries who was imaged 24 h after stroke onset. In addition to loss of NAA, this case also exhibited elevated lactate and choline signals.

Cr

As discussed in Chapter 1, the creatine signal (Cr, 3.02 ppm) is the sum of both creatine and phosphocreatine, which are in exchange via the creatine kinase reaction. Acute ischemia injury causes phosphocreatine to be converted to creatine, but no net change in the total creatine, so it is tempting to use the creatine as a reference signal in the spectrum. However, more recent studies have suggested that creatine may change in both acute and chronic infarction,[14] so it is probably unwise to assume that creatine levels are always normal in human stroke.

Cho

The choline signal (Cho, 3.24 ppm) arises from the trimethylamine groups of glycerophosphocholine (GPC), phosphocholine (PC), and a small amount of free choline,[15] compounds which are involved in membrane synthesis and degradation. Choline has been observed to either be increased or decreased in chronic human stroke.[16,17] Elevation of Cho in stroke may be the result of gliosis or ischemic damage to myelin, while decreases are probably the result of edema, necrosis, and cell loss.

Lactate

In normal human brain, lactate (1.33 ppm) is difficult to detect using conventional MRS techniques, since its concentration is estimated to be about 1 mM concentration or less. When the brain becomes hypoxic or ischemic, the lack of oxygen results in an inability for glucose to be metabolized through the normal tricarboxylic acid (TCA) cycle. The less efficient non-aerobic pathway from pyruvate to lactate is employed instead. Hence, hypoxia or ischemia causes an elevation in brain lactate.[18] Lactate may also be detected in chronic brain infarction, where its presence may be explained by the glycolytic metabolism of macrophages, rather than ongoing chronic ischemia.[19] Lactate may also accumulate in the extracellular space, cysts, and in cerebrospinal fluid.

It should be noted that ischemia and infarction are not the only causes of increased brain lactate. For instance, lactate is quite frequently observed in brain tumors, mitochondrial diseases, and other conditions such as inflammatory demyelinating lesions.[20] Therefore, observation of an increased lactate signal is not necessarily specific for an ischemic lesion.

Metabolic changes during ischemia: animal models

As cerebral blood flow (CBF) decreases, various processes related to cerebral homeostasis gradually fail. [21] Once CBF has decreased below 15–20 ml/100 g tissue/min,[22] the brain becomes ischemic, with the cessation of electrical function, and the accumulation of lactate. Reported CBF thresholds may vary depending on the animal model used, gray or white matter, the type of anesthesia, the type and duration of ischemia, arterial oxygenation and hematocrit, and the method used to measure CBF. However, in complete global ischemia induced by cardiac arrest, lactate levels rise abruptly,[23] and reach a steady state within 10 min of cessation of blood flow.[24,25] As lactate accumulates, the tissue may become acidotic.[22] The final lactate concentration depends on a number of factors, but in particular on the pre-ischemic blood glucose and brain glycogen stores.[23] Under normoglycemic conditions, lactate may typically reach 10–12 mM.[18] Pre-ischemic hyperglycemia may increase final lactate concentrations, and worsen eventual outcome (as seen in human studies). If ischemia is incomplete, or reperfusion occurs, blood flow continues to supply glucose to the tissue, which, if damaged, is unable to metabolize it aerobically, and extremely high lactate concentrations may result.[18,24]

In models of focal ischemia (where presumably CBF reductions are more moderate because of collateral circulation), the increase in lactate may often be significantly slower, increasing over a period of hours. [7,9,10,26,27] For instance, in a permanent middle cerebral artery (MCA) occlusion model, lactate was observed to increase steadily up to 12 h after induction of ischemia (Figure 6.1).[7] In one report, it was also suggested that transient lactate elevations coincided with a burst of cortical spreading depression (CSD) in peri-infarct tissue, which has been postulated to be a mechanism for infarct enlargement into surrounding tissues.[27]

In addition to the increase in lactate, NAA is observed to decrease following the onset of ischemia. It would appear that the rate of NAA decrease (like that of the lactate increase) is dependent on the degree of blood flow reduction to the ischemic tissue, but it is likely that the CBF thresholds for these processes are different, and that they also have different time constants. For instance, in both animal models of ischemia and in human stroke, elevated lactate in peri-infarct regions with near-normal NAA levels has been reported.[16,26] It is tempting to speculate that this dysfunctional tissue with relative neuronal preservation may represent an ischemic metabolic penumbra, [12] although at present this concept is untested.

If the duration and severity of ischemia is short enough (e.g. no more than 10 min in the case of complete ischemia), then most of the metabolic alterations described above are reversible, i.e. establishment of reperfusion will result in restoration of normal metabolite levels and function.[28] Reperfusion after a longer period of ischemia may result in initial restoration of metabolite levels, only to be followed by secondary energy failure over the subsequent 24–48 h.[70,71] As this secondary energy failure continues, or in the case of permanent ischemia, irreversible changes occur and the tissue will progress to neuronal loss, infarction, and gliosis, typically characterized on MRS with low NAA levels.

Metabolic changes in human stroke

In the first two papers reporting MR spectroscopy of human brain infarction, NAA was completely absent from both infarcted tissue at 4 days[2] and at 10 months post-stroke onset.[1] Other metabolic changes have also been reported in the chronic stage of stroke; these include increases of choline containing compounds,[16] and mobile lipid signals.[29] It has also been reported that *myo*-inositol may increase after stroke, perhaps due to glial cell proliferation.[30]

From a clinical viewpoint, the most valuable metabolic information is that acquired during the acute stages of stroke because it may guide treatment decisions and prognosis. Using SV-MRS, it was found that elevated lactate and decreased NAA levels could be detected in cases of acute (< 24 h), [31,32,33], as well as in sub-acute (24 h to 7 days)[2,33,34] and chronic (> 7 days)[1,31,34,35,36] stroke. Lactate was highest in the acute stage, and it was also higher in the most extensive ischemic strokes, whereas it is generally barely detectable by the chronic stage (3 weeks).

Many of these early MRS studies of human stroke used single-voxel localization methods. However, single-voxel techniques do not provide information regarding the spatial distribution and extent of metabolic abnormalities, and require that the location of the ischemic or infarcted region be already known or visible on MRI studies. To address these issues, it is possible to use MR spectroscopic imaging (MRSI) methods for the study of cerebral ischemia, either in one[19,37,38] or two spatial dimensions,[17,39] or using multi-slice 2D MRSI.[12,16]

An example of a serial multi-slice MRSI study of an acute stroke patient is shown in Figure 6.3.[16] The patient presented with a left hemiparesis as the result of a complete occlusion of the right internal carotid artery (ICA), and low flow in the right middle cerebral artery (MCA). Conventional T_2-weighted MR images were normal, while proton MRSI (at 24 h after symptom onset) revealed elevated lactate throughout the right MCA territory, with the highest concentration in the basal ganglia (Figure 6.3A) and a mild reduction in NAA compared to the contralateral hemisphere. Since there were no signs of infarction at this stage, the patient was treated with hypervolemic hypertensive hemodilution therapy and improved clinically. Follow-up imaging performed one week later (Figure 6.3B) revealed that much of the previously ischemic right lateral cortex had resolved (with no elevation of lactate or other abnormality), but that the basal ganglia had progressed to infarction (well-defined T_2 hyperintensity), with almost complete depletion of NAA and increased lactate.

This case shows that, as expected, during the earliest stages of stroke, the main spectroscopic abnormality is an increase in lactate. Elevated lactate in the absence of any other sign of infarction suggests that ischemic tissue at risk of infarction (i.e. a "penumbra") is surrounding an area of documented infarction. Potentially, if emergent MRSI is available in acute stroke patients, this type of information could be useful in making treatment decisions regarding thrombolysis or other interventions. One potential use for MRS might be in refining the penumbral region which is operationally defined by the "mismatch" region in perfusion–diffusion MRI.[40] Even with current state-of-the-art perfusion-weighted imaging (PWI) methodology, it is sometimes difficult to know whether the region of hypoperfusion corresponds to benign oligemia, or ischemic tissue at risk of infarction. One study used MRS in combination with diffusion-weighted imaging (DWI), and found that the combination of lactate and diffusion measurements improved the prediction of stroke outcome compared to either modality alone.[41]

In current practice, however, it is generally unlikely that MRS will be applied to acute stroke cases because of the logistical difficulties involved. Therefore, MRS may be more commonly encountered in a different

A

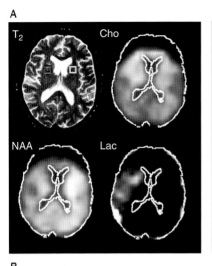
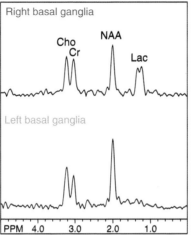

B

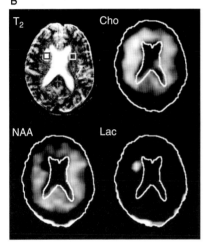
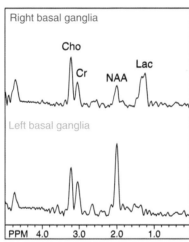

Figure 6.3. (A) T_2 MRI, and proton MRSI in an acute stroke patient with occlusion of the right internal carotid artery and low flow in the right middle cerebral artery 24 h after symptom onset. Minimal changes are visible in the conventional T_2-weighted MRI; however, spectroscopic images show an elevation of lactate through much of the right middle cerebral artery territory, indicating ischemia. NAA is relatively preserved in the right hemisphere, with mild reduction only compared to the left. (B) Follow-up imaging performed one week later shows development of infarction in the right basal ganglia (T_2 hyperintensity, absence of NAA). Lactate is present only in the region of infarction. High Cho is seen in the peri-infarct white matter.

role, namely in identifying lesions of ischemic origin that may mimic other etiologies. As an example, Figure 6.4 shows a patient with a lobular T_2 hyperintense lesion of the corpus callosum, originally suspected to be of neoplastic origin, which was subsequently demonstrated to be a stroke, and which had a high lactate signal on MRS.[42] In this case, MRS helped establish a diagnosis of stroke, although it should be remembered that in a chronic stroke, lactate may not be detected, and there are certainly other pathologies that may exhibit lactate.

MR spectroscopy in hypoxic–ischemic encephalopathy

Hypoxic–ischemic encephalopathy (HIE) results from prolonged oxygen deprivation, and may be caused by reduced oxygen delivery (hypoxemia) or cerebral blood flow (ischemia). HIE is most commonly encountered in perinatal asphyxia in neonates, where hypoxia is the primary insult, and will be the focus of discussion in this section. In addition to perinatal asphyxia, hypoxic–ischemic injury can occur in children and adults. In children, this often results from drowning, hanging, choking, and non-accidental trauma (particularly in young children). In adults, cardiac arrest or cerebrovascular disease with secondary hypoxemia is the more frequent cause,[43,44] as discussed in the previous section.

HIE in neonates is a major cause of death or may be associated with long-term neurological morbidity including seizures, motor deficits, and cognitive impairment. In term infants the estimated frequency is 0.2–0.3% of live births in developed countries, with a

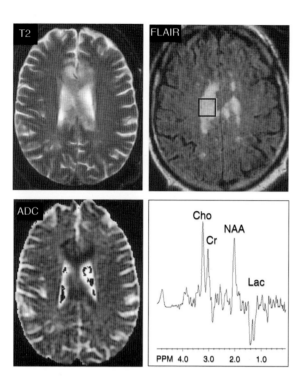

Figure 6.4. Unresponsive 50-year-old male. T_2 and FLAIR MRI show a lobular hyperintense signal abnormality with expansion of the corpus callosum, considered suspicious for neoplasm. ADC maps show reduced diffusion, and MRS shows an elevated lactate and reduced NAA, consistent with acute ischemia disease. Digital subtraction angiography (not shown) revealed an absence of flow in the right A1 segment of the anterior cerebral artery (ACA), but no other vascular abnormalities. Both DWI and MRS in this case were helpful in establishing a diagnosis of stroke. Adapted, with permission, from [42].

10-fold higher frequency in the underdeveloped areas and also at a higher rate of occurrence in the preterm infants.[45] The majority of infants sustain hypoxic–ischemic injury in the immediate perinatal period.[45] Various brain regions are particularly susceptible to injury depending on the maturity of brain, as well as severity and duration of hypoxic–ischemic insult.[43]

During acute insult, there is a conversion of oxidative phosphorylation metabolism to anaerobic glycolysis, resulting in depletion of energy molecules such as ATP and phosphocreatine accompanied by cellular accumulation of lactic acid. Some of these effects may be reversible following resuscitation; however, if the insult is severe and oxygen deprivation is prolonged, additional cytotoxic cascade ensues. The cascade of events includes breakdown of cellular membrane function, passive influx of calcium, cell swelling, release of excitatory neurotransmitters such as glutamate and aspartate from ischemic cells, and activation of free fatty acids and free radicals, leading to acidosis and cellular necrosis occurring at 24–48 h.

Clinically, the initial event is accompanied by umbilical artery pH < 7.1, delayed respiratory effort and poor Apgar scores below 5 at 5 min, with subsequent development of decreased consciousness, cranial nerve and motor deficits.[46,47] Seizure is usually evident at 12–24 h after birth. Together with additional symptoms and signs of hypotonia, poor feeding, and abnormal reflexes, these constitute the clinical syndrome of hypoxic–ischemic (neonatal) encephalopathy. In order to measure severity of HIE and to monitor clinical progress, several grading systems have been devised, of which the most widely used is the Sarnat score.[48] In this system (Sarnat Stages 1–3, ranging from mildest and early syndrome in Stage 1 to the most severe in Stage 3), EEG is used in conjunction with clinical signs and symptoms including level of alertness, feeding, tone, reflex, respiratory and cardiac status, and seizure activity.

Management of HIE is mainly supportive care and prevention of secondary CNS insults such as due to seizures and metabolic derangement. However, various neuroprotective strategies including hypothermia and pharmacologic agents curtailing the effects of excitatory neurotoxicity that offer promising prospects are currently under active investigation. [47,49,50] Prompt and accurate diagnosis is critical for directing neuroprotective measures and therapeutic intervention, as many of these measures have a narrow therapeutic window and should most effectively be targeted to prevent the onset of secondary deterioration (secondary energy failure).

MR imaging is an important part of evaluation of these infants, and may provide important information about the site, extent, severity, and etiology of cerebral injury, as well as prognostic factors.[51] However, conventional (e.g. T_1- or T_2-weighted) MRI generally lacks sensitivity in the acute phase, as the visible MRI abnormalities may not become apparent until several days after onset of injury (at which point neuroprotective therapy is probably too late). Diffusion-weighted imaging (DWI), which is a highly sensitive diagnostic tool in the setting of adult acute ischemia, and classic cases of arterial or venous infarcts in the pediatric population, may also appear unremarkable in early stages of HIE, and frequently underestimates the final extent of injury depending on the time of imaging after acute insult.[52,53] Therefore, there has been interest in developing other MRS-based measurements for the early, quantitative evaluation of neonates with suspected HIE.

^{1}H MR spectroscopy (MRS), particularly by the elevation of lactate (Lac), has been applied to provide the most sensitive and useful diagnosis in acute hypoxic–ischemic injury as early as the first day of life.[54,55] However, corresponding to the physiologic response and akin to DWI, MRS also shows temporal evolution and topological variation of metabolite concentrations.[52]

Normal neonatal brain MRS

Before discussing the metabolic consequences of HIE, it is important to consider the spectral patterns seen in the normal neonatal brain. Proton MRS changes during brain development from (term) birth onwards have been discussed in Chapter 4, so this section largely focuses on the term and preterm neonate.

For ^{31}P MRS in neonates, spectra are usually collected from large regions of brain (sometimes without any form of spatial localization), and visual inspection reveals significant differences from ^{31}P spectra from adults. In particular, the phosphocreatine (PCr) signal is somewhat lower, and phosphomonoesters (PME, primarily phosphocholine (PC) and phosphoethanolamine (PE)) are appreciably higher (Figure 6.5).[56] In terms of ratios, the three resonances from γ-, β-, and α-ATP, and the small signal from inorganic phosphate (Pi), are relatively constant from birth onwards, but absolute quantitation reveals substantial increases in PCr, ATP, Pi, and phosphodiesters (PDE) from birth to adult values (Figure 6.5).[57] These findings most likely reflect (1) high levels of precursors (primarily PE, but also PC[58]) that are needed for membrane synthesis in the immature brain (also seen as a high

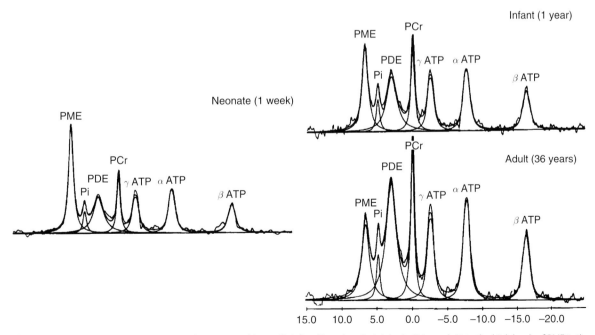

Figure 6.5. Normal phosphorus-31 spectra from neonatal (1 week), infant (1 year) and adult brain (36 years). Note the high levels of PME in the neonatal brain, which decrease in intensity relative to the other metabolites as the brain matures; absolute quantitation indicates this is mainly due to increases in other metabolites (ATP, PCr, etc.) rather than due to a decrease in PME. Spectra are plotted on the same vertical scale, accounting for differences in instrumental factors. Adapted, with permission, from [57].

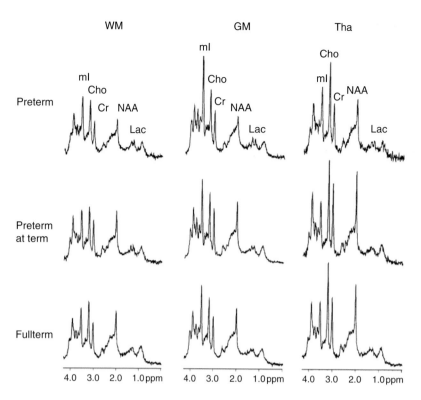

Figure 6.6. Short echo time (TE 20 ms) single voxel (~ 3 cm^3) averaged spectra from centrum semiovale white matter (WM), occipital gray matter (GM), and thalamus (Tha) from normal preterm, preterm at term, and full-term neonates. The less well developed white matter region has lower levels of all metabolites. Preterm gray matter shows very high *myo*-inositol (mI) and low NAA; thalamus has high levels of choline (Cho) and creatine (Cr). Spectra from preterm infants at term are virtually identical to those from term infants. Reproduced with permission from [61].

Cho signal in ^{1}H MRS (see below)), and (2) increasing cellular density occurs as the brain matures (also noted in ^{1}H MRS). Detailed studies in regional ^{31}P spectra in term and preterm neonates as a function of gestational age have not been performed, however.

For ^{1}H MRS, spectra are usually recorded from somewhat smaller and better defined volumes (compared to ^{31}P), and more detailed studies in terms of regional and gestational age-related changes have been performed. For single-voxel studies, 2–8 cm^3 voxels from the thalamus/basal ganglia and frontal or occipital–parietal white matter are commonly used, [59,60,61], while for MRSI, voxel sizes of 1 cm^3 are typical.[62] As described in Chapter 4, at birth in a term infant, *myo*-inositol (mI) and choline (Cho) are higher, and NAA lower than in adult brain.[60] Preterm infants show lower NAA and higher mI (particularly in gray matter and the thalamus) than those at term (Figure 6.6). There are few metabolic differences between (normal) preterm infants at term, and infants delivered at term.[61] At the earliest stages of development (e.g. 25 weeks gestational age (GA)), NAA is barely detectable. The increase in NAA with GA may be interpreted as reflected neuronal development; consistent with this, glutamate (Glu, also believed to be primarily located in neurons) similarly increases with age (Figure 6.7). Detailed metabolic changes as a function of gestational age are available in the literature.[63]

Regional variations in neonate spectra are best evaluated with MRSI, which has shown striking differences between the relatively mature thalamus/basal ganglia and immature peripheral white matter regions. In particular, all basal ganglia metabolite levels (except lactate) are high, presumably because of the higher cellular density of the more mature tissue (Figure 6.8).[62] Lactate (see below) also appears more visible in white matter regions, probably because of the higher water content in the interstitium of unmyelinated white matter, where lactate is more MRS visible because of its very long T_2 (estimated to be on the order of 1400 ms). Higher lactate signals in the CSF than in the brain are also found.

One issue of debate is the detectability of lactate in normal term or preterm neonates. This is an important issue, since elevated lactate is one of the cardinal signs of hypoxic and ischemic injury, so a clear distinction should be made between "normal" and "abnormal" lactate levels. The ability to detect and quantify a lactate signal depends on many factors, in particular signal-to-noise (SNR) ratio, pulse sequence, anatomical region of interest, and analysis software used. Generally, lactate is best detected at long echo

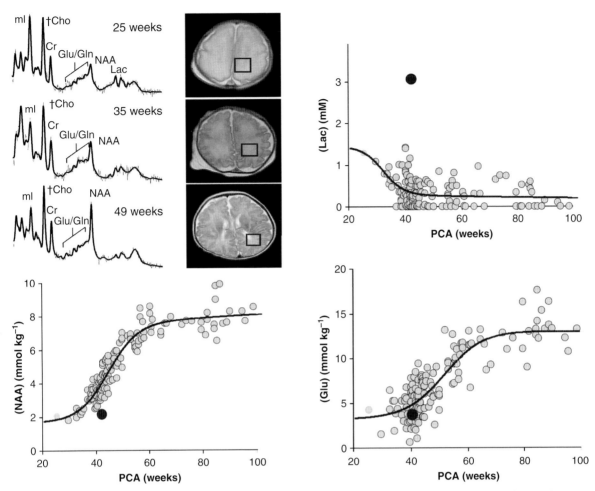

Figure 6.7. Representative short echo time single voxel spectra from neonates of 25, 35, and 49 weeks gestational age. Of note is the decrease of mI and increase in NAA with age. A small lactate peak is also clearly visible in the younger neonates. Normative curves showing evolution of NAA, lactate (Lac) and glutamine (Glu) with gestational age from 20 to 100 weeks are shown. The black dots represent NAA, Lac, and Glu values from one infant with methyl malonic aciduria (MMA), clearly showing the deviation from the normal age-related metabolic pattern (elevated Lac, low NAA). Reproduced with permission from [63].

time where less overlap with lipids occurs, either at 280 ms or as inverted doublet at TE of 140 ms. There are many potential technical pitfalls that may lead to loss of lactate signal (see Chapter 3), so it is important to verify in a phantom that lactate can be detected with good efficiency under the same condition to be used for the neonate study.

Although there have been several studies reporting no detectable MRS lactate signal in normal term neonatal brain,[60,64,65] other studies have reported a lactate signal.[54,59,63] One study reported lactate concentrations of 2.7 ± 0.6 mmol/kg wet weight in thalamus (from 15 infants) and 3.3 ± 1.3 mmol/kg in occipital–parietal regions (9 infants) from normal infants of gestational age 29–41 weeks at mean age of 7–12 days.[59] Another study noted the presence of minimal amounts of lactate in watershed regions, particularly in premature neonates, without evidence of asphyxia or other brain injury.[54] Overall, it appears that premature infants normally have detectable lactate, in association with low NAA peaks, with progressive diminution of Lac and increase of NAA as the infants reach term and develop postnatally.[59,63,66]

Finally, it should be noted that significantly narrower linewidths are obtained in proton spectra of neonates and young children compared to adults. The reasons for this are not totally clear, but appear to most likely be related to higher oxygenation (secondary to higher blood flow), which reduces microscopic susceptibility effects from deoxyhemoglobin

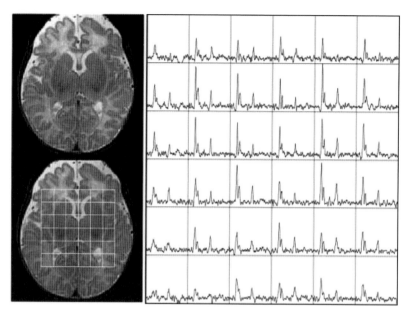

Figure 6.8. Proton MRSI (TR/TE 1000/144 msec) and T_2 MRI of the normal, term neonatal brain. Note that the thalami have the most mature spectra, with lowest choline and highest NAA peaks, and the frontal white matter has the least mature spectra. Reproduced with permission from [80].

and lengthens apparent metabolite T_2 relaxation times, in addition to reduced macroscopic susceptibility effects due to smaller paranasal sinuses, compared to adults. Immature myelination may be another factor that improves shimming. In MRSI, the volume of brain to be covered and hence shimmed is smaller than in adults, which also reduces the field inhomogeneity associated with the main magnetic field.

^{31}P MRS in HIE

Some of the very earliest MRS studies recorded in humans were ^{31}P MRS studies of neonatal brain. [67,68] Using simple, non-localized ^{31}P MRS, it was found that severe HIE gave spectral patterns quite different from those (described above) of normal neonatal brain, with reduced energy metabolites as manifested by low PCr, increased Pi (and hence a very abnormal, low PCr/Pi ratio), and an acidotic shift in pH as measured by the chemical shift of Pi (Figure 6.9). ATP is also typically decreased. Changes of this nature (e.g. a PCr/Pi ratio of more than 2 standard deviations below the mean normal value) have been found to be poor prognostic indicators in terms of survival, or adverse neurological outcome (e.g. measured at 1 or 2 years of age).[68,69]

However, spectroscopic results will depend in some detail on the timing of the MRS exam, and the severity and duration of the HIE event. This has been studied in some detail, both in human neonates as well as appropriate animal models.[70,71] As reported by Cady, in the early MRS studies it was expected that high-energy phosphates would be depleted in MRS studies performed within a few hours of neonatal HIE.[69] However, normal spectroscopic patterns were typically observed at such time points (i.e. less than 1 day) in resuscitated infants, and it was only in fact later on (e.g. 2–4 days, despite the establishment of normal arterial flow, oxygenation, and glucose levels) that the expected adverse spectral changes were observed. This same temporal evolution of ^{31}P spectra has also been observed in a piglet HIE model and has been termed "secondary energy failure".[70] Clearly, the apparent normality of ^{31}P MRS spectra within the first few hours of injury limits the ability of this modality to triage neonates for acute neuroprotective strategies (e.g. such as hypothermia).

^{1}H MRS in HIE

As described above, the cardinal features of HIE on ^{1}H MRS are the loss of NAA and elevation of lactate. [69] The degree of spectral abnormality corresponds to the severity of HIE (Figures 6.10 and 6.11). Other reported spectral changes include increased Glx and lipid,[63,72] most likely due to release of free triglycerides as membrane breakdown occurs, and loss of Cr, probably due to decreased cellularity.[69]

Measurements of lactate, NAA, or the lactate/NAA ratio have been reported to be sensitive to diagnosing HIE, as well as to a prognostic measure.[54,69,73] For

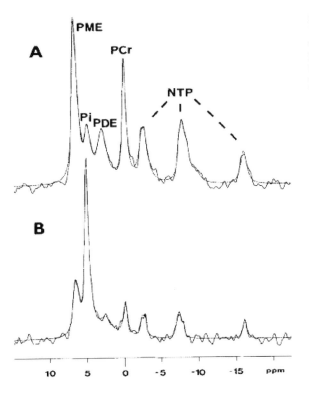

Figure 6.9. ^{31}P spectra (solid lines) from 125 ml voxels (TE 10 ms, TR 12 s, and 160 averages) centered on the thalami in (A) a normal infant of 37 weeks gestational age, and (B) in the infant suffering from perinatal hypoxia–ischemia at age 55 h. (B). The high-energy phosphates PCr and nucleoside triphoshate (NTP – mainly ATP) are conspicuously low in the infant with HIE, and inorganic phosphate (Pi) is markedly increased. The dashed line results from fitting Lorentzian profiles to the spectrum. Reproduced with permission from [81].

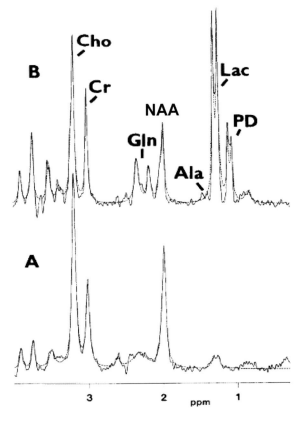

Figure 6.10. Long TE (270 msec) ^{1}H spectra (solid lines) from 8-ml thalamic voxels in (A) a normal infant of 42 weeks gestational age, and (B) in an infant suffering from perinatal hypoxia–ischemia at age 26 h. In the HIE spectrum, lactate (Lac) is conspicuously high and NAA is low, an alanine (Ala) methyl doublet is just visible, and two prominent features, probably mainly due to glutamine (Gln), are present at 2.2 and 2.4 ppm. Note also the prominent peak from propan-1,2-diol (PD) resulting from anti-seizure medications administered to the HIE infant. It is important not to confuse this doublet peak (1.1 ppm) with lactate (1.3 ppm). The dashed line results from fitting Lorentzian profiles to the spectrum. Reproduced with permission from [81].

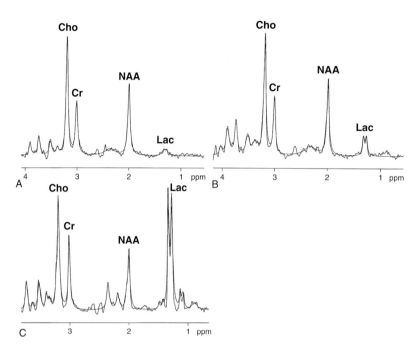

Figure 6.11. Representative thalamic PRESS spectra (8 ml, TR/TE 2000/270 ms) acquired from a control infant and 2 neonates with neonatal encephalopathy. (A) Control. (B) Normal/mild outcome. (C) Severe/fatal outcome. The dashed lines are the spectrum analysis Lorentzian profiles fitted to the peaks. Reproduced with permission from [74].

instance, one study of 17 term HIE infants (divided into normal/mild and severe/fatal outcome groups) and 10 healthy controls at 1 and 2 days after birth, respectively, found that the concentration of NAA was the best determinant to discriminate all three groups. [74] In another study, a high lactate/NAA ratio (> 2 standard deviations above the normal) in neonates examined on day 1 (2 h of age) correctly predicted unfavorable outcome at 1 year.[75] Conversely, an abnormal early neurological exam was not particularly specific for unfavorable outcome, although a normal neurological exam was a good predictor of a favorable outcome.[75] In one study of 21 neonates studied at a mean age of 3.1 years, the lactate/NAA ratio was found to be the best predictor of outcome at 1 year, and performed better than quantitative T_2 relaxometry of brain water (which was less strongly predictive).

As in ^{31}P MRS, the timing of the MRS study relative to the hypoxic event is important, with an initial recovery in most metabolites (once normal flow/oxygenation is established) followed by secondary energy failure at around 24 h (Figure 6.12). Important, however, was the finding that lactate was abnormally elevated within the first 24–48 h, unlike ^{31}P MRS where spectra remained normal at these early stages (Figure 6.13).[66,71]

MRSI studies in HIE have generally shown diffuse metabolic abnormalities; e.g. in patients with grade 3 encephalopathy on the Sarnat scale, 4 of 5 were found to have elevated lactate throughout the brain regions covered by MRSI.[64] Lactate was not observed in neonates with less severe HIE.

Many neonates affected by HIE are given intravenous phenobarbital for seizure control, delivered in a solution of 1,2-propanediol (propylene glycol). This compound readily crosses the blood–brain barrier and has a characteristic pattern on ^{1}H MRS of a doublet centered at 1.1 ppm, which should not be confused with a lactate doublet (1.3 ppm) (Figure 6.10).[63,76]

Diffusion MRI in HIE: comparison to ^{1}H MRS

Since conventional and diffusion-weighted MRI are believed to be sensitive to early HIE, it is worth comparing them to the information available from ^{1}H MRS. The authors of one study performed conventional MRI, DWI, and single-voxel MRS in 7 neonates with encephalopathy following a complicated delivery during the first 24 h of life.[55] Conventional imaging was largely normal, although some cases showed T_2 prolongation, indicating mild edema, in the basal ganglia or cortex. DWI findings were also not dramatically abnormal, but in a few cases revealed reduced diffusion in the posterior limb of the internal capsule (PLIC) and/or lateral thalami. This abnormality was better assessed by measuring apparent diffusion coefficient (ADC) values (rather than visual inspection of

Chapter 6: MRS in stroke and HIE

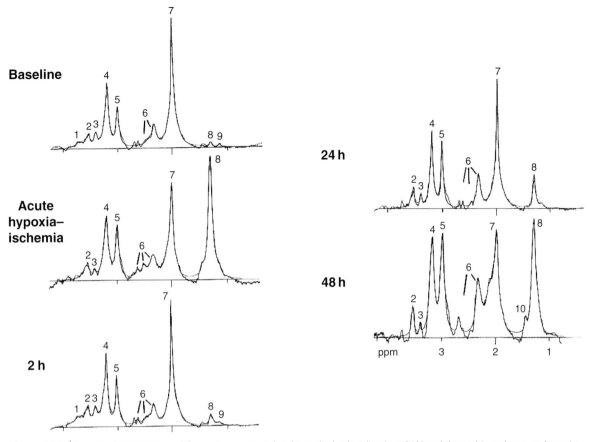

Figure 6.12. ¹H spectra (7 T, TE 270 msec) from a term neonatal piglet studied at baseline (age 24 h) and then subjected to acute hypoxia–ischemia injury (47 min duration). The spectrum during hypoxia was recorded at the end of the insult, and demonstrates a high lactate peak (8). Upon resuscitation, there is an initial recovery to a normal metabolic pattern (2 h), followed by an increase in lactate and decrease in NAA, corresponding to secondary energy failure at 24–48 h later. Resonance identifications: 1 glutamate/glutamine; 2 glycine/myo-inositol; 3 taurine/scyllo-inositol; 4 Cho; 5 Cr; 6 glutamate/glutamine; 7 NAA; 8 Lac; 9, β-hydroxybutyric acid; 10 alanine. Adapted with permission from [71].

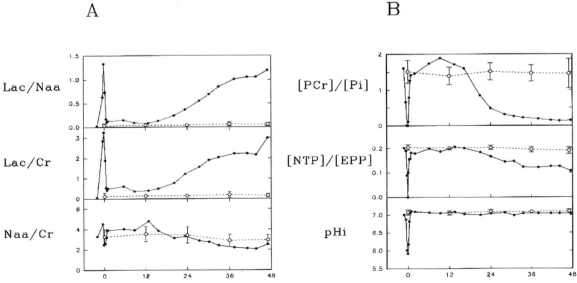

Figure 6.13. Using the same model as in Figure 6.12, summary of (A) Lac/NAA, Lac/Cr, and NAA/Cr, and (B) [PCr]/[Pi], [NTP]/[EPP], and pHi, from an individual piglet subjected to hypoxia–ischemia (solid dots, •). Blood Lac measurements were 1.3, 1.6, 0.9, 0.9, and 1.1 mmol·l⁻¹ at baseline, during the insult, and at 2, 24, and 48 h after resuscitation. The mean control values are shown (open circles, ○) with 95% confidence intervals. Reproduced with permission from [71].

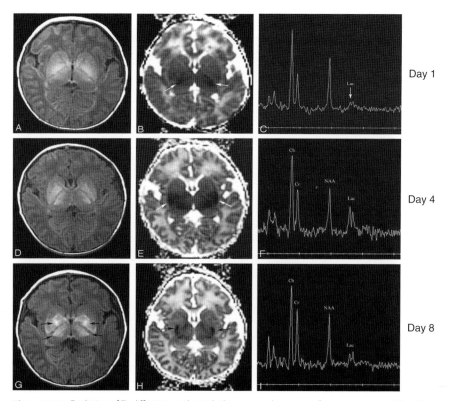

Figure 6.14. Evolution of T_1, diffusivity, and metabolites over 8 days in an infant experiencing HIE. A–C were performed at day 1 (16 h), D–F were performed at 4 days (84 h), and G–I were performed at 8 days (178 h). (A) Axial T_1-weighted image at age 16 h is normal. (B) Axial ADC map at age 16 h shows a small amount of reduced diffusivity on the ventrolateral thalami (arrows). Measurements showed a reduction in ADC of about 10%. (C) Proton MR spectroscopy from the right thalamus at age 16 h shows minimal elevation of lactate (Lac), but is otherwise normal. (D) Axial T_1-weighted image at 84 h shows that the normal hyperintensity in the posterior limb of the internal capsule is no longer seen. Abnormal hyperintensity is seen in the ventrolateral thalami and posterior putamina. (E) Axial ADC map at 84 h shows that reduced diffusivity is now present in the posterior putamina (arrows). Measurements of ADC showed significant reduction since day 1, with values now 50–60% or normal (40–50% reduced) in the thalami and putamina, and dorsal brain stem. Lesser reductions of about 25% were found in the cerebral hemispheric white matter. (F) Proton MR spectroscopy from the right thalamus at 84 h shows an increase in lactate (Lac) and relative reduction of choline and NAA compared with the first study. (G) Axial T_1-weighted image at 8 days shows that the T_1 shortening is becoming less diffuse and more globular (arrows), with the globular regions being located in the globi pallidi, ventrolateral thalami, and at the junction of the anterior globi pallidi and putamina. (H) Axial ADC map at 84 h shows that reduced diffusivity is now almost exclusively seen in the posterior putamina (arrows) with the thalamic abnormality nearly completely gone. Measurements showed that the ADC values of the putamina were still about 30% below normal, but those in the thalami had normalized. (I) Proton MR spectroscopy from the right thalamus at 8 days shows that the lactate peak is significantly smaller. Note that the NAA and choline peaks have continued to decrease in size compared with the creatine peak. Reproduced with permission from [52].

DWI) which showed a 15–20% of decrease throughout the brain compared to healthy controls. In contrast, ^{1}H MRS showed more prominent abnormalities, with elevation of lactate/NAA ratios, and reduction of NAA. Furthermore, follow-up imaging studies suggested that acute MRI and DWI underestimated the topological extent of injury.[55]

Also of note is the temporal evolution of ADC and MRS findings; an example in one patient is shown in Figure 6.14. One study found low ADCs acutely in neonates with HIE, but these values subsequently increased with time, such that only the acute ADC measurements (taken within day 6) showed any prognostic value. In contrast, in the same patients, persistently low NAA concentrations (less than 4 mmol/l in basal ganglia) were found to be a strong indicator of adverse outcome.[77] Conversely, on occasion, focal HIE lesions in the neonatal brain may be too small to be evaluated by MRS, yet nevertheless will show MRI or DWI abnormalities because of the higher resolution of these modalities.[52] It would therefore appear that MRI, DWI, and MRS are complementary techniques for evaluating HIE in neonates.

Some recommendations for MRS protocols for stroke and HIE

Stroke

As discussed above, if MRS is to be attempted in acute stroke, a system needs to be in place for rapidly triaging patients from the emergency department to the MRI suite, with appropriate monitoring and neurology support. Only minimal, essential MRI sequences that can be rapidly performed (e.g. diffusion and perfusion imaging, and a gradient echo sequence to look for hemorrhage) should be included if thrombolysis or other urgent intervention is being considered. Since stroke heterogeneity (infarct vs. penumbra) is important, a rapid MRSI sequence with good brain coverage and spatial resolution should be used if time is available. Generally, this sequence will cover at least the middle cerebral artery, unless clinical suspicion or other imaging modalities suggest involvement of other areas. Since most of the relevant information about ischemia and neuronal viability is contained in the lactate and NAA signals, respectively, use of a long echo time (e.g. TE 140 or 280 ms) is appropriate. Finally, since stroke patients are often confused and uncooperative, fast MRSI techniques which reduce scan time should be considered, as well as techniques which minimize sensitivity to head motion. Ideally, high field (e.g. 3 T) scanners should be used with multi-channel phased-array head coils for good SNR.

HIE

A prerequisite for MRS in neonates is a system in place for transportation to the MRI suite, sedation, and monitoring/support while in the scanner. Dedicated commercial systems are starting to become available for this purpose. Note that sedation is often not needed in neonates, since they will often sleep naturally after feeding and being wrapped comfortably in warm blankets. Full details of all necessary procedures can be found in standard pediatric neuroradiology text books.[78]

Since the neonatal head and brain are much smaller than in adults, poor SNR will be obtained unless appropriate RF coils are used. Ideally, scanning should be performed at 3 T using dedicated pediatric phased-array receiver coils; if these are not available, then coils designed for adult extremity use (such as a knee coil) may yield adequate results.

Many SV-MRS studies in neonates have focused on fairly large volumes of interest in the (relatively speaking) more mature deep gray matter, since this region gives good quality spectra and is often a site of injury in HIE. As in stroke, the compounds of prime interest are usually NAA and lactate, thus long TE protocols (TE 140 or 280 msec) give clean results. If sufficient time and techniques are available, multi-voxel MRSI techniques should also be considered to provide more information on the spatial extent of the injury.

Conclusions

In summary, ischemic and hypoxic injury is associated with profound metabolic alterations in the brain that are detectable by both ^{31}P and ^{1}H MRS, which have important clinical significance. However, because of the long acquisition times and difficulties in scanning both acute stroke patients and neonates with HIE, these techniques are currently not much used in clinical practice. Another issue is the perceived "added-value" compared to conventional MRI techniques, as well as DWI (and PWI), particularly in acute stroke. However, the literature does suggest a positive role for MRS, particularly in neonatal HIE as conventional MRI and even DWI are often unrevealing in the acute stage. If logistical problems can be overcome, MRS may become increasingly used in the future.

References

[1] Berkelbach van der Sprenkel JW, Luyten PR, van Rijen PC, Tulleken CA, den Hollander JA. Cerebral lactate detected by regional proton magnetic resonance spectroscopy in a patient with cerebral infarction. *Stroke* 1988; **19**: 1556–60.

[2] Bruhn H, Frahm J, Gyngell ML, Merboldt KD, Hanicke W, Sauter R. Cerebral metabolism in man after acute stroke: new observations using localized proton NMR spectroscopy. *Magn Reson Med* 1989; **9**: 126–31.

[3] Helpern JA, Vande Linde AMQ, Welch KMA, Levine SR, Schultz LR, Oridge RJ, et al. Acute elevation and recovery of intracellular [Mg^{2+}] following human focal cerebral ischemia. *Neurology* 1993; **43**: 1577–81.

[4] Thulborn KR, Gindin TS, Davis D, Erb P. Comprehensive MR imaging protocol for stroke management: tissue sodium concentration as a measure of tissue viability in nonhuman primate studies and in clinical studies. *Radiology* 1999; **213**: 156–66.

[5] Rothman DL, Howseman AM, Graham GD, Petroff OA, Lantos G, Fayad PB, et al. Localized proton NMR

observation of [3–13C]lactate in stroke after [1–13C] glucose infusion. *Magn Reson Med* 1991; **21**: 302–07.

[6] Higuchi T, Graham SH, Fernandez EJ, Rooney WD, Gaspary HL, Weiner MW, et al. Effects of severe global ischemia on N-acetylaspartate and other metabolites in the rat brain. *Magn Reson Med* 1997; **37**: 851–7.

[7] Monsein LH, Mathews VP, Barker PB, Pardo CA, Blackband SJ, Whitlow WD, et al. Irreversible regional cerebral ischemia: serial MR imaging and proton MR spectroscopy in a nonhuman primate model. *Am J Neuroradiol* 1993; **14**: 963–70.

[8] Sager TN, Laursen H, Fink-Jensen A, Topp S, Stensgaard A, Hedehus M, et al. N-Acetylaspartate distribution in rat brain striatum during acute brain ischemia. *J Cereb Blood Flow Metab* 1999; **19**: 164–72.

[9] Sager TN, Laursen H, Hansen AJ. Changes in N-acetylaspartate content during focal and global brain ischemia of the rat. *J Cereb Blood Flow Metab* 1995; **15**: 639–46.

[10] Higuchi T, Fernandez EJ, Maudsley AA, Shimizu H, Weiner MW, Weinstein PR. Mapping of lactate and N-acetyl-L-aspartate predicts infarction during acute focal ischemia: in vivo 1H magnetic resonance spectroscopy in rats. *Neurosurgery* 1996; **38**: 121–9; discussion 9–30.

[11] van Zijl PCM, Moonen CTW. In situ changes in purine nucleotide and N-acetyl concentrations upon inducing global ischemia in cat brain. *Magn Reson Med* 1993; **29**: 381–5.

[12] Gillard JH, Barker PB, van Zijl PCM, Bryan RN, Oppenheimer SM. Proton MR spectroscopic imaging in acute middle cerebral artery stroke. *Am J Neuroradiol* 1996; **17**: 873–86.

[13] Mathews VP, Barker PB, Blackband SJ, Chatham JC, Bryan RN. Cerebral metabolites in patients with acute and subacute strokes: concentrations determined by quantitative proton MR spectroscopy. *Am J Roentgenol* 1995; **165**: 633–8.

[14] Munoz Maniega S, Cvoro V, Armitage PA, Marshall I, Bastin ME, Wardlaw JM. Choline and creatine are not reliable denominators for calculating metabolite ratios in acute ischemic stroke. *Stroke* 2008; **39**: 2467–9.

[15] Barker P, Breiter S, Soher B, Chatham J, Forder J, Samphilipo M, et al. Quantitative proton spectroscopy of canine brain: in vivo and in vitro correlations. *Magn Reson Med* 1994; **32**: 157–63.

[16] Barker PB, Gillard JH, van Zijl PCM, Soher BJ, Hanley DF, Agildere AM, et al. Acute stroke: evaluation with serial proton magnetic resonance spectroscopic imaging. *Radiology* 1994; **192**: 723–32.

[17] Duijn JH, Matson GB, Maudsley AA, Hugg JW, Weiner MW. Human brain infarction: proton MR spectroscopy. *Radiology* 1992; **183**: 711–8.

[18] Rehncrona S, Rosen I, Siesjo BK. Brain lactic acidosis and ischemic cell damage: 1. Biochemistry and neurophysiology. *J Cereb Blood Flow Metab* 1981; **1**: 297–311.

[19] Petroff OA, Graham GD, Blamire AM, al-Rayess M, Rothman DL, Fayad PB, et al. Spectroscopic imaging of stroke in humans: histopathology correlates of spectral changes. *Neurology* 1992; **42**: 1349–54.

[20] Kruse B, Barker PB, van Zijl PCM, Duyn JH, Moonen CTW, Moser HW. Multislice proton MR spectroscopic imaging in X-linked adrenoleukodystrophy. *Ann Neurol* 1994; **36**: 595–608.

[21] Hossman K-A. Viability thresholds and the penumbra of focal ischemia. *Ann Neurol* 1994; **36**: 557–65.

[22] Crockard HA, Gadian DG, Frackowiak RS, Proctor E, Allen K, Williams SR, et al. Acute cerebral ischaemia: concurrent changes in cerebral blood flow, energy metabolites, pH, and lactate measured with hydrogen clearance and 31P and 1H nuclear magnetic resonance spectroscopy. II. Changes during ischaemia. *J Cereb Blood Flow Metab* 1987; **7**: 394–402.

[23] Petroff OA, Prichard JW, Ogino T, Shulman RG. Proton magnetic resonance spectroscopic studies of agonal carbohydrate metabolism in rabbit brain. *Neurology* 1988; **38**: 1569–74.

[24] Bizzi A, Righini A, Turner R, Le Bihan D, Bockhorst KH, Alger JR. Imaging focal reperfusion injury following global ischemia with diffusion-weighted magnetic resonance imaging and 1H-magnetic resonance spectroscopy. *Magn Reson Imaging* 1996; **14**: 581–92.

[25] Bizzi A, Righini A, Turner R, LeBihan D, DesPres D, Di Chiro G, et al. MR of diffusion slowing in global cerebral ischemia. *Am J Neuroradiol* 1993; **14**: 1347–54.

[26] Dreher W, Kuhn B, Gyngell ML, Busch E, Niendorf T, Hossmann KA, et al. Temporal and regional changes during focal ischemia in rat brain studied by proton spectroscopic imaging and quantitative diffusion NMR imaging. *Magn Reson Med* 1998; **39**: 878–88.

[27] Norris DG, Hoehn-Berlage M, Dreher W, Kohno K, Busch E, Schmitz B. Characterization of middle cerebral artery occlusion infarct development in the rat using fast nuclear magnetic resonance proton spectroscopic imaging and diffusion-weighted imaging. *J Cereb Blood Flow Metab* 1998; **18**: 749–57.

[28] Nagatomo Y, Wick M, Prielmeier F, Frahm J. Dynamic monitoring of cerebral metabolites during and after transient global ischemia in rats by quantitative proton

NMR spectroscopy in vivo. *NMR Biomed* 1995; **8**: 265–70.

[29] Saunders DE, Howe FA, van den Boogart A, McLean MA, Griffiths JR, Brown MM. Continuing ischemic damage after acute middle cerebral artery infarction in humans demonstrated by short-echo time proton spectroscopy. *Stroke* 1995; **26**: 1007–13.

[30] Rumpel H, Lim WE, Chang HM, Chan LL, Ho GL, Wong MC, et al. Is myo-inositol a measure of glial swelling after stroke? A magnetic resonance study. *J Magn Reson Imaging* 2003; **17**: 11–9.

[31] Felber SR, Aichner FT, Sauter R, Gerstenbrand F. Combined magnetic resonance imaging and proton magnetic resonance spectroscopy of patients with acute stroke. *Stroke* 1992; **23**: 1106–10.

[32] Gideon P, Henriksen O, Sperling B, Christiansen P, Olsen TS, Jorgensen HS, et al. Early time course of N-acetylaspartate, creatine and phosphocreatine, and compounds containing choline in the brain after acute stroke. *Stroke* 1992; **23**: 1566–72.

[33] Henriksen O, Gideon P, Sperling B, Olsen TS, Jorgensen HS, Arlien-Soborg P. Cerebral lactate production and blood flow in acute stroke. *J Magn Reson Imaging* 1992; **2**: 511–7.

[34] Fenstermacher MJ, Narayana PA. Serial proton magnetic resonance spectroscopy of ischemic brain injury in humans. *Invest Radiol* 1990; **25**: 1034–9.

[35] Gideon P, Sperling B, Arlien-Soborg P, Olsen TS, Henriksen O. Long-term follow-up of cerebral infarction patients with proton magnetic resonance spectroscopy. *Stroke* 1994; **25**: 967–73.

[36] Sappey-Marinier D, Calabrese G, Hetherington HP, Fisher SN, Deicken R, Van DC, et al. Proton magnetic resonance spectroscopy of human brain: applications to normal white matter, chronic infarction, and MRI white matter signal hyperintensities. *Magn Reson Med* 1992; **26**: 313–27.

[37] Graham GD, Blamire AM, Howseman AM, Rothman DL, Fayad PB, Brass LM, et al. Proton magnetic resonance spectroscopy of cerebral lactate and other metabolites in stroke patients. *Stroke* 1992; **23**: 333–40.

[38] Graham G, Blamire A, Rothman D, Brass L, Fayad P, Petroff O, et al. Early temporal variation of cerebral metabolites after human stroke. *Stroke* 1993; **24**: 1891–6.

[39] Hugg JW, Duijn JH, Matson GB, Maudsley AA, Tsuruda JS, Gelinas DF, et al. Elevated lactate and alkalosis in chronic human brain infarction observed by 1H and 31P MR spectroscopic imaging. *J Cereb Blood Flow Metab* 1992; **12**: 734–44.

[40] Sorensen AG, Buonanno FS, Gonzalez RG, Schwamm LH, Lev MH, Huang-Hellinger FR, et al. Hyperacute stroke: evaluation with combined multisection diffusion-weighted and hemodynamically weighted echo-planar MR imaging. *Radiology* 1996; **199**: 391–401.

[41] Parsons MW, Li T, Barber PA, Yang Q, Darby DG, Desmond PM, et al. Combined (1)H MR spectroscopy and diffusion-weighted MRI improves the prediction of stroke outcome. *Neurology* 2000; **55**: 498–505.

[42] Riedy G, Melhem ER. Acute infarct of the corpus callosum: appearance on diffusion-weighted MR imaging and MR spectroscopy. *J Magn Reson Imaging* 2003; **18**: 255–9.

[43] Huang BY, Castillo M. Hypoxic–ischemic brain injury: imaging findings from birth to adulthood. *Radiographics* 2008; **28**: 417–39; quiz 617.

[44] Biagas K. Hypoxic–ischemic brain injury: advancements in the understanding of mechanisms and potential avenues for therapy. *Curr Opin Pediatr* 1999; **11**: 223–8.

[45] Gunn AJ, Bennet L. Timing of injury in the fetus and neonate. *Curr Opin Obstet Gynecol* 2008; **20**: 175–81.

[46] Ferriero DM. Neonatal brain injury. *N Engl J Med* 2004; **351**: 1985–95.

[47] Shalak L, Perlman JM. Hypoxic–ischemic brain injury in the term infant – current concepts. *Early Hum Dev* 2004; **80**: 125–41.

[48] Sarnat HB, Sarnat MS. Neonatal encephalopathy following fetal distress. A clinical and electroencephalographic study. *Arch Neurol* 1976; **33**: 696–705.

[49] Gluckman PD, Pinal CS, Gunn AJ. Hypoxic–ischemic brain injury in the newborn: pathophysiology and potential strategies for intervention. *Semin Neonatol* 2001; **6**: 109–20.

[50] Vannucci RC, Perlman JM. Interventions for perinatal hypoxic–ischemic encephalopathy. *Pediatrics* 1997; **100**: 1004–14.

[51] Rutherford M, Srinivasan L, Dyet L, Ward P, Allsop J, Counsell S, et al. Magnetic resonance imaging in perinatal brain injury: clinical presentation, lesions and outcome. *Pediatr Radiol* 2006; **36**: 582–92.

[52] Barkovich AJ, Miller SP, Bartha A, Newton N, Hamrick SE, Mukherjee P, et al. MR imaging, MR spectroscopy, and diffusion tensor imaging of sequential studies in neonates with encephalopathy. *Am J Neuroradiol* 2006; **27**: 533–47.

[53] Barkovich AJ, Westmark K, Partridge C, Sola A, Ferriero DM. Perinatal asphyxia: MR findings in the first 10 days. *Am J Neuroradiol* 1995; **16**: 427–38.

[54] Barkovich AJ, Baranski K, Vigneron D, Partridge JC, Hallam DK, Hajnal BL, et al. Proton MR spectroscopy

for the evaluation of brain injury in asphyxiated, term neonates. *Am J Neuroradiol* 1999; **20**: 1399–405.

[55] Barkovich AJ, Westmark KD, Bedi HS, Partridge JC, Ferriero DM, Vigneron DB. Proton spectroscopy and diffusion imaging on the first day of life after perinatal asphyxia: preliminary report. *Am J Neuroradiol* 2001; **22**: 1786–94.

[56] van der Knaap MS, van der Grond J, van Rijen PC, Faber JA, Valk J, Willemse K. Age-dependent changes in localized proton and phosphorus MR spectroscopy of the brain. *Radiology* 1990; **176**: 509–15.

[57] Buchli R, Duc CO, Martin E, Boesiger P. Assessment of absolute metabolite concentrations in human tissue by 31P MRS in vivo. Part I: Cerebrum, cerebellum, cerebral gray and white matter. *Magn Reson Med* 1994; **32**: 447–52.

[58] Bluml S, Seymour KJ, Ross BD. Developmental changes in choline- and ethanolamine-containing compounds measured with proton-decoupled (31)P MRS in in vivo human brain. *Magn Reson Med* 1999; **42**: 643–54.

[59] Cady EB, Penrice J, Amess PN, Lorek A, Wylezinska M, Aldridge RF, et al. Lactate, N-acetylaspartate, choline and creatine concentrations, and spin–spin relaxation in thalamic and occipito-parietal regions of developing human brain. *Magn Reson Med* 1996; **36**: 878–86.

[60] Kreis R, Ernst T, Ross BD. Development of the human brain: in vivo quantification of metabolite and water content with proton magnetic resonance spectroscopy. *Magn Reson Med* 1993; **30**: 424–37.

[61] Kreis R, Hofmann L, Kuhlmann B, Boesch C, Bossi E, Huppi PS. Brain metabolite composition during early human brain development as measured by quantitative in vivo 1H magnetic resonance spectroscopy. *Magn Reson Med* 2002; **48**: 949–58.

[62] Vigneron DB, Barkovich AJ, Noworolski SM, von dem Bussche M, Henry RG, Lu Y, et al. Three-dimensional proton MR spectroscopic imaging of premature and term neonates. *Am J Neuroradiol* 2001; **22**: 1424–33.

[63] Panigrahy A, Bluml S. Advances in magnetic resonance neuroimaging techniques in the evaluation of neonatal encephalopathy. *Top Magn Reson Imaging* 2007; **18**: 3–29.

[64] Groenendaal F, Veenhoven RH, van der Grond J, Jansen GH, Witkamp TD, de Vries LS. Cerebral lactate and N-acetyl-aspartate/choline ratios in asphyxiated full-term neonates demonstrated in vivo using proton magnetic resonance spectroscopy. *Pediatr Res* 1994; **35**: 148–51.

[65] Huppi PS, Posse S, Lazeyras F, Burri R, Bossi E, Herschkowitz N. Magnetic resonance in preterm and term newborns: 1H-spectroscopy in developing human brain. *Pediatr Res* 1991; **30**: 574–8.

[66] Penrice J, Cady EB, Lorek A, Wylezinska M, Amess PN, Aldridge RF, et al. Proton magnetic resonance spectroscopy of the brain in normal preterm and term infants, and early changes after perinatal hypoxia–ischemia. *Pediatr Res* 1996; **40**: 6–14.

[67] Hope PL, Costello AM, Cady EB, Delpy DT, Tofts PS, Chu A, et al. Cerebral energy metabolism studied with phosphorus NMR spectroscopy in normal and birth-asphyxiated infants. *Lancet* 1984; **2**: 366–70.

[68] Azzopardi D, Wyatt JS, Cady EB, Delpy DT, Baudin J, Stewart AL, et al. Prognosis of newborn infants with hypoxic–ischemic brain injury assessed by phosphorus magnetic resonance spectroscopy. *Pediatr Res* 1989; **25**: 445–51.

[69] Cady EB. Magnetic resonance spectroscopy in neonatal hypoxic–ischaemic insults. *Childs Nerv Syst* 2001; **17**: 145–9.

[70] Lorek A, Takei Y, Cady EB, Wyatt JS, Penrice J, Edwards AD, et al. Delayed (secondary) cerebral energy failure after acute hypoxia-ischemia in the newborn piglet: continuous 48-hour studies by phosphorus magnetic resonance spectroscopy. *Pediatr Res* 1994; **36**: 699–706.

[71] Penrice J, Lorek A, Cady EB, Amess PN, Wylezinska M, Cooper CE, et al. Proton magnetic resonance spectroscopy of the brain during acute hypoxia–ischemia and delayed cerebral energy failure in the newborn piglet. *Pediatr Res* 1997; **41**: 795–802.

[72] Malik GK, Pandey M, Kumar R, Chawla S, Rathi B, Gupta RK. MR imaging and in vivo proton spectroscopy of the brain in neonates with hypoxic ischemic encephalopathy. *Eur J Radiol* 2002; **43**: 6–13.

[73] Shanmugalingam S, Thornton JS, Iwata O, Bainbridge A, O'Brien FE, Priest AN, et al. Comparative prognostic utilities of early quantitative magnetic resonance imaging spin–spin relaxometry and proton magnetic resonance spectroscopy in neonatal encephalopathy. *Pediatrics* 2006; **118**: 1467–77.

[74] Cheong JL, Cady EB, Penrice J, Wyatt JS, Cox IJ, Robertson NJ. Proton MR spectroscopy in neonates with perinatal cerebral hypoxic–ischemic injury: metabolite peak-area ratios, relaxation times, and absolute concentrations. *Am J Neuroradiol* 2006; **27**: 1546–54.

[75] Amess PN, Penrice J, Wylezinska M, Lorek A, Townsend J, Wyatt JS, et al. Early brain proton magnetic resonance spectroscopy and neonatal neurology related to neurodevelopmental outcome at

1 year in term infants after presumed hypoxic–ischaemic brain injury. *Dev Med Child Neurol* 1999; **41**: 436–45.

[76] Cady EB, Lorek A, Penrice J, Reynolds EO, Iles RA, Burns SP, *et al*. Detection of propan-1,2-diol in neonatal brain by in vivo proton magnetic resonance spectroscopy. *Magn Reson Med* 1994; **32**: 764–7.

[77] Boichot C, Walker PM, Durand C, Grimaldi M, Chapuis S, Gouyon JB, *et al*. Term neonate prognoses after perinatal asphyxia: contributions of MR imaging, MR spectroscopy, relaxation times, and apparent diffusion coefficients. *Radiology* 2006; **239**: 839–48.

[78] Barkovich AJ. *Pediatric Neuroimaging* (4th edn). Baltimore, MD: Lippincott Williams & Wilkins, 2005.

[79] Monsein LH, Mathews VP, Barker PB, Pardo CA, Blackband SJ, Whitlow WD, *et al*. Irreversible regional cerebral ischemia: serial MR imaging and proton MR spectroscopy in a nonhuman primate model. *Am J Neuroradiol* 1993; **14**: 963–70.

[80] Bartha AI, Yap KR, Miller SP, Jeremy RJ, Nishimoto M, Vigneron DB, *et al*. The normal neonatal brain: MR imaging, diffusion tensor imaging, and 3D MR spectroscopy in healthy term neonates. *Am J Neuroradiol* 2007; **28**: 1015–21.

[81] Cady EB, Amess P, Penrice J, Wylezinska M, Sams V, Wyatt JS. Early cerebral-metabolite quantification in perinatal hypoxic-ischaemic encephalopathy by proton and phosphorus magnetic resonance spectroscopy. *Magn Reson Imaging* 1997; **15**: 605–11.

Chapter 7

MRS in infectious, inflammatory, and demyelinating lesions

Key points

- MRS can provide useful clinical, metabolic information in infection, inflammation, and demyelination.
- Pyogenic abscess have a unique metabolic pattern with decreased levels of all normally observed brain metabolites, and elevation of succinate, alanine, acetate, and amino acids, as well as lipids and lactate. This pattern is quite distinct from that seen in brain tumors.
- Tuberculomas are characterized by elevated lipid and an absence of all other resonances.
- MRS is extensively used in research studies of HIV infection; early changes include elevated choline and *myo*-inositol perhaps associated with microglial proliferation, while later changes (associated with cognitive impairment, and dementia) include reduced NAA (neuronal loss).
- MRS may also be useful in assisting differential diagnosis in HIV-associated lesions.
- MRS shows decreased NAA (suggesting axonal dysfunction and loss) in early multiple sclerosis, as well as increased Cho and *myo*-inositol and lipids (suggesting demyelination). NAA correlates with clinical disability. White matter that appears normal on T_2 MRI may be abnormal metabolically in MS. Lactate may be elevated in acute, inflammatory demyelination.
- Acute disseminated encephalomyelitis (ADEM) may show similar spectral patterns to MS; however, ADEM with good clinical outcome usually only shows mild NAA losses in lesions.

Introduction

Intracranial infection, inflammation, and demyelination include a wide range of disorders of the central nervous system (CNS). Magnetic resonance imaging (MRI) plays a crucial role in the diagnosis and therapeutic decision making in these diseases. In the most recent years, advanced MR methods also have offered their precious contribution and, among these methods, proton MR spectroscopy (^{1}H-MRS) has been shown to provide salient findings for many of these disorders. For instance, ^{1}H-MRS data reported for processes as different as bacterial abscesses, tuberculomas, herpes simplex encephalitis, and HIV-related infections have demonstrated specific metabolic profiles that may be useful in differential diagnosis. It is also noteworthy that the use of ^{1}H-MRS in a demyelinating disorder such as multiple sclerosis (MS) has led to a re-evaluation of the pathogenesis and the natural history of the disease.

Overall, the results reported in the literature support the view that ^{1}H-MRS can be employed to improve diagnosis in single subject cases with infectious, inflammatory, or demyelinating lesions. In chronic conditions such as MS or HIV infections, it can also be used longitudinally to monitor the natural evolution of disease or the response to therapeutic intervention.

Intracranial infections

Infections of the CNS are often life-threatening conditions that may progress rapidly. The prognosis often depends on rapid identification of both the pathogen and site of inflammation to install effective antimicrobial treatment. Significant morbidity or mortality will occur, especially in bacterial infections, if appropriate therapies are not initiated promptly.[1,2]

The clinical presentations of these conditions may vary significantly.[1] The clinical involvement of the CNS usually includes focal deficits of variable entity (due to the focal brain lesion) as well as altered mental status and, possibly, seizures (due to the diffuse cerebritis). Whereas analysis of CSF, laboratory analysis, and, eventually, biopsy remain fundamental to identify the infectious agent, brain MRI is crucial as it clearly shows the inflammatory brain lesions, often allowing a rapid diagnosis and the subsequent therapeutic decisions.

In this context, ^{1}H-MRS has also been shown to be useful in the evaluation of infections, since some brain lesions are difficult to interpret on conventional MRI and, in certain cases, the infected brain tissue can be characterized by specific spectroscopic patterns that are not present in uninfected tissues.[1,3]

Brain abscesses

Brain abscess is a focal suppurative process within the brain parenchyma.[4] The incidence of brain abscesses is variable, higher in developing than in developed countries, accounting for less than 10% of the intracranial space-occupying lesions.[5,6]. They are usually secondary to local extension from a contiguous source of infection (e.g. otitis, sinusitis, or mastoiditis) or to hematogenous dissemination of an extracranial infection. The clinical manifestations are usually fever and signs of raised intracranial pressure. However, in some cases, these signs of infection may be subtle or absent, with serious difficulties in interpreting the clinical picture.

Brain abscess development can be divided into four stages: (i) an early stage, where the focal infection is not well delineated (early cerebritis, 1–4 days); (ii) a subacute stage with focal zones of necrosis, due to the fact that the body begins to react to the infection (late cerebritis, 4–10 days); (iii) the formation of a well-defined capsule, with a central area of necrosis and a surrounding area of inflammation (early capsule formation, 11–14 days); (iv) a progression of the capsule with organization of the surrounding and further necrosis inside (late capsule formation, > 14 days). [1,4,7] In these conditions, imaging features of a brain abscess are strictly related to its stage at the time of imaging.[7]

Conventional MRI is the main noninvasive procedure used to diagnose brain abscesses. This has shown to be very useful in this complicated process and reveals, in general, a ring-enhancing lesion with perifocal edema with characteristics that may vary slightly on the basis of the stage of the abscess (usually, hypointense on T_1-weighted and hyperintense on T_2-weighted images). This MRI pattern, however, is similar to that of other necrotic masses (i.e. glioblastoma and metastasis) that need to be considered in the differential diagnosis.[8,9] In case of hematogenous abscesses, the MRI lesions are usually multiple and enter the differential diagnosis with a metastatic neoplasm. In this context, diffusion-weighted imaging (DWI) is helpful on differentiating both primary and metastatic neoplasms from a brain abscess, as almost all pyogenic abscesses have markedly hyperintense signal on DWI (secondary to restricted water diffusion), whereas non-pyogenic lesions usually show hypointense or mixed signal.[9,10] However, it must be stressed that the diagnosis of pyogenic brain abscesses remains challenging and, in many cases, the biopsy with evacuation represents an inevitable diagnostic step.

^{1}H-MRS

In uncertain cases, additional information can be gathered from ^{1}H-MRS. This complements conventional MRI by enabling better lesion characterization.[11,12] Several in vivo ^{1}H-MRS studies have demonstrated the presence, in the core of the pyogenic lesion, of specific resonances such as succinate (2.40 ppm), acetate (1.92 ppm), alanine (a doublet centered at 1.47 ppm), amino acids (valine, leucine and isoleucine resonating together at 0.90 ppm) as well as lipids (0.90 and 1.30 ppm) and lactate (Lac, a doublet centered at 1.33 ppm) (Figure 7.1). This metabolic pattern has been confirmed in several in vitro studies and might show slight differences in case of aerobic or anerobic infections.[13]

In particular, ^{1}H-MRS has shown to be specifically beneficial in differentiating between brain abscesses and other cystic lesions.[14,15,16] In these cases, the resonances of succinate, acetate, alanine, and amino acids can be found in untreated bacterial abscesses or soon after the initiation of treatment, but are not detected in normal or sterile pathologic human tissue. [1,10,17,18] Spectra of arachnoid cysts, for example, typically show Lac signal and no other metabolites, easily allowing a differential diagnosis with a pyogenic mass.[10] In the latter, the presence of succinate and acetate resonance intensities are probably the results of enhanced glycolysis;[13] the reason for detecting the signals of valine, leucine, and isoleucine should be related to the large breakdown of neutrophils existing in the lesion and resulting in a release of a large amount of proteolytic enzymes that hydrolyze the proteins into amino acids.[19]

In a brain mass, the resonance intensities of acetate, succinate, and amino acids can be considered as markers of infectious involvement, which are not usually found in spectra from intracranial tumors. [12,15,20,21] Moreover, while Lac, alanine, and lipid resonance intensities can be often found in both

Chapter 7: MRS in infectious, inflammatory, and demyelinating lesions

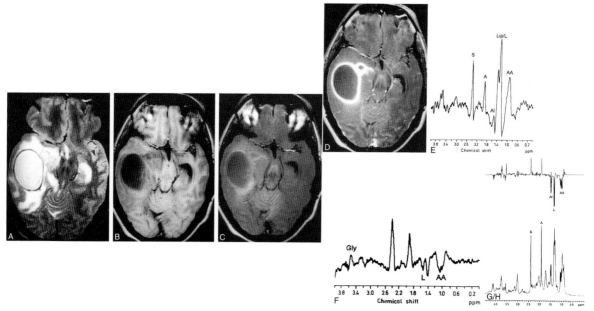

Figure 7.1. Pyogenic brain abscess. A, T_2-weighted MR image through the temporal lobe shows a well-defined hyperintense lesion in the right temporal lobe with peripheral hypointense rim, perifocal edema, and mass effect on the ventricular system. B, T_1-weighted image shows the lesion as hypointense with isointense wall. C, MT T_1-weighted image shows minimal hyperintensity. The MT ratio from the wall is 31.25. D, Post-contrast MT T_1-weighted image shows enhancement of the rim of the lesion. E and F, In vivo MR spectra obtained using STEAM (E) at 3000/20/30/128 (TR/TE/TM/excitations) and spin-echo (F) at 3000/135/128 (TR/TE/excitations) show presence of glycine (Gly) at 3.56 ppm; succinate (S) at 2.4 ppm; acetate (A) at 1.92 ppm; alanine (Al) at 1.5 ppm; lipid/lactate (Lip/L) at 1.3 ppm; and leucine, isoleucine, and valine (AA) at 0.9 ppm. G and H, Ex vivo MR spectroscopy with single-pulse (G) and spin-echo (H) imaging confirm the assignments seen in vivo. [From Gupta et al., Am J Neuroradiol 2001; **22**: 1503–09, figure 1, with permission.]

meningiomas and brain abscesses, an elevated choline (Cho) peak can be found only in conditions of sustained cell proliferation and density such as those of a neoplastic tissue.[1,22,23]

All together, these data suggest that ^{1}H-MRS provides information that could be used for differentiating abscesses from brain masses. However, it should be taken into account that the typical ^{1}H-MRS pattern of the brain abscess changes some days after beginning the antibiotic treatment, with the disappearance of the resonances related to the infective process (i.e. acetate, succinate, and amino acids) in treated patients.[24]

In performing the ^{1}H-MRS exam, due to the characteristics of the peaks that need to be visualized, ideally both short (20–30 ms) and long TE (135–280 ms) should be performed. It is also suggested to center the spectroscopic volume of interest (VOI) over the necrotic core of the lesion to better detect the metabolites of the pyogenic abscess. If the aim is the differential diagnosis with a malignant mass, it might be wise to perform an additional acquisition with the VOI centered on the enhancing rim, or a multi-voxel sequence to rule out choline elevation, a hint for the presence of a tumor.

Tuberculoma

Tuberculosis is still one of the most common and important infections around the world. This can have serious CNS implications, including intracranial tuberculoma.[25] Tuberculosis foci (tubercles) develop in the CNS or adjacent bony structures during the bacillemia that follows primary tuberculosis infection or late reactivation of tuberculosis elsewhere in the body. Intracranial tuberculomas can develop from the tubercles, giving rise to one of the most serious complications of the disease. The clinical course is variable, ranging from complete resolution to the rupture of the tuberculoma with the consequent meningoencephalitis. A prompt diagnosis is necessary to avoid this severe evolution.

On conventional MRI, the lesions can be seen solitary or multiple.[26,27] The non-caseating tuberculomas have high signal on T_2-weighted images, with peripheral enhancement. The immune response can create a granulomatous reaction, with central caseation and a solid center, and in some case, a progression to a liquid center. Caseating tuberculomas with a solid center have low to intermediate signal on T_1-weighted and

Chapter 7: MRS in infectious, inflammatory, and demyelinating lesions

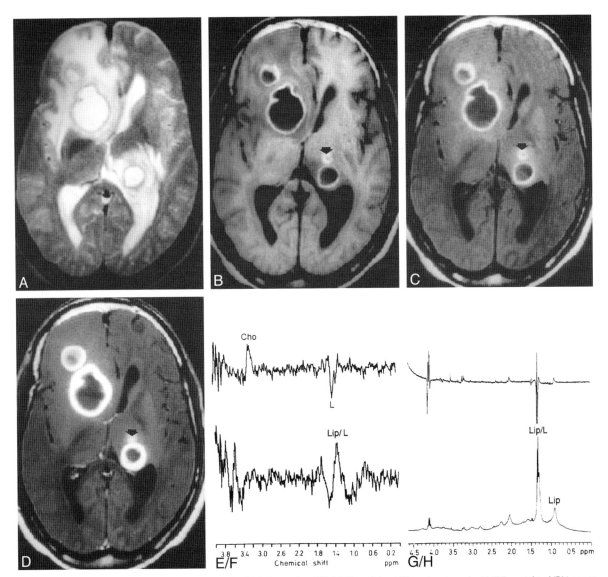

Figure 7.2. Multiple tuberculous abscesses. T_2-weighted (A), T_1-weighted (B), MT T_1-weighted (C), and post-contrast MT T_1-weighted (D) images through the third ventricle show abscesses in the right caudate nucleus, right frontal and left periventricular regions. The lesions show hyperintense core and peripheral hypointense rim along with perifocal edema on T_2-weighted image (A). On T_1-weighted (B) and MT T_1-weighted images (C), these lesions shown hypointense core with hyperintense rim, more clearly visible on MT T_1-weighted images. Post-contrast MT T_1-weighted image shows enhancement of the peripheral rim. Note an additional lesion (arrow) visible on B, C, and D that is not visible in A. The MT ratio from the wall of these three abscesses varied from 18.3 to 20.4. In MR spectroscopy done from the large right caudate nucleus lesion using STEAM at 20 (TE) with a voxel of 1.5 cm^3, only lipid and lactate (Lip/L) at 1.3 ppm are shown (E); on spin-echo spectrum at 135, phase reversal along with reduction in signal is seen (F). The presence of choline (Cho) in F results from the larger size of the voxel (2.0 cm^3) selected on spin-echo sequences that included the wall of the lesion. Ex vivo high-resolution, single-pulse (G) and spin-echo (H) proton MR spectroscopy at 80 ms confirm the presence of Lip/L with no evidence of choline. [Modified from Gupta et al., Am J Neuroradiol 2001; **22**: 1503–09, figure 2, with permission.]

central low signal on T_2-weighted images, with ring-like enhancement. Caseating granulomas with a liquid center have low signal on T_1-weighted and high signal on T_2-weighted images, and can be indistinguishable from pyogenic abscesses.[26]

^{1}H-MRS

^{1}H-MRS can be helpful in differentiating tuberculomas from other cerebral masses (Figure 7.2).[10,18] In contrast to other intracranial infections and brain tumors, the spectroscopic pattern of tuberculomas constantly shows a large lipid peak inside the lesion and no other relevant resonance signals, except the lactate signal on few occasions.[10,18] The ^{1}H-MRS protocols should therefore use a short TE (20–30 ms) acquisition to best detect the lipid signal, with the VOI positioned fully inside the core of the lesion.

Herpes simplex and other viral encephalitis

The most common cause of encephalitis in humans is due to herpes simplex virus (HSV) infection. The virus most often invades the brain after reactivation of latent virus that resides in the ganglia of cranial nerves. Patients present often with hyperpyrexia, altered consciousness, focal neurological deficits, and seizures.

The clinical course and prognosis is usually severe. The gold standard for diagnosis of HSV encephalitis is the detection of HSV DNA in the cerebrospinal fluid by using a polymerase chain reaction.[28] However, as treatment with antiviral therapy should be initiated immediately when HSV encephalitis is suspected, MRI (or CT scan) can be important to support the presumptive diagnosis. High signal is usually found on T_2-weighted and FLAIR images, with specific localization in temporal and inferior frontal lobes.[29,30] Mass effect on the lateral ventricle can sometimes be present. Petechial hemorrhage can also be found on some occasions.[29]

^{1}H-MRS

^{1}H-MRS studies have shown metabolic alterations of the HSV lesion characterized by reduced NAA/Cr, elevated Cho/Cr, and presence of the Lac signal. The decrease of NAA/Cr is usually very severe, probably in relation to the severe neuronal loss caused by the infection.[31] An increase of the Cr signal due to astrocytosis is also likely.[31] The Lac signal (inconstant) is probably due to the activity of macrophages and other cells of inflammation.[32,33] The Cho/Cr ratio is usually lower than that seen in malignant tumors.[10] However, the spectroscopic pattern of HSV infection is not specific and the routine use of ^{1}H-MRS is limited. More useful could be the use of ^{1}H-MRS to monitor the disease evolution. The experience of previous studies suggests that metabolic alterations could be stable,[32] more pronounced,[34] or improved,[33] in parallel with clinical status variations.

Using ^{1}H-MRS to monitor the HSV lesion, serial acquisitions should be performed with accurate repositioning of the VOI on the lesion. As the NAA, Cho, and, eventually, Lac resonances should be monitored, the ^{1}H-MRS protocols should preferably use long TE (135–280 ms) acquisitions.

Other, less frequent viral encephalitis such as infections of West Nile virus, cytomegalovirus, paramyxovirus, Epstein–Barr, and papovavisur do not show specific metabolic alterations (often indistinguishable from those of the HSV infection) and the use of ^{1}H-MRS for diagnosis is only occasional.[1,10]

HIV encephalopathy

The encephalopathy induced by the human immunodeficiency virus (HIV) has been well documented.[35,36,37,38] It generally occurs in the late stages of the disease, when immunodepression becomes more severe, leading to the so-called HIV-associated dementia. The latter is a sub-cortical dementia characterized by progressive loss of cognitive functions, often later accompanied by motor decline.[39] The pathogenesis of the HIV encephalopathy is not fully understood. There is no evidence for a direct viral infection of neurons, but most likely, the neuronal damage found in this condition results from an immune-mediated process leading neurons to death by apoptosis.[37,38]

Cortical/subcortical atrophy is usually present on conventional MRI. This can be found in some degree also in pre-clinical conditions. The primary infection to HIV may also lead to focal abnormalities of the deep white matter ranging from patchy to confluent lesions.[40,41] The former is due to reactive gliosis, macrophages and lymphocytes infiltrates. The latter is due to a diffuse process characterized by tissue loss, water accumulation in the interstitium and demyelination, but little or no inflammation.[42] In severe cases, diffuse symmetric hyperintensity is seen in the supratentorial white matter, predominantly in the frontal and parietal lobes. Mass effect and enhancement are usually absent. On T_1-weighted images, the white matter may appear slightly hypointense.

^{1}H-MRS

There have been many ^{1}H-MRS studies showing metabolic abnormalities in the brains of patients with HIV.[31,43,44,45,46,47,48,49,50] These studies generally showed reductions in NAA and increases in Cho and *myo*-inositol (mI) in both lesional and normal-appearing brain tissues in comparison to normal controls. Decreased NAA signal has been reported in the HIV-related encephalopathy since the earliest studies, in virtually all cases.[43,44,45] The Cho signal likely increases when the immunodepression becomes more relevant and lead to encephalitis.[31] The increase of mI should be put in relation to the glial reaction observed in brain parenchyma in this infection. [31,51,52]

The brain metabolic pattern provided by ^{1}H-MRS is not specific of the disease and does not help significantly in the diagnostic process. However, there is compelling evidence that ^{1}H-MRS reveals changes in neurologically asymptomatic subjects and therefore may be of particular value in documenting early CNS involvement when both neurological examination and conventional MRI are normal.[45,49,52,53,54,55] In addition, several studies have shown that the ^{1}H-MRS patterns have a good correlation with the clinical status, worsening with advancing dementia. [52,56,57,58,59] Furthermore, ^{1}H-MRS has been successfully used in multicenter trials to monitor brain changes after therapy.[57,60,61,62,63,64,65,66] All this suggests that ^{1}H-MRS could be useful for assessing early cerebral involvement and response to therapy in patients with clinical and subclinical HIV infection. [67] This has become particularly important since the implementation of highly active antiretroviral therapy in HIV.[68,69]

Given the importance of NAA and Cho as prognostic markers,[31] the ^{1}H-MRS protocols should preferably use long TE (135–280 ms) acquisitions. Short TE (20–30 ms) acquisitions can be used to measure mI, which may be elevated even at the early stages of HIV-related dementia. The ^{1}H-MRS imaging (^{1}H-MRSI) acquisition can provide a wide spatial coverage and should be preferred. For single-voxel studies, the specific regions to examine should be the white matter of the frontal lobes and the basal ganglia, typically sites of greatest involvement.

HIV-related infections

In HIV-affected patients, the depletion in immunity may be responsible for secondary brain infections. The most frequent are cerebral toxoplasmosis, primary CNS lymphoma, progressive multifocal leukoencephalopathy (PML), and cryptococcosis, and will be illustrated here below. Unlike the primary HIV infection, dementia is not the main feature of these secondary manifestations. Rather, patients show rapidly progressive focal neurologic deficits due to the brain localization of the infective agent. The incidence of opportunistic CNS infections has decreased significantly since the introduction of highly active antiretroviral therapy.[70] However, they are still frequent and life-threatening in HIV-affected individuals and, when they occur, a rapid diagnosis is vital to start specific treatment as early as possible.[153] ^{1}H-MRS can potentially provide additional information to reach the correct diagnosis in all these conditions.[71]

Toxoplasmosis and primary CNS lymphoma

Cerebral toxoplasmosis and primary CNS lymphoma are considered together as they are the two most common secondary manifestations of HIV infection and are characterized by indistinguishable clinical features, with consequently very difficult differential diagnosis.

Cerebral toxoplasmosis is the most common cerebral lesion in HIV-affected patients.[72] Toxoplasmosis results from *Toxoplasma gondii*, an intracellular protozoan parasite diffuse worldwide and, in the majority of cases, asymptomatic. In HIV-affected patients, symptomatic toxoplasmosis is due to reactivation of latent infection as a result of progressive loss of cellular immunity. The frequency of symptomatic cerebral toxoplasmosis in HIV-infected individuals varies from about a quarter to a half of cases.[73] The most frequent clinical manifestations are headache, confusion, fever, and lethargy. Seizures may be an initial manifestation and half of the patients may show focal neurological signs.[73]

Primary CNS lymphoma, mostly due to the Epstein–Barr virus, is an extranodal, non-Hodgkin's neoplasm, which occurs in both immunocompromised and immunocompetent hosts. In HIV-affected individuals, this is the second most frequent space occupying lesion of the brain (after cerebral toxoplasmosis) with an incidence many times higher than in the general population (2–5% of HIV patients).[73] It occurs in severely immunocompromised HIV-infected individuals, with a clinical presentation that is virtually indistinguishable from that of the cerebral toxoplasmosis.[1,73]

In both cerebral toxoplasmosis and lymphoma secondary to HIV infection, the CSF examination may be normal or aspecific. The use (possibly simultaneous) of polymerase chain reaction amplification of *Toxoplasma gondii* and Epstein–Barr virus DNA in CSF has shown to be a sensitive and highly specific assay for the diagnosis.[74,75]

Patients with HIV infection in whom cerebral toxoplasmosis or primary CNS lymphoma is suspected should promptly be evaluated with CT or, preferably, MRI of the brain. Cerebral toxoplasmosis

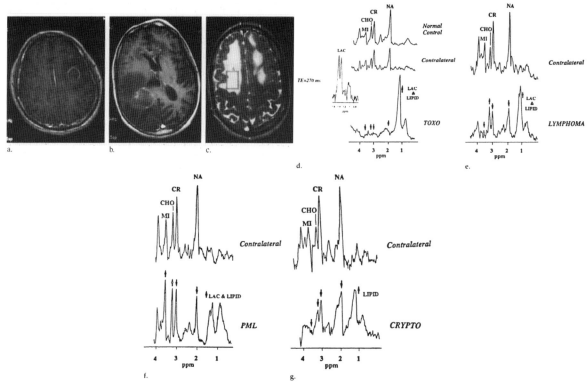

Figure 7.3. (a) Representative T_1-weighted MR image shows a toxoplasmosis lesion with a ring enhancement. (b) T_1-weighted MR image shows a lymphomatous lesion that also has ring enhancement. (c) T_2-weighted MR image obtained in a patient with PML shows a diffuse pattern of hyperintensity in the white matter. The white box shows the region of interest used for proton spectroscopy. (d–g) Representative proton spectra of (d) a toxoplasmosis (TOXO) lesion in the parietal region, (e) a lymphoma in the lateral frontal region, (f) a PML lesion in the subcortical posterior frontal region, and (g) a cryptococcal (CRYPTO) abscess in the subcortical frontal region compared with the contralateral control spectra. [From Chang et al., Radiology 1995; **197**: 525–31, figure 1, with permission.]

and lymphoma often cannot be differentiated on imaging. In both conditions, the multiple lesions may be located anywhere within the brain involving both gray and white matter, seen with high signal on T_2-weighted and FLAIR images, enhancement after contrast administration, and presence of vasogenic edema.[1,73,76] Autopsy studies have revealed that they are always multiple, even when solitary lesions are seen on CT or MRI scans.[73] Unlike toxoplasmosis, lymphoma can involve and cross the corpus callosum.

Given the difficulties in making the correct diagnosis on neuroimaging, the impossibility of waiting for the detection of the DNA in CSF, and the good response to chemotherapy, generally, the empiric anti-toxoplasma therapy represents the first-line approach to most HIV-infected patients with focal brain lesions. In contrast, patients with cerebral lymphoma frequently succumb to other complications such as infection, even when the lymphoma can be controlled (with chemotherapy and/or radiotherapy). Thus, the treatment of cerebral toxoplasma is usually initiated upon presumptive diagnosis and is considered appropriate in patients with low CD4 lymphocytes (below 100 cells/μl), anti-toxoplasma IgG antibody in the serum, consistent clinical and neuroimaging features. The positive response to empiric anti-toxoplasma therapy could confirm the diagnosis. Brain biopsy is recommended for patients who do not fulfill the criteria for presumptive treatment or fail empiric therapy for toxoplasmosis.[73]

^{1}H-MRS

Since brain biopsy is sometimes deferred by both patients and neurosurgeons, there is a need for additional noninvasive methods to make the diagnosis of cerebral toxoplasmosis or primary CNS lymphoma (Figure 7.3).[77] For example, SPECT with radioactive thallium could aid in the differential diagnosis; inflammatory lesions such as toxoplasmosis are negative on

SPECT, while lymphoma uptakes the radioactive thallium. This is, however, not free from false positives and false negatives.[78] In this context, ^{1}H-MRS is another noninvasive modality that has shown promise in distinguishing the etiology of secondary brain lesions in HIV-infected patients.[71]

Studies trying to differentiate between cerebral toxoplasmosis and lymphoma by means of ^{1}H-MRS have provided discrepant results.[79] However, in general, it can be said that ^{1}H-MRS metabolic pattern of cerebral toxoplasmosis shows very large lipid signals and not much else, although Lac may be present. Brain lymphoma may also show a large lipid (generally much smaller than toxoplasmosis) and Lac signals, but does show also very low NAA and, especially, very high Cho levels (that are not detected on toxoplasmosis lesions).[10,71,80]

It must be stressed that, in these conditions, the positioning of the ^{1}H-MRS VOI is crucial.[10] First, the VOI must not contain partial volume from non-lesional tissue (as appearing on conventional MRI), as the latter would contribute with normal or almost normal metabolic signal. Second, the metabolic abnormalities of the necrotic core of the lesion always resemble that of the toxoplasmosis with a very large lipid peak and very low levels of all other metabolites, whereas more correct information is coming from the enhancing edge of the lesions. For these reasons, the best option is to perform a ^{1}H-MRSI study, which guarantees the proper spatial resolution and lesion coverage.

Cryptococcosis

Cryptococcus neoformans causes the most common fungal infection in HIV-infected patients and is the third most frequent neurological complication in patients with AIDS.[73] It can cause meningitis, pseudocysts (encapsulated masses of organisms with little inflammatory reaction), or real abscesses. Pseudocysts or abscesses are preferably localized in the basal ganglia region. In the case of meningitis, patients may present with headache, fever, altered mental status, nausea, and vomiting. Focal neurological signs and seizures occur in about 10% of patients. Elevated intracranial pressure occurs in many cases. The definite diagnosis of cerebral cryptococcosis in HIV-infected patients is based on culture of the organism in CSF.[73]

Neuroimaging studies are useful to confirm the diagnosis. Conventional MRI can show (i) nodular meningeal enhancement; (ii) solid, ring-enhancing masses located in the choroid plexus; (iii) multiple round cystic lesions (hypointense on T_1-weighted images and hyperintense on T_2-weighted and FLAIR images) with no surrounding edema, no or minimal mass effect and no enhancement after contrast agent administration; or (iv) a mix of these findings. The cystic lesions have a predilection for basal ganglia, which poses a differential diagnosis with toxoplasmosis, CNS lymphoma, and abscesses.[10] However, all these brain masses show an enhancement on post-gadolinium T_1-weighted images, which is usually absent in cryptococcal brain lesions.[73]

^{1}H-MRS

^{1}H-MRS can provide additional information useful for the diagnosis and has been shown to be able to discriminate between HIV-related brain masses.[71] More specifically, spectra of cryptococcal infections typically show an increase in Lac (more than lipids) signal, which helps to differentiate this brain mass from that of toxoplasmosis (Figure 7.3). It shows also low NAA signal and, more importantly, low mI signal and unchanged Cho signal. The latter is very helpful for the differential diagnosis with the primary CNS lymphoma.

In order to detect the signals of lipids and mI, short TE (20–30 ms) should be preferred to long TE. As mentioned for the toxoplasmosis and lymphoma lesions, the positioning of the ^{1}H-MRS VOI is a fundamental step and protocols allowing the best spatial resolution should be pursued.

Progressive multifocal leukoencephalopathy

HIV-infected patients can develop progressive multifocal leukoencephalopathy (PML). This is an often fatal demyelinating disease caused by the JC virus, a polyomavirus. Primary infection is usually not associated with clinical symptoms and the virus resides quiescently in the kidney, CNS, and peripheral lymphocytes.[81] The immune deficiency presumably allows the reactivated JC virus to manifest. In these cases, a wide variety of neurologic symptoms can occur, with focal and progressive neurological impairment. Mental status changes may accompany focal neurological deficits. Prior to the introduction of antiretroviral therapy, PML was inevitably fatal in a few months. The administration of specific anti-HIV drugs can cause, sometimes, a stabilization probably by improving the function of the immune system.[73]

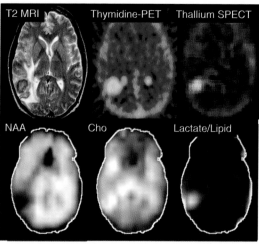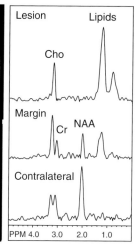

Figure 7.4. Lymphoma. Selected spectra and metabolic images of NAA, Cho and Lactate/Lipid are shown, as well as thymidine-PET and thallium-SPECT. The core of the lymphoma is characterized by an absence of NAA, and elevated Cho and lipids. A similar pattern is also seen in the lesion margin. Low NAA can also be observed in the edematous region of the posterior limb of the internal capsule. Note that the thalamus and insular cortex are normal (i.e. these regions have high Cho in normal subjects). Both thymidine-PET and thallium-SPECT show increased radiotracer uptake. [From Pomper et al., Acad Radiol 2002; **9**: 398–409, with permission.]

Routine analysis of the CSF is usually unhelpful in the diagnosis of PML. However, the detection, by polymerase chain reaction, of JC virus DNA has been shown to be both highly sensitive and specific.[82] Definitive diagnosis of PML is based on pathologic examination of brain tissue, but less-invasive diagnostic methods are more often utilized to make a presumptive diagnosis of PML. Among those, conventional MRI shows multiple, asymmetric foci of hyperintense signal on T_2-weighted and FLAIR images. They are mainly subcortical (with involvement of the U-fibers) and, rarely, periventricular, located almost exclusively in white matter; there is usually no mass effect, with absent or, rarely, very mild contrast enhancement.[83]

¹H-MRS

The JC virus predominantly infects oligodendrocytes and astrocytes, resulting in severe demyelination and cell loss; the inflammatory reaction is usually moderate. [84] The spectra pattern of PML lesions (Figure 7.3) is characterized by increases of Cho signal (in relation to demyelination) and, eventually, increases in Cr signal (in relation to astrocytosis). The Lac signal may be present, probably reflecting necrosis or macrophagic infiltration. The NAA resonance is often decreased, especially in the core of the lesion where neuronal damage and loss secondary to demyelination is more evident. At short TE, large increases of the lipid signal (at 0.9–1.3 ppm) may add to the Lac and, possibly, aminoacid signals. Also, the resonance of mI might be elevated.[80,85,86] All these metabolic alterations are not specific, but this pattern could be useful to differentiate this type of lesion (i.e. demyelinating) from the other brain lesions (i.e. cystic, tumoral, etc.) occurring in HIV-infected individuals.

Examples of MRSI in lymphoma and PML, with comparison to thallium-SPECT and thymidine PET are given in Figures 7.4 and 7.5, respectively.

Subacute sclerosing panencephalitis

Subacute sclerosing panencephalitis (SSPE) is a progressive inflammatory neurological disorder, which is caused by persistent measles virus infection in the CNS.[70] The pathological lesions are characterized by perivascular infiltration by monocytes, astrocytic proliferation, neuronal degeneration, and demyelination. Lesions may occur both in gray and white matter. Symptoms usually include progressive cognitive impairment and myoclonia, with a rapidly progressive disease course. Diagnosis is based on a typical electroencephalographic pattern (with periodic complexes) and on the presence of high measles antibodies in serum and CSF.

On conventional MRI, bilateral hyperintense (on T_2-weighted and FLAIR images) or hypointense (on T_1-weighted images) lesions are seen in the temporoparietal lobes. As the lesions progress, there may be the involvement of the periventricular white matter, corpus callosum, and basal ganglia. Long-term changes include encephalomalacia and atrophy.[10,70]

¹H-MRS

Inside the large white matter lesions, the spectra of subacute sclerosing panencephalitis typically show a low NAA level, high mI and Cho signals and, possibly, the presence of lactate. Interestingly, this pattern,

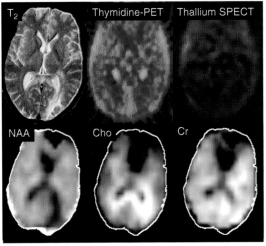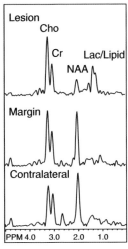

Figure 7.5. PML. Selected spectra and metabolic images of NAA, Cho and Lactate/Lipid are shown, as well as thymidine-PET and thallium-SPECT. The lesion is characterized by elevated Cho and lactate/lipid, and decreased NAA. Note the high Cho signal in the splenium of the corpus callosum and spreading into the right hemisphere, suggesting lesion growth. Both thymidine-PET and thallium-SPECT show no radiotracer uptake above background. [From Pomper et al., Acad Radiol 2002; **9**: 398–409, with permission.]

which is consistent with active inflammation, demyelination, and neurodegeneration, can be seen also at early stages.[87] The MRS protocol should use short TE (20–30 ms) to detect mI. This metabolite provides, with NAA, a much better predictor of clinical severity and outcome than conventional MRI.[88]

Brain demyelination and inflammation

In the last decade or so, MR methods have been extensively used in inflammatory/demyelinating diseases, leading to a re-evaluation of the pathogenesis, the natural history, and the diagnostic approach of many of these disorders. This has been driven to a significant degree by results of ^{1}H-MRS studies. For example, by providing evidence of early neurodegeneration (based on levels of NAA), results of ^{1}H-MRS studies have led to a reconsideration of the role of axonal damage in a primary demyelinating disorder such as MS. By measuring brain changes of metabolites such as choline and *myo*-inositol, ^{1}H-MRS has confirmed the importance of assessing myelin damage and repair.

The next section covers the most relevant applications of ^{1}H-MRS in this area, in particular in diseases such as MS and acute disseminated encephalomyelitis (ADEM). The relevance of the ^{1}H-MRS use in other demyelinating diseases with known infective etiology, such as PML and SSPE, is described above. The ^{1}H-MRS characteristics of other demyelinating disorders due to inherited metabolic abnormalities such as Krabbe disease, adrenoleukodystrophy and other rare leukoencephalopathies is covered in Chapter 11.

Multiple sclerosis

Multiple sclerosis (MS) is the prototype inflammatory autoimmune disorder of the CNS.[89] The etiology of MS is still unknown, but the disease seems to be the result of an interaction between undetermined environmental factors and susceptibility genes.[90] These factors trigger a cascade of pathological events, manifesting in the CNS as acute inflammation, focal demyelination, neurodegeneration, limited remyelination, and ending in the chronic multifocal sclerotic plaques.[91]

Fairly diffuse worldwide (with a northern hemisphere preponderance) and with a lifetime risk of 1 in 400, MS is potentially the commonest cause of neurological disability in young adults.[90] Clinically, it may manifest more frequently with a relapsing–remitting (RR) form, more rarely in a primary progressive (PP) fashion, in both cases with a great variety of symptoms potentially involving any white matter of the CNS (i.e. brain and spinal cord). In the first form, the acute episodes (i.e. relapses) are characterized initially by a complete clinical recovery. However, with time, recurrent relapses are followed by only partial recovery, leaving persistent neurological deficits. This is then followed by a clinical disability progression without relapses (secondary progressive, SP, form). Rarely (10% of cases), the disease manifests with a steady decline in neurologic function from onset, without superimposed attacks (PP form).[89] The overall prognosis of each form is unpredictable, with a high degree of variability in the final outcome, but the patient invariably progresses toward an irrecoverable disability.[92]

Chapter 7: MRS in infectious, inflammatory, and demyelinating lesions

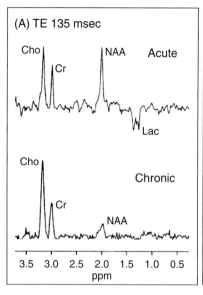 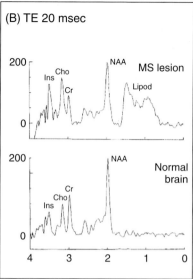

Figure 7.6. (A) Long TE (135 msec) spectra in acute and chronic MS lesions. Both acute and chronic spectra have elevated Cho and reduced NAA (particularly in the chronic lesion); only the acute lesion shows an elevated lactate signal. [Adapted from Miller et al., Lancet 1991; **337**: 58–9, with permission.] (B) Short TE (20 ms) spectra from an acute MS lesion, and normal brain for comparison. The lesion short TE spectra show increased *myo*-inositol, choline, and lipids, as well as slightly decreased Cr and NAA. [Adapted from Koopman et al., Lancet 1993; **341**: 631–2, with permission.]

The widely accepted diagnostic criteria of MS classify individuals in the categories of MS, possible MS, or not MS.[93] This incorporates evidence from conventional MRI, which has a very relevant role in diagnosing and monitoring MS patients. For example, individuals must have a minimum of two attacks (affecting more than one anatomical site) to meet the diagnosis of definite MS ("lesions disseminated in time and space"). Assuming an initial clinical presentation suggestive of MS (the so-called clinically isolated syndrome, or CIS), the second attack does not need to be clinically expressed necessarily, but can be represented by a brain lesion as documented on conventional MRI. Recently, the MRI diagnostic criteria for MS have been further revised, to further improve the final diagnosis of the different forms of MS.[94]

Conventional MRI has a major role in the recently developed diagnostic criteria for MS,[93,94] because of its exquisite sensitivity for detecting MS lesions and their changes over time.[95] Conventional MRI lesions are seen as brain (and spinal cord) multiple foci of various size, irregular shape, and asymmetric distribution of white matter hyperintensity on T_2-weighted images. Abnormalities seen on T_2-weighted images may reflect edema, demyelination, remyelination, gliosis, or axonal loss, with a lack of pathological specificity. [96] A subset of these lesions appear as hypointense on T_1-weighted images, probably more specifically representing areas of axonal loss and severe matrix destruction.[97] On post-gadolinium T_1-weighted images, some MS lesions can appear hyperintense, reflecting intense inflammatory activity and mononuclear cell infiltration.[98]

Despite the sensitivity of conventional MRI for detecting MS lesions, it does have some important limitations. First, there is low pathological specificity of the abnormalities seen on conventional MRI scans. Second, there is the inability of conventional MRI metrics to detect and quantify the extent of damage in normal-appearing brain tissues, which are known to be involved in the pathological process.[99] These limitations probably result in the limited correlation that is found to exist between the conventional MRI metrics and patients' clinical status in MS.[96] These inherent limitations of conventional MRI have prompted the development and application of modern quantitative MR techniques such as ^{1}H-MRS, magnetization transfer (MT) MRI, diffusion-weighted and functional MRI (fMRI) to the study of MS.

^{1}H-MRS

In the last decade, a great number of ^{1}H-MRS studies have provided in vivo accurate chemical–pathological characterization of MR-visible lesions and normal-appearing brain tissues in MS brains.[100–102] In demyelinating lesions large enough to allow spectra to be acquired without substantial partial volume effects (Figure 7.6), ^{1}H-MRS at both short and long echo times reveals increases in Cho and sometimes lactate, resonance intensities from the early phases of the pathological process.[103,104] Changes in the

Chapter 7: MRS in infectious, inflammatory, and demyelinating lesions

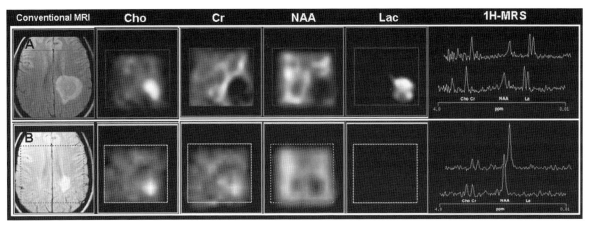

Figure 7.7. From right – to left, conventional brain MRI, spectroscopic images of choline (Cho), creatine (Cr), N-acetylaspartate (NAA), and lactate (LA), and the lesional spectra of a patient with a single giant demyelinating lesion during the acute phase of the disease (A) and 15 months later (B) are shown. During the acute phase of the disease, note the focal increases in Cho and Lac and decreases in NAA and Cr that co-localize with the MRI lesion. The examination performed 15 months later shows a reduction of the MRI lesion and normalization of Cho, Cr, and Lactate metabolic images. NAA shows only a partial recovery.

resonance intensity of Cho can be interpreted as a measure of increases in the steady-state levels of membrane phospholipids released during active myelin breakdown. Increases in Lac may reflect primarily the metabolism of inflammatory cells. In large, acute demyelinating lesions decreases of Cr can also be seen.[105] Short echo time spectra give evidence for transient increases in visible lipids (released during myelin breakdown),[106] and more stable increases in mI.[107] These changes are consistently accompanied by substantial decreases in NAA, interpreted as a measure of axonal injury reflecting metabolic or structural changes.[108] Recently, glutamate levels were found to be elevated in acute lesions suggesting a link between axonal injury in active lesions and glutamate excitotoxicity.[109]

After the acute phase and over a period of days to weeks (Figures 7.6 and 7.7), there is a progressive return of raised Lac resonance intensities to normal levels in focal lesions.[105] Cr also returns to normal within a few days,[105] or may show small residual increases, presumably related to gliosis.[110] Persistent increases in mI signals in chronic lesions may be related to microglial proliferation.[111–113] Resonance intensities of Cho and lipids typically return to normal over months.[105,114] The signal intensity of NAA may remain decreased or show partial recovery, starting soon after the acute phase and lasting for several months.[104] The recovery of NAA may be related in various proportions to reversible metabolic changes in neuronal mitochondria, the resolution of edema, or changes in the relative partial volume of neuronal processes.[104]

Initial ^{1}H-MRS studies were focused mainly on MRI-defined lesions.[115,116] However, more recent studies exploiting the greater coverage and resolution of ^{1}H-MRSI have shown that metabolic abnormalities in MS patients are not restricted to lesions, but are present both adjacent to and distant from the lesions.[106,117–121] The NAA decreases found in the normal-appearing white matter are usually attributed to axonal damage,[122] and, although they can be present at early disease stages,[123] are more pronounced in advanced disease stages.[120,122,124]. The extent of this NAA reduction decreases with the distance from the core of a lesion,[125,126] consistent with the notion that the diffuse changes are at least in part related to dying back of axons transected within plaques.[127] However, decreased levels of NAA also occur without obvious relation to T_2-visible lesions.[128]

Recent ^{1}H-MRS studies have focused on gray matter metabolic changes in MS patients, supporting the notion that the contribution of gray matter pathology is substantial in MS.[129] It has been found that cortical decreases in NAA might be small or absent early, but seem to be considerable in patients with progressive disease.[130–134] In contrast, subcortical gray matter decreases in NAA seem to be more consistently found from early stages.[135–137] In some studies,[135,138] ^{1}H-MRS and histopathological methods have been used in parallel and the amount of ex vivo total loss of thalamic neurons was comparable to the in vivo NAA decrease.

Chapter 7: MRS in infectious, inflammatory, and demyelinating lesions

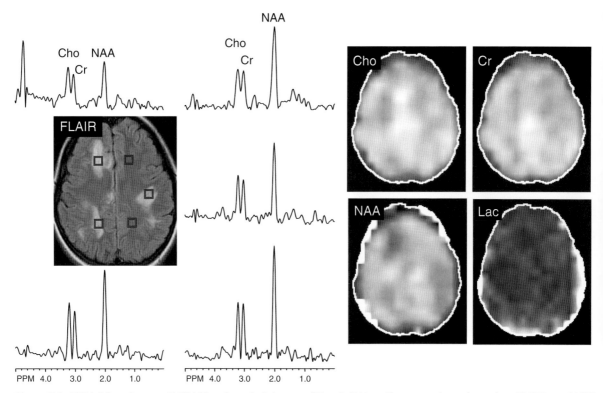

Figure 7.8. ADEM. Selected spectra, FLAIR MRI, and metabolic images of Cho, Cr, NAA, and lactate are shown from a long TE (280 msec) MRSI study. The patient presents acutely with multiple, asymmetric hyperintense lesions in the subcortical white matter of the left and right hemispheres. Note that the only metabolic abnormality associated with the lesions is a mild reduction of NAA, consistent with mild axonal dysfunction and edema. No elevations of lactate or Cho are observed, unlike those typically seen in MS. The ADEM pattern observed here is suggestive of a good clinical outcome; other ADEM cases may have more severe metabolic abnormalities and poor outcome and/or recurrent disease.

A number of spectroscopic studies have demonstrated highly significant correlations between NAA/Cr and clinical disability in patients with isolated acute demyelinating lesions,[105] and in patients with established MS followed through periods of relapse and remission.[139–141] Consistent with other evidence of widespread pathology in MS, a strong correlation also has been found between NAA/Cr decreases and increases in clinical disability in normal-appearing WM.[120] Since changes in Cr could contribute to any changes in NAA/Cr,[113,142,143] it has been suggested that it would be more accurate to interpret decreases of brain NAA/Cr as markers of a less specific disturbance in the "cerebral tissue integrity".[110]

Despite its potential to monitor the temporal evolution of metabolite changes reflecting tissue integrity in demyelinating lesions and normal-appearing brain tissue, the use of ^{1}H-MRS in longitudinal studies to monitoring the response to drug therapies are uncommon,[144–148] and its large-scale use as a primary or secondary endpoint in clinical trials has not been attempted. However, recently recommendations for a standardized use ^{1}H-MRS protocol in MS multicenter clinical studies have been provided.[149]

Acute disseminated encephalomyelitis

Acute disseminated encephalomyelitis (ADEM) is an acute, autoimmune, demyelinating disorder of the CNS, often associated with a previous viral exposure or vaccination.[150] Usual symptoms include decreased consciousness and multiple focal neurological deficits. Diagnosis generally is made on the basis of clinical characteristics, analysis of CSF, and neuroimaging studies. A differential diagnosis should be made with MS and, sometimes, with other less frequent conditions such as vasculitis and leukoencephalopaties. Clinical symptoms may guide diagnosis, as ADEM usually is monophasic and presents days after a viral illness or immunization.

On conventional MRI, lesions of ADEM are markedly similar to other demyelinating pathologies such as MS. Lesions include foci with variable low T_1-weighted

Table 7.1. ^{1}H-MRS brain changes in infectious, inflammatory, and demyelinating disorders.

		NAA	Lactate	Creatine	Choline	Lipid	Myo-inositol	Glutamate	Succinate	Acetate	Alanine	Amino acids
Abscess			↑						↑	↑	↑	↑
Tuberculoma			↑ sometimes									
Herpes Simplex Encephalitis		↓↓ (NAA/Cr)	↑		↑ (Cho/Cr)	↑						
HIV encephalopathy		↓	↑		↑		↑					
Toxoplasmosis		↓↓		↓	↓	↑↑						
Lymphoma		↓↓	↑		↑↑	↑						
Cryptococcus		↓	↑		↔	↔	↓					
Progressive Multifocal Leukoencephalopathy		↓	↔ to ↑	↑	↑	↑	↑					
Subacute sclerosing Pan Encephalitis		↓	↔ to ↑		↑		↑					
Multiple Sclerosis[1]	Early	↓	↑		↑	↑	↑	↑				
	Late	↓	↔	↔ to ↑	↔	↔ to ↑	↔ to ↑					
Acute Disseminated Encephalomyelitis[2]		↓	↑ acute		↑ acute							

↑/↓ – elevated/reduced; ↓↓/↑↑ – much elevated/reduced; ↔ – no change; NL – normal
[1] see reduced NAA in normal-appearing brain.
[2] do not see reduced NAA in normal-appearing brain.

signal intensity, increased T_2-weighted signal intensity, and inconstant contrast enhancement after gadolinium administration. The white matter is primarily affected, but deep gray structures and brain stem may be involved as well.

¹H-MRS

Early diagnosis might have implications for treatment, particularly in distinguishing ADEM from MS.[151] By providing additional metabolic information on MRI lesions and normal-appearing tissues, ¹H-MRS can be useful for the diagnosis and for monitoring clinical outcome in patients with ADEM. As in other demyelinating conditions, spectra of ADEM lesions show decrease in NAA, variable changes in Cho, and presence of Lac signal during the acute phase.[152] Low NAA can be the only abnormality in the acute or chronic phase; in few cases, a partial recovery of NAA has been reported.[153] The lack of Cho elevation may favor the diagnosis of ADEM versus MS (Figure 7.8). Interestingly, the metabolic pattern of the MRI normal-appearing brain is usually normal,[153] helping in distinguishing ADEM from MS. In this context, a ¹H-MRSI acquisition, which allows wide spatial coverage and resolution, is preferable. A long TE (135–280 ms) protocol is recommended.

Summary

¹H-MRS can be used on a routine basis to characterize the metabolic pattern of infectious, inflammatory, and demyelinating disease, and provides an important contribution to the intricate diagnostic process of these complex disorders. Table 7.1 provides a summary of the metabolic changes that are typically associated with each disorder.

The information provided by ¹H-MRS often offers a better understanding of the pathology underlying these disorders and has led, in some occasions, to a re-evaluation of the pathogenesis of these diseases. Finally, in some disorders such as HIV-related disease and MS, ¹H-MRS has shown to be useful to follow disease evolution, and may be used to monitor therapeutic response.

References

[1] Foerster BR, Thurnher MM, Malani PN, Petrou M, Carets-Zumelzu F, Sundgren PC. Intracranial infections: clinical and imaging characteristics. *Acta Radiol* 2007; **48**: 875–93.

[2] Garvey G. Current concepts of bacterial infections of the central nervous system. Bacterial meningitis and bacterial brain abscess. *J Neurosurg* 1983; **59**: 735–44.

[3] Cecil KM, Lenkinski RE. Proton MR spectroscopy in inflammatory and infectious brain disorders. *Neuroimaging Clin N Am* 1998; **8**: 863–80.

[4] Habib AA, Mozaffar T. Brain abscess. *Arch Neurol* 2001; **58**: 1302–04.

[5] Krcmery V, Fedor-Freybergh PG. Neuroinfections in developed versus developing countries. *Neuro Endocrinol Lett* 2007; **28**(Suppl 2): 5–6.

[6] Osenbach RK, Loftus CM. Diagnosis and management of brain abscess. *Neurosurg Clin N Am* 1992; **3**: 403–20.

[7] Calfee DP, Wispelwey B. Brain abscess. *Semin Neurol* 2000; **20**: 353–60.

[8] Haimes AB, Zimmerman RD, Morgello S, Weingarten K, Becker RD, Jennis R, et al. MR imaging of brain abscesses. *Am J Roentgenol* 1989; **152**: 1073–85.

[9] Kastrup O, Wanke I, Maschke M. Neuroimaging of infections. *Neurorx* 2005; **2**: 324–32.

[10] Kingsley PB, Shah TC, Woldenberg R. Identification of diffuse and focal brain lesions by clinical magnetic resonance spectroscopy. *NMR Biomed* 2006; **19**: 435–62.

[11] Poptani H, Gupta RK, Gupta K, Roy R, Pandey R, Jain VK, et al. Characterization of intracranial mass lesions with in vivo proton MR spectroscopy. *Am J Neuroradiol* 1995; **16**: 1593–603.

[12] Poptani H, Kaartinen J, Gupta RK, Niemitz M, Hiltunen Y, Kauppinen RA. Diagnostic assessment of brain tumours and non-neoplastic brain disorders in vivo using proton nuclear magnetic resonance spectroscopy and artificial neural networks. *J Cancer Res Clin Oncol* 1999; **125**: 343–9.

[13] Garg M, Gupta RK, Husain M, Chawla S, Chawla J, Kumar R, et al. Brain abscesses: etiologic categorization with in vivo proton MR spectroscopy. *Radiology* 2004; **230**: 519–27.

[14] Poptani H, Gupta RK, Jain VK, Roy R, Pandey R. Cystic intracranial mass lesions: possible role of in vivo MR spectroscopy in its differential diagnosis. *Magn Reson Imaging* 1995; **13**: 1019–29.

[15] Dev R, Gupta RK, Poptani H, Roy R, Sharma S, Husain M. Role of in vivo proton magnetic resonance spectroscopy in the diagnosis and management of brain abscesses. *Neurosurgery* 1998; **42**: 37–42.

[16] Shukla-Dave A, Gupta RK, Roy R, Husain N, Paul L, Venkatesh SK, et al. Prospective evaluation of in vivo proton MR spectroscopy in differentiation of similar appearing intracranial cystic lesions. *Magn Reson Imaging* 2001; **19**: 103–10.

[17] Lai PH, Hsu SS, Ding SW, Ko CW, Fu JH, Weng MJ, et al. Proton magnetic resonance spectroscopy and diffusion-weighted imaging in intracranial cystic mass lesions. *Surg Neurol* 2007; **68**(Suppl 1): S25–36.

[18] Gupta RK, Vatsal DK, Husain N, Chawla S, Prasad KN, Roy R, et al. Differentiation of tuberculous from pyogenic brain abscesses with in vivo proton MR spectroscopy and magnetization transfer MR imaging. *Am J Neuroradiol* 2001; **22**: 1503–09.

[19] Mendz GL, McCall MN, Kuchel PW. Identification of methyl resonances in the 1H NMR spectrum of incubated blood cell lysates. *J Biol Chem* 1989; **264**: 2100–07.

[20] Grand S, Lai ES, Esteve F, Rubin C, Hoffmann D, Remy C, et al. In vivo 1H MRS of brain abscesses versus necrotic brain tumors. *Neurology* 1996; **47**: 846–8.

[21] Lai PH, Ho JT, Chen WL, Hsu SS, Wang JS, Pan HB, et al. Brain abscess and necrotic brain tumor: discrimination with proton MR spectroscopy and diffusion-weighted imaging. *Am J Neuroradiol* 2002; **23**: 1369–77.

[22] Tedeschi G, Lundbom N, Raman R, Bonavita S, Duyn JH, Alger JR, et al. Increased choline signal coinciding with malignant degeneration of cerebral gliomas: a serial proton magnetic resonance spectroscopy imaging study. *J Neurosurg* 1997; **87**: 516–24.

[23] McKnight TR, Lamborn KR, Love TD, Berger MS, Chang S, Dillon WP, et al. Correlation of magnetic resonance spectroscopic and growth characteristics within Grades II and III gliomas. *J Neurosurg* 2007; **106**: 660–6.

[24] Burtscher IM, Holtas S. In vivo proton MR spectroscopy of untreated and treated brain abscesses. *Am J Neuroradiol* 1999; **20**: 1049–53.

[25] Bayindir C, Mete O, Bilgic B. Retrospective study of 23 pathologically proven cases of central nervous system tuberculomas. *Clin Neurol Neurosurg* 2006; **108**: 353–7.

[26] Bernaerts A, Vanhoenacker FM, Parizel PM, Van Goethem JW, Van AR, Laridon A, et al. Tuberculosis of the central nervous system: overview of neuroradiological findings. *Eur Radiol* 2003; **13**: 1876–90.

[27] Shah GV. Central nervous system tuberculosis: imaging manifestations. *Neuroimaging Clin N Am* 2000; **10**: 355–74.

[28] Cinque P, Cleator GM, Weber T, Monteyne P, Sindic CJ, van Loon AM. The role of laboratory investigation in the diagnosis and management of patients with suspected herpes simplex encephalitis: a consensus report. The EU Concerted Action on Virus Meningitis and Encephalitis. *J Neurol Neurosurg Psychiatry* 1996; **61**: 339–45.

[29] Tien RD, Felsberg GJ, Osumi AK. Herpesvirus infections of the CNS: MR findings. *Am J Roentgenol* 1993; **161**: 167–76.

[30] Schroth G, Gawehn J, Thron A, Vallbracht A, Voigt K. Early diagnosis of herpes simplex encephalitis by MRI. *Neurology* 1987; **37**: 179–83.

[31] Salvan AM, Confort-Gouny S, Cozzone PJ, Vion-Dury J. Atlas of brain proton magnetic resonance spectra. Part III: Viral infections. *J Neuroradiol* 1999; **26**: 154–61.

[32] Menon DK, Sargentoni J, Peden CJ, Bell JD, Cox IJ, Coutts GA, et al. Proton MR spectroscopy in herpes simplex encephalitis: assessment of neuronal loss. *J Comput Assist Tomogr* 1990; **14**: 449–52.

[33] Takanashi J, Sugita K, Ishii M, Aoyagi M, Niimi H. Longitudinal MR imaging and proton MR spectroscopy in herpes simplex encephalitis. *J Neurol Sci* 1997; **149**: 99–102.

[34] Demaerel P, Wilms G, Robberecht W, Johannik K, Van HP, Carton H, et al. MRI of herpes simplex encephalitis. *Neuroradiology* 1992; **34**: 490–3.

[35] Price RW. Neurological complications of HIV infection. *Lancet* 1996; **348**: 445–52.

[36] Price RW, Epstein LG, Becker JT, Cinque P, Gisslen M, Pulliam L, et al. Biomarkers of HIV-1 CNS infection and injury. *Neurology* 2007; **69**: 1781–8.

[37] Anthony IC, Bell JE. The neuropathology of HIV/AIDS. *Int Rev Psychiatry* 2008; **20**: 15–24.

[38] Hult B, Chana G, Masliah E, Everall I. Neurobiology of HIV. *Int Rev Psychiatry* 2008; **20**: 3–13.

[39] Nath A, Schiess N, Venkatesan A, Rumbaugh J, Sacktor N, McArthur J. Evolution of HIV dementia with HIV infection. *Int Rev Psychiatry* 2008; **20**: 25–31.

[40] Trotot PM, Gray F. Diagnostic imaging contribution in the early stages of HIV infection of the brain. *Neuroimaging Clin N Am* 1997; **7**: 243–60.

[41] Offiah CE, Turnbull IW. The imaging appearances of intracranial CNS infections in adult HIV and AIDS patients. *Clin Radiol* 2006; **61**: 393–401.

[42] Dal Canto MC. Mechanisms of HIV infection of the central nervous system and pathogenesis of AIDS–dementia complex. *Neuroimaging Clin N Am* 1997; **7**: 231–41.

[43] Menon DK, Baudouin CJ, Tomlinson D, Hoyle C. Proton MR spectroscopy and imaging of the brain in AIDS: evidence of neuronal loss in regions that appear normal with imaging. *J Comput Assist Tomogr* 1990; **14**: 882–5.

[44] Meyerhoff DJ, MacKay S, Bachman L, Poole N, Dillon WP, Weiner MW, et al. Reduced brain N-acetylaspartate suggests neuronal loss in

cognitively impaired human immunodeficiency virus-seropositive individuals: in vivo ^{1}H magnetic resonance spectroscopic imaging. *Neurology* 1993; **43**: 509–15.

[45] Jarvik JG, Lenkinski RE, Grossman RI, Gomori JM, Schnall MD, Frank I. Proton MR spectroscopy of HIV-infected patients: characterization of abnormalities with imaging and clinical correlation. *Radiology* 1993; **186**: 739–44.

[46] Chong WK, Paley M, Wilkinson ID, Hall-Craggs MA, Sweeney B, Harrison MJ, et al. Localized cerebral proton MR spectroscopy in HIV infection and AIDS. *Am J Neuroradiol* 1994; **15**: 21–5.

[47] Barker PB, Lee RR, McArthur JC. AIDS dementia complex: evaluation with proton MR spectroscopic imaging. *Radiology* 1995; **195**: 58–64.

[48] Wilkinson ID, Miller RF, Miszkiel KA, Paley MN, Hall-Craggs MA, Baldeweg T, et al. Cerebral proton magnetic resonance spectroscopy in asymptomatic HIV infection. *AIDS* 1997; **11**: 289–95.

[49] Tarasow E, Wiercinska-Drapalo A, Kubas B, Dzienis W, Orzechowska-Bobkiewicz A, Prokopowicz D, et al. Cerebral MR spectroscopy in neurologically asymptomatic HIV-infected patients. *Acta Radiol* 2003; **44**: 206–12.

[50] Yiannoutsos CT, Ernst T, Chang L, Lee PL, Richards T, Marra CM, et al. Regional patterns of brain metabolites in AIDS dementia complex. *Neuroimage* 2004; **23**: 928–35.

[51] Moller HE, Vermathen P, Lentschig MG, Schuierer G, Schwarz S, Wiedermann D, et al. Metabolic characterization of AIDS dementia complex by spectroscopic imaging. *J Magn Reson Imaging* 1999; **9**: 10–8.

[52] Meyerhoff DJ, Bloomer C, Cardenas V, Norman D, Weiner MW, Fein G. Elevated subcortical choline metabolites in cognitively and clinically asymptomatic HIV+ patients. *Neurology* 1999; **52**: 995–1003.

[53] Skolnick AA. Magnetic resonance spectroscopy may offer early look at HIV disease-mediated changes in brain [News]. *J Am Med Assoc* 1993; **269**: 1084.

[54] Tracey I, Carr CA, Guimaraes AR, Worth JL, Navia BA, Gonzalez RG. Brain choline-containing compounds are elevated in HIV-positive patients before the onset of AIDS dementia complex: a proton magnetic resonance spectroscopic study. *Neurology* 1996; **46**: 783–8. [Erratum appears in *Neurology* 1996; **46**: 1787.]

[55] Suwanwelaa N, Phanuphak P, Phanthumchinda K, Suwanwela NC, Tantivatana J, Ruxrungtham K, et al. Magnetic resonance spectroscopy of the brain in neurologically asymptomatic HIV-infected patients. *Magn Reson Imaging* 2000; **18**: 859–65.

[56] Chong WK, Sweeney B, Wilkinson ID, Paley M, Hall-Craggs MA, Kendall BE, et al. Proton spectroscopy of the brain in HIV infection: correlation with clinical, immunologic, and MR imaging findings. *Radiology* 1993; **188**: 119–24.

[57] Salvan AM, Vion-Dury J, Confort-Gouny S, Nicoli F, Lamoureux S, Cozzone PJ. Brain proton magnetic resonance spectroscopy in HIV-related encephalopathy: identification of evolving metabolic patterns in relation to dementia and therapy. *AIDS Res Hum Retrovir* 1997; **13**: 1055–66.

[58] Chang L, Ernst T, Witt MD, Ames N, Gaiefsky M, Miller E. Relationships among brain metabolites, cognitive function, and viral loads in antiretroviral-naive HIV patients. *Neuroimage* 2002; **17**: 1638–48.

[59] Paul RH, Yiannoutsos CT, Miller EN, Chang L, Marra CM, Schifitto G, et al. Proton MRS and neuropsychological correlates in AIDS dementia complex: evidence of subcortical specificity. *J Neuropsychiatry Clin Neurosci* 2007; **19**: 283–92.

[60] Vion-Dury J, Nicoli F, Salvan AM, Confort-Gouny S, Dhiver C, et al. Reversal of brain metabolic alterations with zidovudine detected by proton localised magnetic resonance spectroscopy [Letter]. *Lancet* 1995; **345**: 60–1.

[61] Sacktor N, Skolasky RL, Ernst T, Mao X, Selnes O, Pomper MG, et al. A multicenter study of two magnetic resonance spectroscopy techniques in individuals with HIV dementia. *J Magn Reson Imaging* 2005; **21**: 325–33.

[62] Chang L, Lee PL, Yiannoutsos CT, Ernst T, Marra CM, Richards T, et al. A multicenter in vivo proton-MRS study of HIV-associated dementia and its relationship to age. *Neuroimage* 2004; **23**: 1336–47.

[63] Lee PL, Yiannoutsos CT, Ernst T, Chang L, Marra CM, Jarvik JG, et al. A multi-center 1H MRS study of the AIDS dementia complex: validation and preliminary analysis. *J Magn Reson Imaging* 2003; **17**: 625–33.

[64] Paley M, Cozzone PJ, Alonso J, Vion-Dury J, Confort-Gouny S, Wilkinson ID, et al. A multicenter proton magnetic resonance spectroscopy study of neurological complications of AIDS. *AIDS Res Hum Retrovir* 1996; **12**: 213–22.

[65] Chang L, Ernst T, Leonido-Yee M, Witt M, Speck O, Walot I, et al. Highly active antiretroviral therapy reverses brain metabolite abnormalities in mild HIV dementia. *Neurology* 1999; **53**: 782–9.

[66] Schifitto G, Navia BA, Yiannoutsos CT, Marra CM, Chang L, Ernst T, et al. Memantine and HIV-associated cognitive impairment: a neuropsychological and proton magnetic resonance spectroscopy study. *AIDS* 2007; **21**: 1877–86.

[67] Avison MJ, Nath A, Berger JR. Understanding pathogenesis and treatment of HIV dementia: a role for magnetic resonance? *Trends Neurosci* 2002; **25**: 468–73.

[68] Navia BA, Rostasy K. The AIDS dementia complex: clinical and basic neuroscience with implications for novel molecular therapies. *Neurotox Res* 2005; **8**: 3–24.

[69] Pomper MG, Sacktor N. New techniques for imaging Human Immunodeficiency Virus associated cognitive impairment in the era of highly active antiretroviral therapy. *Arch Neurol* 2007; **64**: 1233–5.

[70] Maschke M, Kastrup O, Forsting M, Diener HC. Update on neuroimaging in infectious central nervous system disease. *Curr Opin Neurol* 2004; **17**: 475–80.

[71] Chang L, Miller BL, McBride D, Cornford M, Oropilla G, Buchthal S, et al. Brain lesions in patients with AIDS: H-1 MR spectroscopy. *Radiology* 1995; **197**: 525–31.

[72] Simpson DM, Tagliati M, Ramcharitar S. Neurologic complications of AIDS: new concepts and treatments. *Mt Sinai J Med* 1994; **61**: 484–91.

[73] Mamidi A, DeSimone JA, Pomerantz RJ. Central nervous system infections in individuals with HIV-1 infection. *J Neurovirol* 2002; **8**: 158–67.

[74] Antinori A, Ammassari A, De Luca A, Cingolani A, Murri R, Scoppettuolo G, et al. Diagnosis of AIDS-related focal brain lesions: a decision-making analysis based on clinical and neuroradiologic characteristics combined with polymerase chain reaction assays in CSF. *Neurology* 1997; **48**: 687–94.

[75] Roberts TC, Storch GA. Multiplex PCR for diagnosis of AIDS-related central nervous system lymphoma and toxoplasmosis. *J Clin Microbiol* 1997; **35**: 268–9.

[76] Sadler M, Brink NS, Gazzard BG. Management of intracerebral lesions in patients with HIV: a retrospective study with discussion of diagnostic problems. *Q J Med* 1998; **91**: 205–17.

[77] Cingolani A, De Luca A, Larocca LM, Ammassari A, Scerrati M, Antinori A, et al. Minimally invasive diagnosis of acquired immunodeficiency syndrome-related primary central nervous system lymphoma. *J Natl Cancer Inst* 1998; **90**: 364–9.

[78] Skiest DJ, Erdman W, Chang WE, Oz OK, Ware A, Fleckenstein J. SPECT thallium-201 combined with *Toxoplasma* serology for the presumptive diagnosis of focal central nervous system mass lesions in patients with AIDS. *J Infect* 2000; **40**: 274–81.

[79] Pomper MG, Constantinides CD, Barker PB, Bizzi A, Dogan S, Yokoi F, et al. Quantitative MR spectroscopic imaging of brain lesions in patients with AIDS: correlation with [11C-methyl]thymidine PET and thallium-201 SPECT. *Acad Radiol* 2002; **9**: 398–409.

[80] Simone IL, Federico F, Tortorella C, Andreula CF, Zimatore GB, Giannini P, et al. Localised 1H-MR spectroscopy for metabolic characterisation of diffuse and focal brain lesions in patients infected with HIV. *J Neurol Neurosurg Psychiatry* 1998; **64**: 516–23.

[81] Ferrante P, Caldarelli-Stefano R, Omodeo-Zorini E, Vago L, Boldorini R, Costanzi G. PCR detection of JC virus DNA in brain tissue from patients with and without progressive multifocal leukoencephalopathy. *J Med Virol* 1995; **47**: 219–25.

[82] Marzocchetti A, Sanguinetti M, Giambenedetto SD, Cingolani A, Fadda G, Cauda R, et al. Characterization of JC virus in cerebrospinal fluid from HIV-1 infected patients with progressive multifocal leukoencephalopathy: insights into viral pathogenesis and disease prognosis. *J Neurovirol* 2007; **13**: 338–46.

[83] Yousry TA, Major EO, Ryschkewitsch C, Fahle G, Fischer S, Hou J, et al. Evaluation of patients treated with natalizumab for progressive multifocal leukoencephalopathy. *N Engl J Med* 2006; **354**: 924–33.

[84] von Einsiedel RW, Fife TD, Aksamit AJ, Cornford ME, Secor DL, Tomiyasu U. Progressive multifocal leukoencephalopathy in AIDS: a clinicopathologic study and review of the literature. *J Neurol* 1993; **240**: 391–406.

[85] Chang L, Ernst T, Tornatore C, Aronow H, Melchor R, Walot I, et al. Metabolite abnormalities in progressive multifocal leukoencephalopathy by proton magnetic resonance spectroscopy. *Neurology* 1997; **48**: 836–45.

[86] Iranzo A, Moreno A, Pujol J, Marti-Fabregas J, Domingo P, Molet J, et al. Proton magnetic resonance spectroscopy pattern of progressive multifocal leukoencephalopathy in AIDS. *J Neurol Neurosurg Psychiatry* 1999; **66**: 520–3.

[87] Teksam M, Cakir B, Agildere AM. Proton MR spectroscopy in the diagnosis of early-stage subacute sclerosing panencephalitis. *Diagn Interv Radiol* 2006; **12**: 61–3.

[88] Aydin K, Tatli B, Ozkan M, Ciftci K, Unal Z, Sani S, et al. Quantification of neurometabolites in subacute sclerosing panencephalitis by 1H-MRS. *Neurology* 2006; **67**: 911–3.

[89] McDonald WI, Ron MA. Multiple sclerosis: the disease and its manifestations. *Phil Trans R. Soc Lond B Biol Sci* 1999; **354**: 1615–22.

[90] Compston A, Coles A. Multiple sclerosis. *Lancet* 2002; **359**: 1221–31.

[91] McDonald WI, Miller DH, Barnes D. The pathological evolution of multiple sclerosis. *Neuropathol Appl Neurobiol* 1992; **18**: 319–34.

[92] Vukusic S, Confavreux C. Natural history of multiple sclerosis: risk factors and prognostic indicators. *Curr Opin Neurol* 2007; **20**: 269–74.

[93] McDonald WI, Compston A, Edan G, Goodkin D, Hartung HP, Lublin FD, *et al*. Recommended diagnostic criteria for multiple sclerosis: guidelines from the International Panel on the diagnosis of multiple sclerosis. *Ann Neurol* 2001; **50**: 121–7.

[94] Polman CH, Reingold SC, Edan G, Filippi M, Hartung HP, Kappos L, *et al*. Diagnostic criteria for multiple sclerosis: 2005 revisions to the "McDonald Criteria". *Ann Neurol* 2005; **58**: 840–6.

[95] Filippi M, Rocca MA, Comi G. The use of quantitative magnetic-resonance-based techniques to monitor the evolution of multiple sclerosis. *Lancet Neurol* 2003; **2**: 337–46.

[96] Filippi M, Rocca MA. Conventional MRI in multiple sclerosis. *J Neuroimaging* 2007; **17**(Suppl 1): 3S–9S.

[97] van Waesberghe JH, van Walderveen MA, Castelijns JA, Scheltens P, Nijeholt GJ, Polman CH, *et al*. Patterns of lesion development in multiple sclerosis: longitudinal observations with T1-weighted spin-echo and magnetization transfer MR. *Am J Neuroradiol* 1998; **19**: 675–83.

[98] Katz D, Taubenberger JK, Cannella B, McFarlin DE, Raine CS, McFarland HF. Correlation between magnetic resonance imaging findings and lesion development in chronic, active multiple sclerosis. *Ann Neurol* 1993; **34**: 661–9.

[99] Peterson JW, Trapp BD. Neuropathobiology of multiple sclerosis. *Neurol Clin* 2005; **23**: 107–29.

[100] Arnold DL, De Stefano N, Narayanan S, Matthews PM. Proton MR spectroscopy in multiple sclerosis. *Neuroimaging Clin N Am* 2000; **10**: 789–98.

[101] Filippi M, Arnold DL, Comi G. *Magnetic Resonance Spectroscopy in Multiple Sclerosis*. Milan: Springer, 2007.

[102] Narayana PA. Magnetic resonance spectroscopy in the monitoring of multiple sclerosis. *J Neuroimaging* 2005; **15**: 46S–57S.

[103] Davie CA, Hawkins CP, Barker GJ, Brennan A, Tofts PS, Miller DH, *et al*. Serial proton magnetic resonance spectroscopy in acute multiple sclerosis lesions. *Brain* 1994; **117**: 49–58.

[104] De Stefano N, Matthews PM, Arnold DL. Reversible decreases in N-acetylaspartate after acute brain injury. *Magn Reson Med* 1995; **34**: 721–7.

[105] De Stefano N, Matthews PM, Antel JP, Preul M, Francis G, Arnold DL. Chemical pathology of acute demyelinating lesions and its correlation with disability. *Ann Neurol* 1995; **38**: 901–09.

[106] Narayana PA, Doyle TJ, Lai D, Wolinsky JS. Serial proton magnetic resonance spectroscopic imaging, contrast-enhanced magnetic resonance imaging, and quantitative lesion volumetry in multiple sclerosis. *Ann Neurol* 1998; **43**: 56–71.

[107] Fernando KT, McLean MA, Chard DT, MacManus DG, Dalton CM, Miszkiel KA, *et al*. Elevated white matter *myo*-inositol in clinically isolated syndromes suggestive of multiple sclerosis. *Brain* 2004; **127**: 1361–9.

[108] Matthews PM, De Stefano N, Narayanan S, Francis GS, Wolinsky JS, Antel JP, *et al*. Putting magnetic resonance spectroscopy studies in context: axonal damage and disability in multiple sclerosis. *Semin Neurol* 1998; **18**: 327–36.

[109] Srinivasan R, Sailasuta N, Hurd R, Nelson S, Pelletier D. Evidence of elevated glutamate in multiple sclerosis using magnetic resonance spectroscopy at 3 T. *Brain* 2005; **128**: 1016–25.

[110] Caramanos Z, Narayanan S, Arnold DL. 1H-MRS quantification of tNA and tCr in patients with multiple sclerosis: a meta-analytic review. *Brain* 2005; **128**: 2483–506.

[111] Brex PA, Parker GJ, Leary SM, Molyneux PD, Barker GJ, Davie CA, *et al*. Lesion heterogeneity in multiple sclerosis: a study of the relations between appearances on T1 weighted images, T1 relaxation times, and metabolite concentrations. *J Neurol Neurosurg Psychiatry* 2000; **68**: 627–32.

[112] Kapeller P, Brex PA, Chard D, Dalton C, Griffin CM, McLean MA, *et al*. Quantitative 1H MRS imaging 14 years after presenting with a clinically isolated syndrome suggestive of multiple sclerosis. *Mult Scler* 2002; **8**: 207–10.

[113] Helms G, Stawiarz L, Kivisakk P, Link H. Regression analysis of metabolite concentrations estimated from localized proton MR spectra of active and chronic multiple sclerosis lesions. *Magn Reson Med* 2000; **43**: 102–10.

[114] Brenner RE, Munro PM, Williams SC, Bell JD, Barker GJ, Hawkins CP, *et al*. The proton NMR spectrum in acute EAE: the significance of the change in the Cho: Cr ratio. *Magn Reson Med* 1993; **29**: 737–45.

[115] Arnold DL, Matthews PM, Francis G, Antel J. Proton magnetic resonance spectroscopy of human brain in vivo in the evaluation of multiple sclerosis: assessment of the load of disease. *Magn Reson Med* 1990; **14**: 154–9.

[116] Wolinsky JS, Narayana PA, Fenstermacher MJ. Proton magnetic resonance spectroscopy in multiple sclerosis. *Neurology* 1990; **40**: 1764–9.

[117] Husted CA, Goodin DS, Hugg JW, Maudsley AA, Tsuruda JS, de Bie SH, et al. Biochemical alterations in multiple sclerosis lesions and normal-appearing white matter detected by in vivo 31P and 1H spectroscopic imaging. *Ann Neurol* 1994; **36**: 157–65.

[118] Narayanan S, Fu L, Pioro E, De Stefano N, Collins DL, Francis GS, et al. Imaging of axonal damage in multiple sclerosis: spatial distribution of magnetic resonance imaging lesions. *Ann Neurol* 1997; **41**: 385–91.

[119] Davie CA, Barker GJ, Thompson AJ, Tofts PS, McDonald WI, Miller DH. 1H magnetic resonance spectroscopy of chronic cerebral white matter lesions and normal appearing white matter in multiple sclerosis. *J Neurol Neurosurg Psychiatry* 1997; **63**: 736–42.

[120] Fu L, Matthews PM, De Stefano N, Worsley KJ, Narayanan S, Francis GS, et al. Imaging axonal damage of normal-appearing white matter in multiple sclerosis. *Brain* 1998; **121**: 103–13.

[121] Sarchielli P, Presciutti O, Pelliccioli GP, Tarducci R, Gobbi G, Chiarini P, et al. Absolute quantification of brain metabolites by proton magnetic resonance spectroscopy in normal-appearing white matter of multiple sclerosis patients. *Brain* 1999; **122**: 513–21.

[122] Matthews PM, De Stefano N, Narayanan S, Francis GS, Wolinsky JS, Antel JP, et al. Putting magnetic resonance spectroscopy studies in context: axonal damage and disability in multiple sclerosis. *Semin Neurol* 1998; **18**: 327–36.

[123] De Stefano N, Narayanan S, Francis GS, Arnaoutelis R, Tartaglia MC, Antel JP, et al. Evidence of axonal damage in the early stages of multiple sclerosis and its relevance to disability. *Arch Neurol* 2001; **58**: 65–70.

[124] Falini A, Calabrese G, Filippi M, Origgi D, Lipari S, Colombo B, et al. Benign versus secondary-progressive multiple sclerosis: the potential role of proton MR spectroscopy in defining the nature of disability. *Am J Neuroradiol* 1998; **19**: 223–9.

[125] De Stefano N, Narayanan S, Matthews PM, Francis GS, Antel JP, Arnold DL. In vivo evidence for axonal dysfunction remote from focal cerebral demyelination of the type seen in multiple sclerosis. *Brain* 1999; **122**: 1933–9.

[126] Arnold DL. Changes observed in multiple sclerosis using magnetic resonance imaging reflect a focal pathology distributed along axonal pathways. *J Neurol* 2005; **252**(Suppl 5): v25–v29.

[127] Trapp BD, Peterson J, Ransohoff RM, Rudick R, Mork S, Bo L. Axonal transection in the lesions of multiple sclerosis. *N Engl J Med* 1998; **338**: 278–85.

[128] De Stefano N, Narayanan S, Francis SJ, Smith S, Mortilla M, Tartaglia MC, et al. Diffuse axonal and tissue injury in patients with multiple sclerosis with low cerebral lesion load and no disability. *Arch Neurol* 2002; **59**: 1565–71.

[129] Filippi M. Multiple sclerosis: a white matter disease with associated gray matter damage. *J Neurol Sci* 2001; **185**: 3–4.

[130] Sharma R, Narayana PA, Wolinsky JS. Grey matter abnormalities in multiple sclerosis: proton magnetic resonance spectroscopic imaging. *Mult Scler* 2001; **7**: 221–6.

[131] Sarchielli P, Presciutti O, Tarducci R, Gobbi G, Alberti A, Pelliccioli GP, et al. Localized (1)H magnetic resonance spectroscopy in mainly cortical gray matter of patients with multiple sclerosis. *J Neurol* 2002; **249**: 902–10.

[132] Filippi M, Bozzali M, Rovaris M, Gonen O, Kesavadas C, Ghezzi A, et al. Evidence for widespread axonal damage at the earliest clinical stage of multiple sclerosis. *Brain* 2003; **126**: 433–7.

[133] Adalsteinsson E, Langer-Gould A, Homer RJ, Rao A, Sullivan EV, Lima CA, et al. Gray matter N-acetyl aspartate deficits in secondary progressive but not relapsing-remitting multiple sclerosis. *Am J Neuroradiol* 2003; **24**: 1941–5.

[134] Sastre-Garriga J, Ingle GT, Chard DT, Ramio-Torrenta L, McLean MA, Miller DH, et al. Metabolite changes in normal-appearing gray and white matter are linked with disability in early primary progressive multiple sclerosis. *Arch Neurol* 2005; **62**: 569–73.

[135] Wylezinska M, Cifelli A, Jezzard P, Palace J, Alecci M, Matthews PM. Thalamic neurodegeneration in relapsing–remitting multiple sclerosis. *Neurology* 2003; **60**: 1949–54.

[136] Inglese M, Liu S, Babb JS, Mannon LJ, Grossman RI, Gonen O. Three-dimensional proton spectroscopy of deep gray matter nuclei in relapsing–remitting MS. *Neurology* 2004; **63**: 170–2.

[137] Geurts JJ, Reuling IE, Vrenken H, Uitdehaag BM, Polman CH, Castelijns JA, et al. MR spectroscopic evidence for thalamic and hippocampal, but not cortical, damage in multiple sclerosis. *Magn Reson Med* 2006; **5**: 478–83.

[138] Cifelli A, Arridge M, Jezzard P, Esiri MM, Palace J, Matthews PM. Thalamic neurodegeneration in multiple sclerosis. *Ann Neurol* 2002; **52**: 650–3.

[139] Davie CA, Barker GJ, Webb S, Tofts PS, Thompson AJ, Harding AE, et al. Persistent functional deficit in multiple sclerosis and autosomal dominant cerebellar ataxia is associated with axon loss. *Brain* 1995; **118**: 1583–92.

[140] De Stefano N, Matthews PM, Narayanan S, Francis GS, Antel JP, Arnold DL. Axonal dysfunction and disability in a relapse of multiple sclerosis: longitudinal study of a patient. *Neurology* 1997; **49**: 1138–41.

[141] De Stefano N, Matthews PM, Fu L, Narayanan S, Stanley J, Francis GS, *et al.* Axonal damage correlates with disability in patients with relapsing–remitting multiple sclerosis. Results of a longitudinal magnetic resonance spectroscopy study. *Brain* 1998; **121**: 1469–77.

[142] Vrenken H, Barkhof F, Uitdehaag BM, Castelijns JA, Polman CH, Pouwels PJ. MR spectroscopic evidence for glial increase but not for neuro-axonal damage in MS normal-appearing white matter. *Magn Reson Med* 2005; **53**: 256–66.

[143] Filippi M, Falini A, Arnold DL, Fazekas F, Gonen O, Simon JH, *et al.* Magnetic resonance techniques for the in vivo assessment of multiple sclerosis pathology: consensus report of the white matter study group. *J Magn Reson Imaging* 2005; **21**: 669–75.

[144] Narayanan S, De Stefano N, Francis GS, Arnaoutelis R, Caramanos Z, Collins DL, *et al.* Axonal metabolic recovery in multiple sclerosis patients treated with interferon beta-1b. *J Neurol* 2001; **248**: 979–86.

[145] Schubert F, Seifert F, Elster C, Link A, Walzel M, Mientus S, *et al.* Serial 1H-MRS in relapsing-remitting multiple sclerosis: effects of interferon-beta therapy on absolute metabolite concentrations. *MAGMA* 2002; **14**: 213–22.

[146] Parry A, Corkill R, Blamire AM, Palace J, Narayanan S, Arnold D, *et al.* Beta-Interferon treatment does not always slow the progression of axonal injury in multiple sclerosis. *J Neurol* 2003; **250**: 171–8.

[147] Khan O, Shen Y, Caon C, Bao F, Ching W, Reznar M, *et al.* Axonal metabolic recovery and potential neuroprotective effect of glatiramer acetate in relapsing-remitting multiple sclerosis. *Mult Scler* 2005; **11**: 646–51.

[148] Mostert JP, Sijens PE, Oudkerk M, De KJ. Fluoxetine increases cerebral white matter NAA/Cr ratio in patients with multiple sclerosis. *Neurosci Lett* 2006; **402**: 22–4.

[149] De Stefano N, Filippi M, Miller D, Pouwels PJ, Rovira A, Gass A, *et al.* Guidelines for using proton MR spectroscopy in multicenter clinical MS studies. *Neurology* 2007; **69**: 1942–52.

[150] Talbot PJ, Arnold D, Antel JP. Virus-induced autoimmune reactions in the CNS. *Curr Top Microbiol Immunol* 2001; **253**: 247–71.

[151] Kesselring J, Miller DH, Robb SA, Kendall BE, Moseley IF, Kingsley D, *et al.* Acute disseminated encephalomyelitis. MRI findings and the distinction from multiple sclerosis. *Brain* 1990; **113**: 291–302.

[152] Gabis LV, Panasci DJ, Andriola MR, Huang W. Acute disseminated encephalomyelitis: an MRI/MRS longitudinal study. *Pediatr Neurol* 2004; **30**: 324–9.

[153] Bizzi A, Ulug AM, Crawford TO, Passe T, Bugiani M, Bryan RN, *et al.* Quantitative proton MR spectroscopic imaging in acute disseminated encephalomyelitis. *Am J Neuroradiol* 2001; **22**: 1125–30.

Chapter 8
MRS in epilepsy

Key points

- MRS is principally used as an adjunct diagnostic technique for evaluating patients with medically intractable epilepsy (in order to identify the seizure focus).
- Most commonly, NAA is reduced in epileptogenic tissue; metabolic abnormalities are often subtle.
- Metabolic abnormalities may be more widespread than seen on MRI, and present in the contralateral hemisphere.
- MRS may occasionally be helpful when other techniques (e.g. MRI) are either normal or non-specific.
- MRS measures of the inhibitory neurotransmitter GABA using spectral editing may help determine optimal drug regimen.
- MRS may also be a useful research tool for determining epileptogenic networks in the brain.

Introduction

Epilepsy, the condition of recurrent seizures, is a relatively common neurological disorder, estimated to affect between 1 and 2 million people in the US alone. A multitude of etiologies cause epilepsy, including tumors, developmental abnormalities, febrile illness, trauma, or infection. However, not infrequently, the cause is unknown. Many patients with epilepsy can be successfully treated pharmacologically, but when medical management fails to adequately control seizure activity, surgical resection of the epileptogenic tissue may be considered. For surgery to be successful, seizures must be of focal onset from a well-defined location. It has been estimated that up to 10% of patients with epilepsy are medically intractable, of whom approximately 20% may be candidates for surgical treatment. Traditionally, scalp electroencephalography (EEG) and often invasive (subdural grid or depth electrode) EEG are used to identify the epileptogenic regions of the brain, but increasingly magnetic resonance imaging (MRI), positron emission tomography (PET), ictal single photon emission computed tomography (SPECT), and, more recently, magnetoencephalography (MEG) are also used.

MRI is the modality of choice for identifying brain tumors, cortical malformations, infectious and other causes of epilepsy (Figure 8.1). In mesial temporal sclerosis (MTS), the most common abnormality in patients with temporal lobe epilepsy, MRI typically shows hippocampus volume loss,[1] with abnormal signal intensity on T_2-weighted images (Figure 8.1A),[2] which corresponds histologically to neuronal loss and gliosis. Sensitivity of MTS detection may be increased by performing careful, quantitative T_2 measurements from multiple echo data acquisitions,[3] or by using the CSF-suppressed FLAIR sequence.[4] Quantitative volume measurements more reliably detect small changes in hippocampal volume and are generally preferable, particularly when atrophy may be subtle.[5,6] Lateralization of seizure focus in patients with temporal lobe epilepsy has been reported to be over 90% efficient with volumetric analysis of hippocampal and amygdaloid formations using high-resolution 3D scans.[5,7]

While these studies show that MRI is a sensitive tool for the detection of MTS, the clinical significance of these findings should be carefully considered.[8] First, many published studies have been performed in retrospectively selected patients who were already candidates for epilepsy surgery by other criteria, such as EEG. This may increase sensitivity and specificity by excluding patients who might have negative MRI findings, or who are "complicated" cases. Second, a significant number of patients will have symmetric hippocampi (either no atrophy or bilateral atrophy), and yet still have successful seizure control after surgery,[9] indicating that bilateral sclerosis is not necessarily a contra-indication for

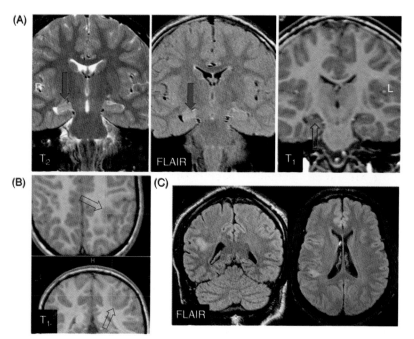

Figure 8.1. MRI findings in patients with epilepsy: (A) 8-year-old female with right-sided mesial temporal sclerosis (MTS) visible as hyperintensity in both T_2 and FLAIR coronal MRI (closed arrows), as well as reduced hippocampal volume on T_1-weighted coronal MRI (open arrow). (B), (C) Malformations of cortical development; (B) left parietal focal cortical dysplasia (FCD type I) in a patient with intractable seizures, and (C) right posterior frontal subcortical region of T_2 hyperintensity due to type II FCD in a patient with seizures and psychosis.

surgery. Third, longer-term follow-up post-surgery is often not reported; one study found seizure-free outcome in 70–80% of patients 1 year after surgery, but by 5 years this number had fallen to 50–60%.[10] Interestingly, relapse only occurred in the patients who originally presented with hippocampal atrophy. Collectively, these studies demonstrate that MRI is a valuable tool for the evaluation of patients with epilepsy, but that it also has limitations. For this reason, other imaging studies are often considered for the evaluation of epilepsy patients, particularly "functional" techniques that measure blood flow and metabolism, such as SPECT and PET.[11]

Epilepsy has been extensively studied by PET since the early 1980s, the majority of the studies using ^{18}F-fluorodeoxyglucose (FDG) to measure glucose metabolism.[12] Interictally, glucose uptake is reduced compared to normal brain, while ictally increases in uptake may be observed.[13] The hypometabolic region is usually larger on interictal PET scans than the electrically defined volume of pathology.[13] Seizure foci also have been found to be associated with changes in cerebral perfusion,[14] which can be monitored with oxygen-15 PET,[15] SPECT,[16] or perfusion MRI.[17,18] In one study,[20] PET and MRI were determined to be comparable in terms of ability to detect abnormalities in patients with temporal lobe epilepsy, but PET had better concordance than MRI with the EEG localization. Worse surgical outcome was also associated with hypometabolism which extended beyond the region of the temporal lobe. MRI and PET imaging are usually done interictally for logistical reasons.[19] The main advantage of SPECT, despite having lower resolution and being less quantitative than PET, is that SPECT can be performed ictally, i.e. the SPECT tracer is injected while (or just after) the patient experiences a seizure. Ictal SPECT may provide unique information about the location of the epileptic focus.

All together, these studies demonstrate the additional value of flow and metabolic based imaging studies, in addition to structural anatomic scans, for the evaluation of patients with non-lesional epilepsy.

MR spectroscopy in epilepsy

^{31}P MRS

The use of MR spectroscopy for the metabolic evaluation of animal models of epilepsy was first investigated by the Yale group in the early 1980s.[21,22] At that time, ^{31}P was the most widely used nucleus for in vivo MR studies of the brain, although some proton MRS studies were also performed. Generally, status epilepticus is associated with reductions of high-energy phosphates (nucleotide triphosphates (NTP) and phosphocreatine (PCr)), increases in low-energy phosphates (inorganic phosphate (Pi)), and cerebral acidosis as determined by the chemical shift of the Pi peak. However, not all studies report these findings; in an interictal cortical spike focus in the rat, no significant ^{31}P MRS or MRI changes were reported.[23]

After these initial results in animal models, the use of ^{31}P MRS in humans with seizure disorders was reported in the late 1980s and early 1990s.[24,25,26,27] Interictally, seizure foci were found by ^{31}P MRS to be alkaline, and this was proposed as a means of lateralization of the seizure foci,[25,26] although this finding was not reproduced subsequently.[28] ^{31}P MR spectroscopy of infants experiencing status epilepticus showed a decrease in the PCr/Pi ratio,[24] and in general, in most forms of adult epilepsy, the most common finding (ictally or interictally) appears to be bioenergetic impairment (i.e. reduced ratios of PCr/Pi and/or PCr/ATP).[29] In one study using ^{31}P MRS at high field (4.1 T) in a group of 30 patients, ^{31}P MRS successfully lateralized temporal lobe epilepsy in 70–73% using either PCr/Pi or ATP/Pi ratios,[30] a rate that was actually higher than that achieved with MRI in the same study. An example of a ^{31}P MRS of a 2-year-old child with Lennox–Gastaut syndrome, before and after initiation of the "ketogenic diet" for seizure control, is shown in Figure 8.2; a small but noticeable increase in PCr can be determined.

However, the relatively coarse spatial resolution and low sensitivity of MRS ($\approx$30 cm^3 voxel size for human brain studies at 1.5 T) limit the application of this technique to rather large focal abnormalities or diffuse brain pathologies. The technique does not appear to be able to map the extent of epileptogenic tissue because of its low spatial resolution. Finally, it is not particularly widely available since appreciable hardware modifications are required on most MRI scanners as a result of the lower resonant frequency of the ^{31}P nucleus. For all of these reasons, there have been many more proton MRS studies of epilepsy than ^{31}P, although with high-field (i.e. 3 T and above) MRI systems becoming increasingly available, there is still some interest in using ^{31}P to investigate the biochemical processes occurring in patients with epilepsy. [31,32] High-field ^{31}P MRS offers higher sensitivity which results in better spatial resolution ($\sim$ 6–12 cm^3 in 40–50 min scan time)[29] compared to lower fields.

Proton MRS

Proton spectroscopy has a considerable sensitivity advantage compared to ^{31}P which allows significantly better spatial resolution, and can be used on most MRI scanners without hardware modifications. Over the last several years, therefore, most spectroscopy studies of human epilepsy have utilized the proton nucleus. The first published study involved two patients with Rasmussen's syndrome (both of whom had abnormal MRI scans). Both patients showed decreased N-acetyl aspartate (NAA), and the single patient who had seizures during spectral acquisition showed increased lactate.[33] Since the NAA signal is believed to originate from neuronal cells,[34] the reduction in NAA has been attributed to neuronal loss within the seizure focus, which is also a common histological finding.[35] Increased lactate in patients who are experiencing active seizures is consistent with the hypermetabolism observed in ictal FDG-PET scans, indicating that the increased glucose uptake is at least partly metabolized anaerobically to lactate, as opposed to the normal path through pyruvate to the TCA cycle. Ictal, or early post-ictal (up to about 6 h) elevations of lactate have therefore been found to be useful in the identification of seizure foci.[36] However, most spectroscopy studies of epilepsy are performed interictally (where lactate is not normally observed), and the most universal finding associated with seizure foci is a decrease in NAA, either measured quantitatively[37] or as a ratio to creatine, ratio to choline, or both.[38,39,40] In some

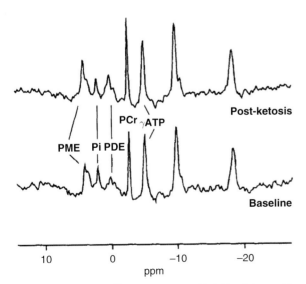

Figure 8.2. Example of ^{31}P MRS in a 2-year-old child with intractable epilepsy due to Lennox–Gastaut syndrome, before and after initiation of the ketogenic diet. A small increase in PCr can be seen after treatment. Reproduced with permission from [31].

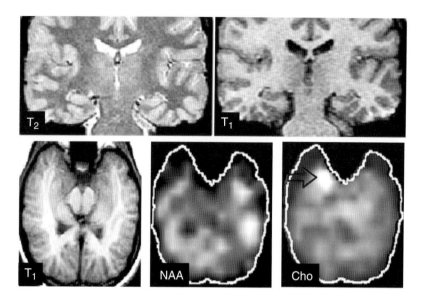

Figure 8.3. Twenty-five-year-old female with intractable seizures of right anterior temporal lobe origin (right-sided spikes on sphenoidal EEG). Conventional MRI is unremarkable; spectroscopic images of Cho show markedly increased signal in the right anterior mesial temporal lobe (arrow); NAA is also mildly reduced. This patient was seizure-free after right anterior temporal lobectomy.

cases, increases in choline may also be observed,[41] perhaps as the result of gliosis or neoplastic proliferation, since glial cells are believed to have high choline levels.[42] An example of a patient with intractable seizures of right anterior temporal lobe origin with high Cho seen on MRSI is shown in Figure 8.3.

In short echo time spectroscopy, lower levels of glutamate and glutamine (Glx) than in control subjects have also been reported in patients with hippocampal sclerosis.[43] This is consistent with lower glutamate (the major component of the "Glx" peak) in association with neuronal loss – in an in vitro study of temporal lobe specimens from patients undergoing epilepsy surgery, it was found that Glu gave an excellent correlation with NAA, with both Glu and NAA showing trends for negative correlations with hippocampal neuronal counts.[44] Consistent with this, an MRSI study in temporal lobe epilepsy also found lower levels of Glu in patients with temporal lobe epilepsy: lower in the ipsilateral temporal lobe, but lower than healthy controls in the contralateral temporal lobe as well.[45] Finally, in vitro NMR spectroscopy studies of perchloric acid extracts of gliotic hippocampal tissue have also shown increased *myo*-inositol and decreased glutamate,[46] consistent with gliosis and neuronal loss.

Temporal lobe epilepsy

In early published studies of patients with temporal lobe epilepsy (TLE), reductions of NAA in the affected hippocampus were found in 100%,[47] 100%,[37] 88% (note 40% of the cases had bilateral reductions, so that lateralization was obtained in 60% of cases [38]), 90%, [40] 100%,[48] and 100%[39] of cases studied. An example of proton MRS in a patient with unilateral mesial temporal sclerosis, showing decreased NAA, is depicted in Figure 8.4. Generally, sample sizes in these studies were in the range between 10 and 25 subjects, and they most likely contained carefully pre-selected cases with clear-cut abnormalities on other modalities. Larger and more recent studies have reported more variable success rates in terms of lateralizing abnormalities in patients with temporal lobe epilepsy; in one MRSI study in 50 patients, success of localization (based on neuroradiological interpretation of spectra) varied from 62% to 76%,[49] while in another the NAA/Cr ratio only successfully lateralized 18 of 40 cases (45%); however, cases which were lateralized by MRS had excellent surgical outcomes.[50] In a study of 100 cases (using MRSI and volumetric MRI), MRSI was found to correctly localize 86% of cases, very similar to volumetric MRI and EEG localization rates.[51] Despite published studies such as these with high success rates, MRS and/or MRSI have had relatively little clinical impact over the last few years for presurgical evaluation of epilepsy patients. There are probably several reasons for this.

(1) Research studies typically pre-select well-characterized patients to study (see above), who are somewhat different from the general epileptic population, and not typical of the "difficult" cases that may be referred for special MRS studies.

(2) The spectroscopic changes are subtle, so while group studies may show statistically significant differences,

Chapter 8: MRS in epilepsy

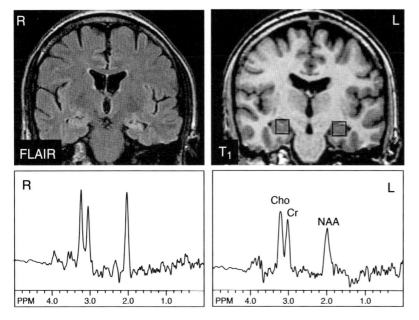

Figure 8.4. Fifty-one-year-old male with nocturnal seizures and left-sided mesial temporal sclerosis visible on coronal FLAIR MRI. PRESS spectra (1.7 × 1.7 × 2.0 cm, TR/TE 1500/144 msec), voxel locations depicted on coronal T_1 MRI, show reduced NAA in the left hippocampus compared to the right. Note that the field homogeneity is worse in the left hemisphere than the right, leading to broader peak height and lower signal intensities for all resonances in the spectrum.

decision-making confidence in individual patients may be low.

(3) The site of abnormality in many patients, the anterior mesial temporal lobe, is often in a region of poor field homogeneity because of magnetic susceptibility effects from adjacent paranasal sinuses and mastoids, leading to poor quality spectra that are difficult to interpret (see Chapter 3).

(4) Metabolic abnormalities may be bilateral, even in patients with unilateral MTS or who have good surgical outcome following unilateral temporal lobectomy (i.e. seizure free after 1 year – class 1 on the Engel surgical outcome scale).

(5) In many patients, MRI and MRS may be concordant, which, while improving confidence in the diagnosis, may not warrant the performance of the MRS study in addition to conventional MRI.

Points (4) and (5) above warrant some extra discussion below.

Diffuse metabolic abnormalities in temporal lobe epilepsy

As already indicated,[38] MRS may frequently show bilateral hippocampal abnormalities (i.e. NAA is lower than normal control values in both left and right hippocampi in patients with seizures).[52] This could reflect bilateral sclerosis, or it could represent the effect of seizure propagation from the epileptogenic hippocampus to the other side, causing metabolic impairment. In this regard, it is interesting to note that the metabolism of the contralateral hippocampus typically "improves" after ipsilateral surgery and seizure control, suggesting neuronal dysfunction rather than irreversible neuronal loss.[53,54] This effect seems to occur over a time period of months following surgery.[55] Conversely, untreated patients may show progressive worsening of NAA/Cr ratios over time.[56]

More recent studies using MRSI with 1 ml nominal spatial resolution have suggested that the network of brain regions affected by seizures originating in the mesial temporal lobe can be mapped by looking for correlations between metabolite levels in different regions of the brain.[57] Figure 8.5 shows an example of network connection derived from MRSI; in the TLE patients in this study, in addition to low NAA/Cr in ipsi- and contra-lateral hippocampi, NAA/Cr was also lower than controls in both ipsi- and contra-lateral thalami. Furthermore, ipsi-lateral hippocampal NAA/Cr values were correlated with contralateral hippocampi, and ipsi- and contra-lateral thalami and putamina, suggesting these structures are all functionally linked and metabolically affected by seizure activity.[57]

Seizure activity may also result in more widespread metabolic abnormalities; for instance, it has been found that frontal lobe NAA levels are lower in TLE patients than in controls in both gray and white matter

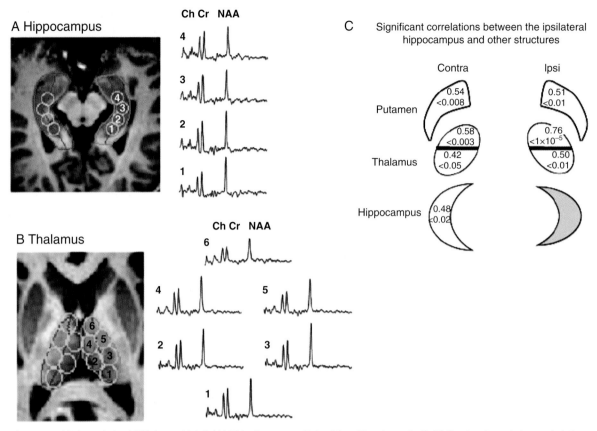

Figure 8.5. High-resolution MRSI data at high field (4 T) in the temporal lobe (A) and basal ganglia (B). (C) Results of correlation analysis for NAA/Cr between different structures in patients with TLE. Correlations are seen between the ipsilateral hippocampus and (in order of R value) ipsi-lateral and contralateral anterior thalamus, contra- and ipsi-lateral putamen, ipsilateral posterior thalamus, and contralateral hippocampus. Reproduced with permission from [57].

regions,[58] as well as in other lobes.[59] Conversely, in patients with epilepsy in the neocortex, hippocampal NAA reductions have also been reported.[60,61] These factors should be kept in mind when using MRSI to evaluate whether seizures are of temporal, neocortical, or extratemporal origin; however, in making this distinction, the largest metabolic abnormality is generally reported to be in the site of seizure onset.[62]

Finally, it is interesting to note that many published MRS studies are apparently successful in identifying the epileptogenic temporal lobe, despite the fact that large regions-of-interest are often used for spectroscopic analysis (e.g. 8 cm^3 for single-voxel studies). Voxels are even larger for ^{31}P studies (e.g. ≈ 30 cm^3 or larger). Since the hippocampus occupied only a small fraction of these localized volumes, these results might indicate that there are diffuse spectroscopic abnormalities in the temporal lobe, even when the only MRI finding is that of hippocampal atrophy. These results are therefore consistent with the common observation by PET of extensive hypometabolism throughout the temporal lobe. Since seizure control is often obtained by selective amygdalohippocampalectomy, clearly not all of the metabolically abnormal tissue is epileptogenic.

MRS in TLE cases where MRI is normal, or symmetrically abnormal

Arguably, the most useful scenario for MRS is when results from other modalities (particularly MRI) are either normal, ambiguous, or bilaterally abnormal. In these TLE patients, MRS has the potential of lateralizing the epileptic focus. Presence of bilateral MRI abnormalities does not necessarily indicate a poor surgical outcome,[9] but does increase the difficulty in correctly lateralizing the epileptogenic source. In an MRSI study of 21 patients with bilateral hippocampal atrophy who were operated on the side of greatest

EEG abnormality, it was found that factors in favor of good surgical outcome were: (1) concordant MRSI-EEG localization; (2) greater asymmetry of NAA/Cr between hippocampi; and (3) an absence of contralateral posterior NAA/Cr abnormalities.[63]

MRS may also play a role when MRI is normal – in a study of 7 patients with intractable epilepsy but completely normal MRI findings, it was found that 5 of 7 cases had abnormal NAA/(Cr+Cho) ratios, 2 of which were bilateral.[64] Although this study did not report detailed EEG correlation or surgical outcome, it did suggest that MRS may provide additional information when MRI is normal. NAA has also been found to be lower than normal control values in the ipsilateral hippocampus (to EEG) in another study of MRI negative patients.[65] However, another study found that metabolic abnormalities (in particular, well-localized NAA asymmetry) surprisingly did not predict seizure-free outcome after surgery, although the presence of contralateral abnormality did predict poor outcome in this group.[66]

Cortical malformations

There have been a number of reports of MRS in patients with malformations of cortical development (MCD).[67,68,69,70] Despite their frequent epileptogenic nature, MCDs typically show only subtle (or sometimes no) metabolic abnormalities;[68] when metabolic abnormalities are observed, most commonly NAA is reduced and Cho increased, particularly for focal cortical dysplasias. As with other types of epilepsy, metabolic changes remote from the presumed seizure focus (e.g. in the contralateral hemisphere) may be different (typically lower NAA/Cr) from healthy controls, and in fact were not significantly different from the ipsilateral side.[71]

Frontal lobe epilepsy

There have been fewer reports of MR spectroscopy in extratemporal epilepsy than in TLE. Garcia *et al.* have studied frontal lobe epilepsy using both ^{31}P[27] and ^{1}H[41] MR spectroscopy. In the proton study, all eight cases exhibited a reduced NAA/Cr ratio in the epileptogenic tissue compared to an anatomical similar contralateral location. Stanley *et al.* also reported the results of proton spectroscopy imaging in 20 cases with frontal lobe epilepsy.[72] As in TLE, it was found that the ratio of NAA/(Cho+Cr) successfully lateralized the epileptogenic tissue as defined by EEG.

Widespread NAA reductions were also noted (i.e. the contralateral NAA was also lower than control values), indicating extensive neuronal loss not confined to just the side of seizure onset, as commonly observed in TLE.

Childhood epilepsies, Rasmussen's encephalitis

A number of epilepsies of childhood have been studied by MRS.[73,74,75] MRS has also been used to investigate cerebral metabolism in the "ketogenic diet", which is becoming increasingly popular as an alternative to pharmacological means of seizure control – using ^{31}P MRS, improvements in bioenergetic status have been reported (Figure 8.2),[31] while proton MRS has shown that the ketone bodies such as β-hydroxy-butyrate and acetone may be detected.[76,77]

Rasmussen's encephalitis (RE) is a rare, chronic, and progressive epilepsy of childhood involving one hemisphere of the brain. While the cause is largely unknown, currently the only treatment that can reliably provide effective seizure relief is hemispherectomy. Definitive diagnosis is usually made on the basis of clinical, electroencephalographic, and neuroimaging findings; however, in the early stages diagnosis may not be straightforward. Breiter *et al.*[37] found hemispheric NAA reductions in 5 cases of Rasmussen's syndrome, an example of which can be seen in Figure 8.6, where on MRSI the whole hemisphere shows low NAA, consistent with neuronal loss. It can also be seen that choline is elevated in the affected hemisphere, particularly the white matter, consistent with the microglial proliferation typically seen on pathology. The hemispheric nature of the metabolic involvement confirms the clinical observation that complete hemispherectomy is necessary for effective seizure control in most cases.

An alternative pattern of involvement can be seen in Figure 8.7, which shows a more regional distribution, with high Cho and low NAA primarily occurring in the insular cortex, putamen, and frontal lobe.

Neurotransmitters: brain GABA levels

While the majority of MRS studies of epilepsy have studied the most readily observed metabolites Cho, Cr, and NAA, by the use of spectral editing methods it is also possible to measure the inhibitory neurotransmitter, γ-aminobutyric acid (GABA) (see Chapter 2).[78] Typically, macromolecules and the

Chapter 8: MRS in epilepsy

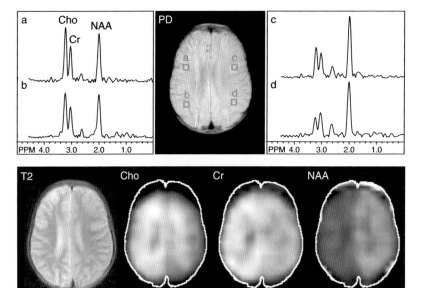

Figure 8.6. Four-year-old female with Rasmussen's encephalitis. Proton density (PD) localizer MRI is unremarkable, while T_2 MRI shows mild right-sided atrophy. MRSI (TR/TE 2300/280 ms) shows hemispheric right-sided reduction in NAA, and increased Cho, particularly in white matter regions, suggesting neuronal loss and microglial proliferation, respectively. Cr is also mildly elevated on the right.

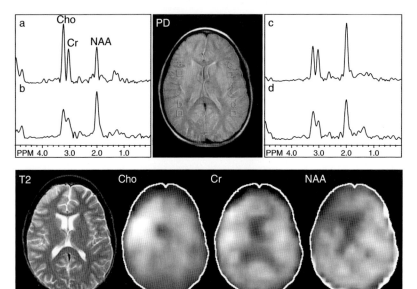

Figure 8.7. Ten-year-old female with Rasmussen's encephalitis. Proton density (PD) and T_2 MRI show mild right-sided atrophy. MRSI (TR/TE 2300/280 ms) shows focal increase in Cho and decreased NAA in right frontal lobe only.

dipeptide homocarnosine also co-edit with GABA, so that the peak observed by MRS is sometimes labeled GABA+ to distinguish it from "pure" GABA. It has been reported that GABA may be (globally) decreased in the brains of epilepsy patients with poor seizure control,[79] and that GABA levels can be increased (and seizure control obtained) using the antiepileptic drugs vigabatrin,[80] topiramate,[81] and gabapentin.[82] Techniques of this type are promising for monitoring the effects of therapy and establishing optimal drug dosages,[80] and will be more widely used as the MEGA-PRESS editing technique becomes commercially available. It should be noted that other compounds of potential clinical significance, such as glutamate and N-acetyl-aspartyl-glutamate (NAAG) can also be measured using MEGA-PRESS.[83]

Some recommendations for MRS protocols for patients with epilepsy

Since metabolic changes in interictal epilepsy patients are often subtle, high quality MRS with good SNR is essential. Ideally, high-field field scanners (e.g. 3 T) with multiple phased-array receiver coils should be used.

For temporal lobe epilepsy, the simplest protocol is to compare the body of the left and right hippocampi at intermediate TE (typically 140 msec) using single-voxel PRESS, or PRESS-MRSI, angulated along the long axis of the hippocampus, as described in Chapter 1 (Figures 1.8 and 1.9). While this may seem relatively simple, in fact considerable care has to be taken with these protocols. Voxels should be positioned carefully (both because of the small structures to be observed, and also because of metabolic and field homogeneity changes along the hippocampus[84]), preferably using full 3-view localizers (axial T_1-weighted 3D "MP-RAGE" scan reconstructed in both sagittal and coronal views works well (Figure 1.9)). High-bandwidth slice-selective RF pulses are preferable to minimize left–right asymmetries due to chemical shift displacement effects. Second-order shimming is also important (especially at higher fields such as 3 T), particularly for MRSI protocols, since significant non-linear field inhomogeneities occur in the temporal lobes which cannot be corrected using linear shims alone. At the anterior tip of the temporal lobe (pes or head of the hippocampus), field homogeneity is usually particularly poor because of susceptibility effects from the nearby paranasal sinuses. Poor field homogeneity results in insufficient quality spectra for analysis in most adult subjects, even with high order shimming corrections. This is unfortunate, since this is often the target of surgical resection when anterior mesial temporal lobectomies are performed, and is presumably the primary site of pathology. Because of the poor field homogeneity in anterior regions, as well as potential lipid contamination from retro-orbital fat and other skull base structures, it is helpful to apply saturation bands in these anterior regions (Figure 1.8).

For extratemporal lobe epilepsy, the site of the epileptogenic focus may be unknown, or even when known (e.g. MCDs) the spectroscopic findings may be heterogeneous, therefore in these cases the best approach is to use MRSI with high spatial coverage (e.g. multi-slice or 3D). MRSI is also important since the observation of abnormal metabolism remote from (e.g. surrounding, connected via fiber pathways, or contralateral to) the primary focus is also commonly reported, and may be of clinical significance.

Since MRS changes in epilepsy are generally subtle, demands are placed on the accuracy of both the acquisition technique and spectral analysis software methods. Overviews of these are provided in Chapters 2 and 3. For single-voxel epilepsy MRS, the LC model software is particularly recommended, since it can provide metabolite concentrations and yield uncertainty estimates.[85] In addition, it is advisable to have matched control data using the same scanner, brain region, and MRS technique for comparison; more advanced studies use statistical tests (such as a z-score) to estimate how abnormal any particular spectrum may be.[86] Precision will also be improved by applying corrections to metabolite concentrations according to the gray matter, white matter, and CSF composition within the localized voxel.[87]

Conclusions

In summary, MRS of epilepsy is now a relatively mature field, with the reduction of NAA in abnormal tissue the most common finding. Despite this observation, and the relatively high reported sensitivity and specificity of MRS for seizure focus lateralization in TLE in most research studies, the technique has not found widespread application in clinical practice. This is due to the reasons that are listed in the introduction, of which the most likely is the relatively subtle metabolic changes that are found in most patients. Such subtle changes, along with data that are susceptible to minor instrumental imperfections, make interpretation of individual studies challenging. The "added-value" of MRS compared to other diagnostic techniques remains questionable, although it is apparent that it may be helpful in at least some of the cases that are MRI negative or symmetrically abnormal. As high-field scanners and proton MRS techniques improve (i.e. with improved SNR and accuracy of quantitation), and as the use of editing techniques for neurotransmitters such as GABA or glutamate increases, it is hoped that the clinical utilization of MRS in the evaluation of patients with epilepsy will increase. In the long run, MRS may help obviate the need for invasive EEG procedures and expensive alternative imaging procedures such as PET.

References

[1] Watson C, Jack CR, Jr., Cendes F. Volumetric magnetic resonance imaging. Clinical applications and contributions to the understanding of temporal lobe epilepsy. *Arch Neurol* 1997; **54**: 1521–31.

[2] Jack CR, Jr. Hippocampal T2 relaxometry in epilepsy: past, present, and future. *Am J Neuroradiol* 1996; **17**: 1811–4.

[3] Jackson GD, Connelly A, Duncan JS, Grunewald RA, Gadian DG. Detection of hippocampal pathology in intractable partial epilepsy: increased sensitivity with quantitative magnetic resonance T2 relaxometry. *Neurology* 1993; **43**: 1793–9.

[4] Jack CR, Jr., Rydberg CH, Krecke KN, Trenerry MR, Parisi JE, Rydberg JN, et al. Mesial temporal sclerosis: diagnosis with fluid-attenuated inversion – recovery versus spin–echo MR imaging. *Radiology* 1996; **199**: 367–73.

[5] Cendes F, Andermann F, Gloor P, Evans A, Jones-Gotman M, Watson C, et al. MRI volumetric measurement of amygdala and hippocampus in temporal lobe epilepsy. *Neurology* 1993; **43**: 719–25.

[6] Jack CR, Jr., Sharbrough FW, Twomey CK, Cascino GD, Hirschorn KA, Marsh WR, et al. Temporal lobe seizures: lateralization with MR volume measurements of the hippocampal formation. *Radiology* 1990; **175**: 423–9.

[7] Jack CR, Jr. MRI-based hippocampal volume measurements in epilepsy. *Epilepsia* 1994; **35**(Suppl 6): S21–9.

[8] Spencer SS. MRI and epilepsy surgery. *Neurology* 1995; **45**: 1248–50.

[9] Jack CR, Jr., Trenerry MR, Cascino GD, Sharbrough FW, So EL, O'Brien PC. Bilaterally symmetric hippocampi and surgical outcome. *Neurology* 1995; **45**: 1353–8.

[10] Berkovic SF, McIntosh AM, Kalnins RM, Jackson GD, Fabinyi GC, Brazenor GA, et al. Preoperative MRI predicts outcome of temporal lobectomy: an actuarial analysis. *Neurology* 1995; **45**: 1358–63.

[11] Duncan JS. Imaging and epilepsy. *Brain* 1997; **120**(Pt 2): 339–77.

[12] Fisher RS, Frost JJ. Epilepsy. *J Nucl Med* 1991; **32**: 651–9.

[13] Engel J, Jr. The use of positron emission tomographic scanning in epilepsy. *Ann Neurol* 1984; **15**(Suppl): S180–91.

[14] Heiss WD, Turnheim M, Vollmer R, Rappelsberger P. Coupling between neuronal activity and focal blood flow in experimental seizures. *Electroencephalogr Clin Neurophysiol* 1979; **47**: 396–403.

[15] Franck G, Salmon E, Sadzot B, Maquet P. Epilepsy: the use of oxygen-15-labeled gases. *Semin Neurol* 1989; **9**: 307–16.

[16] Duncan R, Patterson J, Hadley DM, Wyper DJ, McGeorge AP, Bone I. Tc99m HM-PAO single photon emission computed tomography in temporal lobe epilepsy. *Acta Neurol Scand* 1990; **81**: 287–93.

[17] Pizzini F, Farace P, Zanoni T, Magon S, Beltramello A, Sbarbati A, et al. Pulsed-arterial-spin-labeling perfusion 3 T MRI following single seizure: a first case report study. *Epilepsy Res* 2008; **81**: 225–7.

[18] Warach S, Levin JM, Schomer DL, Holman BL, Edelman RR. Hyperperfusion of ictal seizure focus demonstrated by MR perfusion imaging. *Am J Neuroradiol* 1994; **15**: 965–8.

[19] Van Paesschen W. Ictal SPECT. *Epilepsia* 2004; **45** (Suppl 4): 35–40.

[20] Swartz BE, Tomiyasu U, Delgado-Escueta AV, Mandelkern M, Khonsari A. Neuroimaging in temporal lobe epilepsy: test sensitivity and relationships to pathology and postoperative outcome. *Epilepsia* 1992; **33**: 624–34.

[21] Petroff OA, Prichard JW, Behar KL, Alger JR, Shulman RG. In vivo phosphorus nuclear magnetic resonance spectroscopy in status epilepticus. *Ann Neurol* 1984; **16**: 169–77.

[22] Young RS, Chen B, Petroff OA, Gore JC, Cowan BE, Novotny EJ, Jr., et al. The effect of diazepam on neonatal seizure: in vivo 31P and 1H NMR study. *Pediatr Res* 1989; **25**: 27–31.

[23] Karlik SJ, Stavraky RT, Taylor AW, Fox AJ, McLachlan RS. Magnetic resonance imaging and 31P spectroscopy of an interictal cortical spike focus in the rat. *Epilepsia* 1991; **32**: 446–53.

[24] Younkin DP, Delivoria-Papadopoulos M, Maris J, Donlon E, Clancy R, Chance B. Cerebral metabolic effects of neonatal seizures measured with in vivo 31P NMR spectroscopy. *Ann Neurol* 1986; **20**: 513–9.

[25] Hugg JW, Laxer KD, Matson GB, Maudsley AA, Husted CA, Weiner MW. Lateralization of human focal epilepsy by 31P magnetic resonance spectroscopic imaging. *Neurology* 1992; **42**: 2011–8.

[26] Laxer KD, Hubesch B, Sappey-Marinier D, Weiner MW. Increased pH and inorganic phosphate in temporal seizure foci demonstrated by [31P]MRS. *Epilepsia* 1992; **33**: 618–23.

[27] Garcia PA, Laxer KD, van der Grond J, Hugg JW, Matson GB, Weiner MW. Phosphorus magnetic resonance spectroscopic imaging in patients with frontal lobe epilepsy. *Ann Neurol* 1994; **35**: 217–21.

[28] Chu WJ, Hetherington HP, Kuzniecky RJ, Vaughan JT, Twieg DB, Faught RE, et al. Is the intracellular pH different from normal in the epileptic focus of patients with temporal lobe epilepsy? A 31P NMR study. *Neurology* 1996; **47**: 756–60.

[29] Hetherington HP, Kim JH, Pan JW, Spencer DD. 1H and 31P spectroscopic imaging of epilepsy: spectroscopic and histologic correlations. *Epilepsia* 2004; **45**(Suppl 4): 17–23.

[30] Chu WJ, Hetherington HP, Kuzniecky RI, Simor T, Mason GF, Elgavish GA. Lateralization of human temporal lobe epilepsy by 31P NMR spectroscopic imaging at 4.1 T. *Neurology* 1998; **51**: 472–9.

[31] Pan JW, Bebin EM, Chu WJ, Hetherington HP. Ketosis and epilepsy: 31P spectroscopic imaging at 4.1 T. *Epilepsia* 1999; **40**: 703–07.

[32] Pan JW, Williamson A, Cavus I, Hetherington HP, Zaveri H, Petroff OA, et al. Neurometabolism in human epilepsy. *Epilepsia* 2008; **49**(Suppl 3): 31–41.

[33] Matthews PM, Andermann F, Arnold DL. A proton magnetic resonance spectroscopy study of focal epilepsy in humans. *Neurology* 1990; **40**: 985–9.

[34] Birken DL, Oldendorf WH. N-acetyl-L-aspartic acid: a literature review of a compound prominent in 1H-NMR spectroscopic studies of brain. *Neurosci Biobehav Rev* 1989; **13**: 23–31.

[35] Dam AM. Epilepsy and neuron loss in the hippocampus. *Epilepsia* 1980; **21**: 617–29.

[36] Cendes F, Stanley JA, Dubeau F, Andermann F, Arnold DL. Proton magnetic resonance spectroscopic imaging for discrimination of absence and complex partial seizures. *Ann Neurol* 1997; **41**: 74–81.

[37] Breiter SN, Arroyo S, Mathews VP, Lesser RP, Bryan RN, Barker PB. Proton MR spectroscopy in patients with seizure disorders. *Am J Neuroradiol* 1994; **15**: 373–84.

[38] Connelly A, Jackson GD, Duncan JS, King MD, Gadian DG. Magnetic resonance spectroscopy in temporal lobe epilepsy. *Neurology* 1994; **44**: 1411–7.

[39] Hetherington HP, Kuzniecky RI, Pan JW, Vaughan JT, Twieg DB, Pohost GM. Application of high field spectroscopic imaging in the evaluation of temporal lobe epilepsy. *Magn Reson Imaging* 1995; **13**: 1175–80.

[40] Ng TC, Comair YG, Xue M, So N, Majors A, Kolem H, et al. Temporal lobe epilepsy: presurgical localization with proton chemical shift imaging. *Radiology* 1994; **193**: 465–72.

[41] Garcia PA, Laxer KD, van der Grond J, Hugg JW, Matson GB, Weiner MW. Proton magnetic resonance spectroscopic imaging in patients with frontal lobe epilepsy. *Ann Neurol* 1995; **37**: 279–81.

[42] Urenjak J, Williams SR, Gadian DG, Noble M. Proton nuclear magnetic resonance spectroscopy unambiguously identifies different neural cell types. *J Neurosci* 1993; **13**: 981–9.

[43] Simister RJ, Woermann FG, McLean MA, Bartlett PA, Barker GJ, Duncan JS. A short-echo-time proton magnetic resonance spectroscopic imaging study of temporal lobe epilepsy. *Epilepsia* 2002; **43**: 1021–31.

[44] Petroff OA, Errante LD, Rothman DL, Kim JH, Spencer DD. Neuronal and glial metabolite content of the epileptogenic human hippocampus. *Ann Neurol* 2002; **52**: 635–42.

[45] Pan JW, Venkatraman T, Vives K, Spencer DD. Quantitative glutamate spectroscopic imaging of the human hippocampus. *NMR Biomed* 2006; **19**: 209–16.

[46] Peeling J, Sutherland G. 1H magnetic resonance spectroscopy of extracts of human epileptic neocortex and hippocampus. *Neurology* 1993; **43**(3 Pt 1): 589–94.

[47] Cendes F, Andermann F, Preul MC, Arnold DL. Lateralization of temporal lobe epilepsy based on regional metabolic abnormalities in proton magnetic resonance spectroscopic images. *Ann Neurol* 1994; **35**: 211–6.

[48] Hugg JW, Laxer KD, Matson GB, Maudsley AA, Weiner MW. Neuron loss localizes human temporal lobe epilepsy by in vivo proton magnetic resonance spectroscopic imaging. *Ann Neurol* 1993; **34**: 788–94.

[49] Capizzano AA, Vermathen P, Laxer KD, Ende GR, Norman D, Rowley H, et al. Temporal lobe epilepsy: qualitative reading of 1H MR spectroscopic images for presurgical evaluation. *Radiology* 2001; **218**: 144–51.

[50] Kantarci K, Shin C, Britton JW, So EL, Cascino GD, Jack CR, Jr. Comparative diagnostic utility of 1H MRS and DWI in evaluation of temporal lobe epilepsy. *Neurology* 2002; **58**: 1745–53.

[51] Cendes F, Caramanos Z, Andermann F, Dubeau F, Arnold DL. Proton magnetic resonance spectroscopic imaging and magnetic resonance imaging volumetry in the lateralization of temporal lobe epilepsy: a series of 100 patients. *Ann Neurol* 1997; **42**: 737–46.

[52] Ende GR, Laxer KD, Knowlton RC, Matson GB, Schuff N, Fein G, et al. Temporal lobe epilepsy: bilateral hippocampal metabolite changes revealed at proton MR spectroscopic imaging. *Radiology* 1997; **202**: 809–17.

[53] Cendes F, Andermann F, Dubeau F, Matthews PM, Arnold DL. Normalization of neuronal metabolic dysfunction after surgery for temporal lobe epilepsy. Evidence from proton MR spectroscopic imaging. *Neurology* 1997; **49**: 1525–33.

[54] Vermathen P, Ende G, Laxer KD, Walker JA, Knowlton RC, Barbaro NM, et al. Temporal lobectomy for epilepsy: recovery of the contralateral hippocampus measured by (1)H MRS. *Neurology* 2002; **59**: 633–6.

[55] Serles W, Li LM, Antel SB, Cendes F, Gotman J, Olivier A, et al. Time course of postoperative recovery of N-acetyl-aspartate in temporal lobe epilepsy. *Epilepsia* 2001; **42**: 190–7.

[56] Bernasconi A, Tasch E, Cendes F, Li LM, Arnold DL. Proton magnetic resonance spectroscopic imaging suggests progressive neuronal damage in human temporal lobe epilepsy. *Prog Brain Res* 2002; **135**: 297–304.

[57] Hetherington HP, Kuzniecky RI, Vives K, Devinsky O, Pacia S, Luciano D, et al. A subcortical network of dysfunction in TLE measured by magnetic resonance spectroscopy. *Neurology* 2007; **69**: 2256–65.

[58] Mueller SG, Suhy J, Laxer KD, Flenniken DL, Axelrad J, Capizzano AA, et al. Reduced extrahippocampal NAA in mesial temporal lobe epilepsy. *Epilepsia* 2002; **43**: 1210–6.

[59] Capizzano AA, Vermathen P, Laxer KD, Matson GB, Maudsley AA, Soher BJ, et al. Multisection proton MR spectroscopy for mesial temporal lobe epilepsy. *Am J Neuroradiol* 2002; **23**: 1359–68.

[60] Mueller SG, Laxer KD, Cashdollar N, Lopez RC, Weiner MW. Spectroscopic evidence of hippocampal abnormalities in neocortical epilepsy. *Eur J Neurol* 2006; **13**: 256–60.

[61] Mueller SG, Laxer KD, Barakos JA, Cashdollar N, Flenniken DL, Vermathen P, et al. Identification of the epileptogenic lobe in neocortical epilepsy with proton MR spectroscopic imaging. *Epilepsia* 2004; **45**: 1580–9.

[62] Li LM, Caramanos Z, Cendes F, Andermann F, Antel SB, Dubeau F, et al. Lateralization of temporal lobe epilepsy (TLE) and discrimination of TLE from extra-TLE using pattern analysis of magnetic resonance spectroscopic and volumetric data. *Epilepsia* 2000; **41**: 832–42.

[63] Li LM, Cendes F, Antel SB, Andermann F, Serles W, Dubeau F, et al. Prognostic value of proton magnetic resonance spectroscopic imaging for surgical outcome in patients with intractable temporal lobe epilepsy and bilateral hippocampal atrophy. *Ann Neurol* 2000; **47**: 195–200.

[64] Connelly A, Van Paesschen W, Porter DA, Johnson CL, Duncan JS, Gadian DG. Proton magnetic resonance spectroscopy in MRI-negative temporal lobe epilepsy. *Neurology* 1998; **51**: 61–6.

[65] Doelken MT, Stefan H, Pauli E, Stadlbauer A, Struffert T, Engelhorn T, et al. (1)H-MRS profile in MRI-positive versus MRI-negative patients with temporal lobe epilepsy. *Seizure* 2008; **17**: 490–7.

[66] Suhy J, Laxer KD, Capizzano AA, Vermathen P, Matson GB, Barbaro NM, et al. 1H MRSI predicts surgical outcome in MRI-negative temporal lobe epilepsy. *Neurology* 2002; **58**: 821–3.

[67] Kuzniecky R, Hetherington H, Pan J, Hugg J, Palmer C, Gilliam F, et al. Proton spectroscopic imaging at 4.1 tesla in patients with malformations of cortical development and epilepsy. *Neurology* 1997; **48**: 1018–24.

[68] Mueller SG, Laxer KD, Barakos JA, Cashdollar N, Flenniken DL, Vermathen P, et al. Metabolic characteristics of cortical malformations causing epilepsy. *J Neurol* 2005; **252**: 1082–92.

[69] Simister RJ, McLean MA, Barker GJ, Duncan JS. Proton magnetic resonance spectroscopy of malformations of cortical development causing epilepsy. *Epilepsy Res* 2007; **74**: 107–15.

[70] Li LM, Cendes F, Bastos AC, Andermann F, Dubeau F, Arnold DL. Neuronal metabolic dysfunction in patients with cortical developmental malformations: a proton magnetic resonance spectroscopic imaging study. *Neurology* 1998; **50**: 755–9.

[71] Leite CC, Lucato LT, Sato JR, Valente KD, Otaduy MC. Multivoxel proton MR spectroscopy in malformations of cortical development. *Am J Neuroradiol* 2007; **28**: 1071–5; discussion 6–7.

[72] Stanley JA, Cendes F, Dubeau F, Andermann F, Arnold DL. Proton magnetic resonance spectroscopic imaging in patients with extratemporal epilepsy. *Epilepsia* 1998; **39**: 267–73.

[73] Holopainen IE, Valtonen ME, Komu ME, Sonninen PH, Manner TE, Lundbom NM, et al. Proton spectroscopy in children with epilepsy and febrile convulsions. *Pediatr Neurol* 1998; **19**: 93–9.

[74] Cross JH, Connelly A, Jackson GD, Johnson CL, Neville BG, Gadian DG. Proton magnetic resonance spectroscopy in children with temporal lobe epilepsy. *Ann Neurol* 1996; **39**: 107–13.

[75] Hanefeld F, Kruse B, Holzbach U, Christen HJ, Merboldt KD, Hanicke W, et al. Hemimegalencephaly: localized proton magnetic resonance spectroscopy in vivo. *Epilepsia* 1995; **36**: 1215–24.

[76] Pan JW, Telang FW, Lee JH, de Graaf RA, Rothman DL, Stein DT, et al. Measurement of beta-hydroxybutyrate in acute hyperketonemia in human brain. *J Neurochem* 2001; **79**: 539–44.

[77] Seymour KJ, Bluml S, Sutherling J, Sutherling W, Ross BD. Identification of cerebral acetone by 1H-MRS in patients with epilepsy controlled by ketogenic diet. *Magma* 1999; **8**: 33–42.

[78] Rothman DL, Petroff OA, Behar KL, Mattson RH. Localized 1H NMR measurements of gamma-aminobutyric acid in human brain in vivo. *Proc Natl Acad Sci USA* 1993; **90**: 5662–6.

[79] Petroff OA, Rothman DL, Behar KL, Mattson RH. Low brain GABA level is associated with poor seizure control. *Ann Neurol* 1996; **40**: 908–11.

[80] Petroff OA, Rothman DL, Behar KL, Mattson RH. Human brain GABA levels rise after initiation of vigabatrin therapy but fail to rise further with increasing dose. *Neurology* 1996; **46**: 1459–63.

[81] Petroff OA, Hyder F, Mattson RH, Rothman DL. Topiramate increases brain GABA, homocarnosine, and pyrrolidinone in patients with epilepsy. *Neurology* 1999; **52**: 473–8.

[82] Petroff OA, Rothman DL, Behar KL, Lamoureux D, Mattson RH. The effect of gabapentin on brain gamma-aminobutyric acid in patients with epilepsy. *Ann Neurol* 1996; **39**: 95–9.

[83] Edden RA, Pomper MG, Barker PB. In vivo differentiation of *N*-acetyl aspartyl glutamate from *N*-acetyl aspartate at 3 Tesla. *Magn Reson Med* 2007; **57**: 977–82.

[84] Vermathen P, Laxer KD, Matson GB, Weiner MW. Hippocampal structures: anteroposterior *N*-acetylaspartate differences in patients with epilepsy and control subjects as shown with proton MR spectroscopic imaging. *Radiology* 2000; **214**: 403–10.

[85] Provencher SW. Estimation of metabolite concentrations from localized in vivo proton NMR spectra. *Magn Reson Med* 1993; **30**: 672–9.

[86] Hetherington H, Kuzniecky R, Pan J, Mason G, Morawetz R, Harris C, *et al*. Proton nuclear magnetic resonance spectroscopic imaging of human temporal lobe epilepsy at 4.1 T. *Ann Neurol* 1995; **38**: 396–404.

[87] Bonekamp D, Horska A, Jacobs MA, Arslanoglu A, Barker PB. Fast method for brain image segmentation: application to proton magnetic resonance spectroscopic imaging. *Magn Reson Med* 2005; **54**: 1268–72

Chapter 9
MRS in neurodegenerative disease

Key points

- Despite the relatively common occurrence of neurodegenerative diseases, MRS is lightly used in these conditions, most likely because of lack of sensitivity and overlap of spectral findings in different disorders.
- MRS usually shows decreased levels of NAA in dementia.
- Dementias associated with gliosis (e.g. Alzheimer's) also have increased *myo*-inositol (mI).
- mI/NAA ratios correlate with clinical severity and histopathological involvement in Alzheimer's disease.
- mI/NAA ratios, and regional variations in metabolite levels, may be helpful in the differential diagnosis of different dementias (Alzheimer, vascular, frontotemporal, Lewy body).
- Parkinson's disease does not seem to be associated with any metabolic disorders, although other Parkinsonian disorders (e.g. multiple system atrophy) may show reduced NAA in the basal ganglia.
- Metabolic changes in Huntington's disease are unclear; some studies have reported elevated lactate levels in the basal ganglia, but others have not.
- Prion diseases are characterized by decreased NAA levels.
- In amyotrophic lateral sclerosis (ALS), upper motor neuron NAA decreases may be helpful in establishing a diagnosis.

Introduction

Neurodegenerative diseases include a very wide group of disorders affecting the central nervous system (CNS). Many of these disorders arise from the combined effects of genetic predisposition and environmental factors. This results in reduced cognition (e.g. Alzheimer's disease, dementia with Lewy bodies, and vascular dementia), motor system performance (e.g. amyotrophic lateral sclerosis), or both (e.g. Parkinson's disease and prion diseases). In general, neurodegenerative diseases show a wide diversity of etiology and a broad clinical phenotype spectrum, but all have in common the decrease in neuronal function and neuronal cell death due to activation of apoptotic pathways, programmed cell death, or other mechanisms.

The evolution of magnetic resonance (MR) techniques for imaging the CNS has led to significant advances in the understanding of brain changes associated with neurodegenerative disorders. Conventional MR imaging (MRI) provides detailed anatomic information with excellent tissue contrast and spatial resolution. Additional MR sequences can provide information concerning tissue metabolism, water diffusion, and perfusion. All together, these MR modalities have been demonstrated to be relevant in the clinical evaluation of neurodegenerative disorders for early diagnosis, differential diagnosis, and monitoring of disease activity.[1,2,3,4] MR spectroscopy (MRS), in particular, has provided insights into some of the metabolic abnormalities associated with neurodegeneration, thereby helping to elucidate the underlying pathophysiology of these disorders, although it has yet to have significant impact in routine clinical usage.[5]

The goal here is to describe the common findings of MR spectroscopy (MRS) in this field and its limitations. We will focus on proton MRS (^{1}H-MRS), as the majority of MRS studies in neurodegenerative disorders utilize this technique and it can readily be performed as part of a routine MR study. It should be stressed, however, that other nuclei (such as ^{31}P or ^{13}C) have been used in research studies of various neurodegenerative disorders.[5]

Dementia

Dementia is a clinical diagnosis defined as a decline in memory and other cognitive functions that affect the daily life in an alert patient. Major causes of dementia include Alzheimer's disease (AD) and vascular dementia (VD) and, less commonly, frontotemporal dementia and dementia with Lewy bodies.[6] Consensus clinical criteria have been applied for diagnosis of different dementias, but the sensitivity and specificity of these criteria are variable.[7] The diagnosis of dementing disorders by means of the current clinical criteria remains difficult and, occasionally, due to the mixed pathology existing in the brain of demented patients, the underlying cause of dementia cannot be definitely determined even after the histopathologic examination of the brain.[8]

The difficulty in making diagnoses exclusively based on the current clinical criteria has generated the incentive for identifying specific neuroimaging markers for various dementing pathologies. Thus, in vivo changes of demented brains have been studied with different MR modalities with increasing frequency. In this context, ^{1}H-MRS has been shown to be useful in the differential diagnosis of dementing illnesses, as well as in monitoring early disease progression and effectiveness of therapies.[9]

Alzheimer's disease

Alzheimer's disease (AD) is the most common cause of dementia. Several million people worldwide are afflicted by the disease, and the number of individuals affected is expected to grow with the increasing life expectancy.[10] Generally, it is diagnosed in people over 65 years of age, although the less-prevalent early-onset AD can occur much earlier.[10]

The first symptoms of the disease are often mistaken as related to aging or stress.[11] At this stage, neuropsychological testing can reveal mild cognitive impairment (MCI).[12] These early symptoms can have an effect on the most complex daily living activities.[13] The most noticeable deficits observed at this stage are memory loss (short-term memory loss), problems with executive functions, and apathy. In patients with AD, the increasing impairment of learning and memory will lead to a definitive diagnosis.[14] Subsequently, language problems and apraxia will appear. At this stage, AD patients can still perform tasks independently, but may need assistance or supervision with the most complicated activities. As the disease progresses further, progressive deterioration hinders independence. Reading and writing skills are progressively lost, urinary incontinence can develop, motor and memory problems worsen, and behavioral changes become evident. During the latest stage of AD, the patient is completely dependent upon caregivers, language is reduced to simple phrases or single words, and mobility deteriorates to the point where the patient is bedridden. Death occurs from complications such as pressure ulcers and pneumonia, and not from the disease itself.

The brains of individuals with AD are characterized by the presence of abundant neurofibrillary tangles (pathological protein aggregates found within neurons) and neuritic plaques (deposits of the beta-amyloid in the extracellular spaces). These deposits ultimately cause neuron disintegration, collapsing the neuron's transport system.[15] Although many individuals develop some plaques and tangles as a consequence of aging, the brains of AD patients usually have a greater number of them in specific brain regions such as the limbic system and temporal neocortex, with a tendency to spread to other brain regions as the disease progresses.[16] However, typically this pathology manifests as clinical AD only after a certain quantitative threshold is reached, and by the time the individual is diagnosed with AD, a significant loss of synapses and neurons has already occurred.[17]

In agreement with pathologic findings, most MRI studies have reported global or focal signs of brain atrophy in the brains of AD patients. Gray-matter atrophy was consistently found in the frontal, temporal, and parietal lobes and limbic system of patients with AD, presumably reflecting neuron loss in these regions.[1] In this context, accurate volumetric measures of regional brain volumes on MRI images have demonstrated to support diagnostic decision-making and differential diagnosis.[18,19,20] This is particularly useful for detection of cerebral changes in MCI, as these are usually too subtle to be detected by visual inspection of MRI scans alone. Compared with elderly controls, patients with MCI seem to have significant hippocampal and entorhinal cortex volume losses. However, as expected, gray matter volume loss is in general less severe and diffuse in MCI than in clinically evident AD patients.[18,19,21,22,23]

^{1}H-MRS

Large decreases in brain NAA have been observed since the earliest proton MRS and MR spectroscopic imaging (MRSI) studies of AD patients.[1] Reduction of NAA concentration, or the ratio of NAA to other metabolites such as Cr, has been consistently found in the mesial temporal lobe, posterior cingulate gyrus,

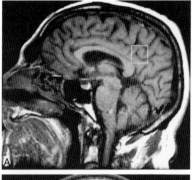
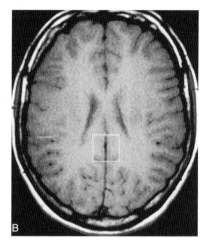
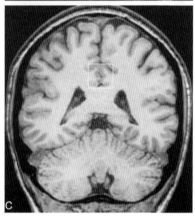

Figure 9.1. (A) Midsagittal, (B) axial, and C, coronal T_1-weighted images (700/14) show the location of the 8 cm³ (2 × 2 × 2 cm) posterior cingulate spectroscopy voxel typically used for MRS studies of Alzheimer's disease. This ROI is chosen because involvement of the posterior cingulate is common in AD, and this region gives excellent quality spectra. [From Kantarci et al., Am J Neuroradiol 2004; **24**: 834–49, figure 1, with permission.]

parietotemporal region, frontal lobe, occipital lobe, and hippocampus.[9,24,25,26,27,28,29,30,31] These NAA decreases are much less evident in the white matter, probably due to the fact that AD affects the cortical regions primarily.[30] Measurements that account for atrophy in acquired voxels show that the decreases in NAA are independent of CSF content. [30,32,33] NAA is a sensitive marker for neuronal density or viability,[34] but its loss is certainly not specific for AD. However, it is interesting to note that decreases in NAA seem more pronounced in anatomic locations that show higher severity of neuropathologic findings (e.g. amyloid plaques and neurofibrillary tangles) on postmortem studies.[35,36] This suggests that in AD patients, brain NAA changes do correlate with neuronal loss or dysfunction in patients with AD.

Proton MRS studies have reported conflicting results on choline (Cho) metabolite levels in brains of AD patients, with increases, no changes or decreases of this metabolite in different brain locations.[9,30] In contrast, increases in *myo*-inositol (mI) have consistently been reported in several anatomic locations of brains of AD patients, with again an effect more pronounced in gray matter (e.g. mesial temporal lobe, anterior and posterior cingulate gyrus, and parietal lobe) than in white matter.[24,28,37,38,39] As most of the mI in the brain is present in glial cells,[40] it is likely that persistent elevation of mI levels reflects microglial proliferation in AD.[38] In patients with MCI, brain regional (e.g. posterior cingulate – see Figure 9.1) levels of mI are often increased without a large decrease in NAA, suggesting that MRS may be sensitive to the biochemical changes in the pathological progression of prodromal AD even before there is a significant loss of neuronal integrity (Figure 9.2). [41,42,43]

Owing to the consistent coexistence of high mI and low NAA brain levels in patients with AD, the ratio of NAA/mI has been proposed by several authors as the most accurate ^{1}H-MRS measurement in this disease. [1,9,31,44] Whereas the clinical specificity of the NAA decline in AD is poor, the addition of mI information increases accuracy. Thus, the NAA/mI ratio is able to distinguish clinically diagnosed patients with AD from normal elderly people, with good sensitivity and specificity.[9] For the more challenging task of discriminating AD cases from other possible dementia diagnoses, the NAA/mI ratio can still be useful. Whereas NAA levels are almost constantly decreased in the different dementia types, levels of mI are

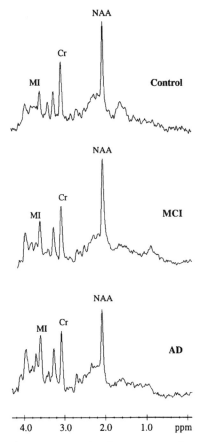

Figure 9.2. Examples of proton spectra obtained from the posterior cingulate VOI with a TE of 30 ms in a control subject (top), in a patient with MCI (middle), and in a patient with AD (bottom). NAA/Cr ratio is lower in the patient with AD than both the patient with MCI and the control subject. mI/Cr ratios are higher in patients with MCI and AD than the control subject. The mI/Cr ratio is also higher in the patient with AD than the patient with MCI. [Adapted from Kantarci et al., Neuroimag Clin N Am 2003; **13**: 197–209, figure 3, with permission.]

elevated predominantly in dementias that are pathologically characterized by gliosis, such as AD. Thus, higher levels of NAA/mI are expected in patients with, for example, vascular dementia or in dementia with Lewy bodies where gliosis is little or absent.[28,29,32] Furthermore, in patients with AD, neuropsychological measures of cognitive function may correlate with levels of the NAA/mI ratio, with region-specific association between neuropsychological performance and MRS metabolite changes depending on the cognitive domain being studied.[45,46] The intriguing suggestion that ^{1}H-MRS may have a useful role in prognosis of mental function and tracking of disease progression seems to be confirmed by the close correlation recently reported between antemortem metabolite changes found on ^{1}H-MRS examination and AD-type pathology subsequently seen in the same brains at autopsy (e.g. strong association of NAA/mI and Braak stage, a histopathological estimate of extent of neurofibrillary tangle involvement).[47]

Despite its potential to monitor the temporal evolution of metabolite changes, longitudinal ^{1}H-MRS studies are uncommon.[48,49,49,50,51] There may be several reasons for this, including technical difficulties. For example, it is technically demanding to obtain reproducible measurements from the anteromedial temporal lobe (a region which is known to be affected earlier and more severely than any other brain regions in AD patients), as this brain region is in proximity to the tissue–air interface near the petrous bone and, consequently, difficult to achieve a homogenous magnetic field and an optimal water suppression within that VOI. In addition, the $\sim 8\,cm^3$ voxel size generally used to obtain the sufficient SNR is much larger than the volume of the cortical gray matter structures, with the consequent lack in the anatomic specificity of the measurements. Since short TE (30–35 ms) ^{1}H-MRSI acquisition has become easier to use and routinely available on clinical MR systems, this should be the preferred method for quantification of spectroscopic metabolites in AD brains. Furthermore, more sophisticated post-processing using segmented MRIs to calculate percentages of gray and white matter in each ^{1}H-MRSI voxel can allow estimation of "pure" cortical changes, improving the reliability of the spectroscopic measure.[52,53] Alternatively, short TE single voxel (usually 8 cm^3 size) ^{1}H-MRS acquisitions from the posterior cingulate region and mesial occipitoparietal cortex provide good quality data and seem to show changes specific enough to AD.[30]

Finally, it must be stressed that the diagnosis of AD as well as other forms of dementia is difficult and relies mainly on clinical information. According to the most recent guidelines by the American Academy of Neurology, quantitative MR techniques are not recommended for routine use of dementia evaluation at this time because superiority to clinical criteria has not been demonstrated.[54] However, the early diagnosis of AD using a combination of quantitative MR methodologies,[165] including ^{1}H-MRSI, seems possible in the near future.[55]

Vascular dementia

Vascular dementia (VD) or multi-infarct dementia is one of the most common causes of cognitive decline in adult humans, secondary only to AD.[10] This is an increasingly diagnosed condition, where diffuse microvascular ischemic disease of the brain leads to progressive motor impairment and dementia. The wide range of clinical phenotypes has inspired numerous classification schemes. However, despite the efforts of many researchers,[10,56,57,58,59,60], there are no universal clinical diagnostic criteria for VD. In addition, histopathologic data make this picture even more complex, showing that in many cases vascular pathology coexists with the pathology of AD. Although "pure" vascular pathology is relatively uncommon,[61,62] it is widely accepted that VD can be split into cortical versus subcortical dementias: the former has a prevalent involvement of the cortical gray matter, whereas the latter is mostly characterized by lacunar infarcts and deep white matter changes.[63]

The sensitivity for detecting vascular brain injury has widened significantly with the advent of modern neuroimaging.[64,65,66] In patients with VD, quantitative MRI studies have consistently identified general brain atrophy and loss of cortical gray matter, enlargement of the ventricular CSF spaces, and even atrophy of the hippocampus and amygdala. Since all of these findings can also be seen in AD patients, the MRI comparison between these two types of dementia is usually inconclusive, especially at the early stages of each disease.[67] Compared with normal subjects, an increased load of white matter lesions on T_2-weighted images has been identified consistently in VD.[67] The term *leukoaraiosis* has been suggested to describe the white matter changes associated with cerebrovascular disease, but the etiology and clinical relevance of these changes remain to be determined in most cases.[68] The severity and frequency of leukoaraiosis increase with advancing age, risk factors for stroke, previous strokes (particularly of the lacunar type), and dementia. On the other hand, however, patients who meet neuroimaging criteria for leukoariosis may also be free of clinical signs of dementia, suggesting that the imaging changes are pathophysiologically nonspecific.[69,70,71]

Although the lack of pathological specificity of conventional MRI does not allow for the differentiation of the heterogeneous pathological processes associated with VD, neuroimaging still plays an important role in the diagnostic process of this disease. One of the most used diagnostic criteria, the so-called NINDS-AIREN criteria,[60] includes brain imaging to support clinical findings, on the basis of lesion location and pattern typically associated with VD.[64] Neuroimaging confirmation of cerebrovascular disease in VD provides information about the topography and severity of vascular lesions. It may also assist with the differential diagnosis of dementia associated with normal pressure hydrocephalus, chronic subdural hematoma, arteriovenous malformation, or tumoral diseases.[64] In this context, MRI is preferred over CT scan because of its superior soft-tissue contrast.[65] New MR modalities such as diffusion tensor imaging (DTI) have been reported to detect abnormalities not shown with conventional acquisition sequences and even attempt differential diagnosis, for example between AD and VD, on the basis of the brain regional differences in diffusion abnormalities.[72,73,74]

^{1}H-MRS

As mentioned above, the differentiation between AD and VD is challenging. In particular, since the patient may have mixed pathology, one may want to identify how much, if any, of the two pathologies are contributing to AD or VD, so that appropriate therapies can be planned. In this difficult context, the metabolic information provided by ^{1}H-MRS can have some value. Levels of NAA and NAA/Cr are similarly diffusely reduced in the gray matter regions of patients with VD and AD.[9] However, cortical levels of mI/Cr are normal in VD patients and elevated in AD patients (Figure 9.3).[9,28,29] This can help in the differential diagnosis and, perhaps, to even identify concomitant AD in a demented patient with cerebrovascular disease.[9] In addition, white matter levels of NAA/Cr are generally lower in patients with VD than in those with AD, reflecting the WM ischemic damage in VD compared to the cortical degenerative pathology in AD.[9,75,76] This might be particularly useful for early diagnosis and in the context of specific familial conditions such as cerebral autosomal dominant arteriopathy with subcortical infarcts and leukoencephalopathy (CADASIL).[77]

The technical difficulties and limitations in performing an MRS examination described in AD are also relevant in this context. The use of short TE (~20–35 ms) is recommended. Finally, given the importance of assessing metabolic changes in cortical gray matter structures without contamination from other brain tissues, the use of MRI segmentation to allow tissue specific spectroscopic measures is strongly suggested.[52,53]

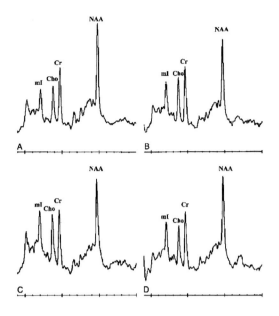
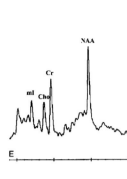

Figure 9.3. Examples of proton spectra obtained from the posterior cingulate VOI with a TE of 30 ms in a normal subject (A), a patient with Alzheimer's disease (AD) (B), a patient with frontotemporal dementia (FTD) (C), a patient with dementia with Lewy bodies (DLB) (D), and a patient with vascular dementia (VD) (E). The NAA/Cr ratio is lower in patients with AD, FTLD, and VD than both the DLB and the normal subjects. mI/Cr ratio is higher in the patients with AD and FTLD than both the VD and the normal subjects. All spectra are scaled to the height of the reference peak Cr, which is assumed to be constant across (A) through (E) (from Kantarci et al., Neurology 2004; **63**: 1393–8, figure 2, with permission.]

Frontotemporal dementia

Frontotemporal dementia (FTD) describes a group of syndromes caused by pathological processes predominantly affecting the frontal and temporal lobes.[78,79] It represents the most frequent cause of non-AD degenerative dementia and has been increasingly recognized as an important cause of early-onset dementia.[79] At least three major anatomic variants of FTD can be described: a bifrontal, slightly asymmetric subtype with more involvement of the right frontotemporal region (*frontal variant*), a temporal-predominant subtype (*temporal variant* or *semantic dementia*), and a left frontal-predominant subtype (*progressive nonfluent aphasia*).[79,80] There is clinical and pathological overlap between the syndromes and the influence of genetic factors varies substantially across the syndromes.[80] Symptoms can be generally related to impairment of functions of the affected brain regions, with behavioral symptoms and deficits in executive functions.[79] Extensive loss of pyramidal neurons in the frontotemporal cortex, severe gliosis within the gray and white matter, and presence of Pick bodies are the most common histologic findings.[78]

Structural MRI scans frequently reveal frontal lobe and/or anterior temporal lobe atrophy, often with significant asymmetry.[1] When advanced volumetric techniques such as voxel-based morphometry are used, brain atrophy in the different types of FTD seem to be closely correlated to the cerebral location responsible for the clinical syndrome.[81,82,83,84] It has been suggested that it may be possible to differentiate autopsy-proven FTD from AD on the basis of the brain atrophy pattern.[85] However, at the earliest disease stages, the differential diagnosis between dementia types is difficult, even using quantitative neuroimaging techniques.

^{1}H-MRS

The metabolite changes of FTD patients are very similar to those seen in AD (Figure 9.3). Usually, NAA/Cr is lower than in normal controls and mI/Cr is higher.[28] However, accurate regional ^{1}H-MRS measurements may help to differentiate dementias that display regionally specific involvement such as FTD.[9,86] This may be particularly important at the early stages of disease progression, as these regional differences may disappear at later stages when the neurodegenerative pathology becomes more widespread and involves the majority of the cerebral cortex.

Dementia with Lewy bodies

Dementia with Lewy bodies (DLB) is recognized as a distinct neurodegenerative disease with established clinical criteria.[87] Lewy bodies, which are the characteristic features of Parkinson's disease in the substantia nigra, are abundant in the neocortex. They are commonly found in people with other types of dementia and LDB by itself is less common than the mixed (AD and Lewy body) type.[88] However, pathologic studies have revealed a consistent pattern of vulnerability to Lewy body formation across subcortical, paralimbic, and neocortical structures that was not related to the amount of AD changes, indicating that dementia with Lewy bodies is a distinct pathology rather than a variant of AD.[88,89] Clinically, patients with DLB exhibit symptoms close to those of AD patients, but with relevant motor features of Parkinsonism.[90,91] Generally, DLB is diagnosed when cognitive symptoms develop within a year or two of movement disorder/Parkinsonian symptoms.[91]

Based on in vivo cerebral MRI studies, patients with DLB have various morphologic changes, including global brain volume loss, regional atrophy in the frontal and temporal lobes, hippocampus, and amygdala.[1] Differences in the severity of atrophy between AD and DLB have been investigated, usually reporting similar losses in the total brain volume, but less severe atrophy in the hippocampus and temporoparietal cortex in DLB patients than in those with AD.[1,92,93]

^{1}H-MRS

There are few MRS studies on patients with DLB, and the technique is not generally used for diagnostic purposes in this disease. It is interesting to note, however, that in comparison with AD patients, those with DLB show less (or no) decrease in NAA/Cr levels in brain regions such as the posterior cingulate gyri (Figure 9.3).[9] This may reflect the relative sparing of neurons in that region seen at autopsy,[89] and again could be useful at early disease stages to distinguish patients with DLB from other dementia syndromes.[9] Increases in the Cho/Cr ratio have been reported in DLB, which have been suggested to be due to increased membrane turnover in this disease.[9]

Parkinson's disease and related disorders

Parkinson's disease (PD) is a progressive neurological disorder characterized by a variable degree of impairment in motor skills, speech, and other CNS functions.[94] Rest tremor, bradykinesia, rigidity, and loss of postural reflexes are generally considered the cardinal signs of PD. Other clinical features include secondary motor symptoms (e.g. dysphagia, sialorrhoea, micrographia, shuffling gait, and festination) and non-motor symptoms (e.g. autonomic dysfunction, cognitive/neurobehavioral abnormalities, sleep disorders, and sensory abnormalities). The symptoms are the results of decreased stimulation of the motor cortex by the basal ganglia, normally caused by the insufficient formation and action of dopamine due to an idiopathic degeneration of the brain dopaminergic system.[95] The mechanism by which brain cells are lost may consist of an abnormal protein accumulation (alpha-synuclein to ubiquitin) in the damaged cells, which leads to the accumulation of the characteristic inclusions called Lewy bodies.[96] Excessive accumulation of iron, which is toxic to nerve cells, are also typically observed in conjunction with the protein inclusions.[97] Recently, genetic mutations, protein mishandling, increased oxidative stress, mitochondrial dysfunction, inflammation, and other pathogenic mechanisms have been identified as contributing factors in the death of dopaminergic and non-dopaminergic cells in the brains of PD patients.[94,98,99] There are no definitive diagnostic tests for the diagnosis of PD. Thus the disease must be diagnosed based on clinical criteria, which are typically based on the presence of a combination of cardinal motor features, associated and exclusionary symptoms, and response to levodopa.[94,100] Pathological confirmation of the hallmark Lewy bodies on autopsy is still considered the criterion for definite PD diagnosis.[94] The disease is not fatal, but it progresses with time, dramatically worsening the subject's quality of life and decreasing his/her average life expectancy. The treatment includes drug therapy (e.g. levodopa, dopamine agonists, and monoamine oxidase-B inhibitors),[101] as well as surgery and deep brain stimulation (in advanced PD patients for whom drug therapy is no longer sufficient).[102]

Although the diagnosis of PD is straightforward when patients have a classical presentation, differentiating PD from other forms of PD related disorders is difficult.[103] These affections include secondary (acquired) Parkinsonism, progressive supranuclear palsy (PSP), multiple system degeneration (MSA), and corticobasal degeneration (CBD). The absence of rest tremor, early occurrence of gait difficulty, postural instability, dementia, and the presence of dysautonomia, ophthalmoparesis, ataxia, and other atypical features, coupled with poor or no response to levodopa, can help in the differential diagnosis of these disorders.[94] However, at early disease stages, when signs and symptoms overlap, this can be very challenging, leading to a significant number of misdiagnoses.[104] Due to the very different natural histories of these diseases, an early differentiation between PD and related disorders is important for correct prognosis and treatment strategy.[103,104,105]

Conventional MRI is normal in PD patients and this investigation is usually performed to exclude a structural cause for the development of Parkinsonism.[106] At late disease stages, the atrophy of the substantia nigra may become evident.[107] Conventional MRI, however, may be useful in the differentiation of the various Parkinsonian syndromes, as it frequently shows abnormalities in these patients.[106] Qualitative and quantitative studies have shown that atrophy and signal changes in putamen and infratentorial structures can differentiate patients with MSA from PD patients with high specificity at late stages, although sensitivity is suboptimal especially at early stages.[107,108,109,110] In a recent study, an MR index based on the midbrain area and the width of superior cerebellar peduncles was able to distinguish patients with PSP from those with PD and MSA with high sensitivity and specificity.[111] On CBD, the midbrain is generally not atrophic, which may help to distinguish this condition from other atypical Parkinsonisms.[106] In some cases, an asymmetrical atrophy (more marked on the side opposite to the clinically involved side of the body and prevalently involving the frontoparietal cortex) can be found.[107]

¹H-MRS

Studies of MRS in PD have produced conflicting results, showing either no difference in the basal ganglia metabolite levels between PD patients and normal controls, or decreases of NAA/Cr levels in PD.[5,112,113] The variability of the results is probably related to the difficulty of reliably assessing metabolic abnormalities in the substantia nigra, due to its small size and high iron content.[5] Interestingly, by using high field MR strength (4 T), recent work has shown the ability to measure multiple metabolites (including GABA) very accurately in a small volume (2.2 ml) including the substantia nigra, but did not find any differences between patients and controls in a relatively small study (10 PD, 11 controls) (Figure 9.4).[114]

In contrast to patients with PD, the reduction of NAA in basal ganglia and other brain regions seems to be consistent in patients with related Parkinsonian disorders.[5,112,113] Decreases in basal ganglia levels of NAA have been reported in patients with MSA.[115,116,117] In an MRS study at high field strength (3 T), NAA/Cr decreases were confirmed to be significantly reduced in the putamen and in the pontine base of MSA patients, suggesting that these measurements may be of diagnostic value early in the disease course.[118] Another study was performed in groups of

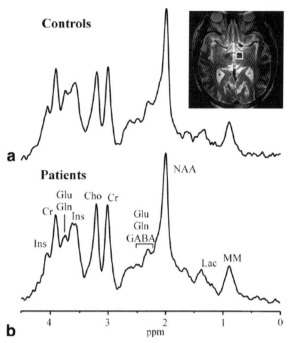

Figure 9.4. Short echo time group-averaged proton spectra (TE 5 ms, TR 4.5 s, 400 averages for each subject) acquired with a 4 T magnet from 2.2-ml volumes that encompass the substantia nigra of (a) 11 healthy volunteers and (b) 10 patients with PD. The VOI is shown in the T_2-weighted image. [From Oz et al., Magn Reson Med 2006; **55**: 296–301, figure 1, with permission.]

patients with PD, PSP, and CBD using a multi-slice MRSI approach,[119] which allowed the assessment of the metabolic profile of several brain regions with good spatial resolution. Decreases of NAA/Cr were observed in PSP patients in brainstem, centrum semiovale, frontal lobe, and precentral cortex, as well as a reduction of NAA/Cho in the lentiform nucleus. However, the largest decrease in NAA/Cho was in the lentiform nucleus of CBD patients, exactly where one would expect the most prominent neuropathological abnormality in this disease. Again, this study confirmed that the PD patient group showed no metabolic abnormalities in any of the brain regions studied.

Collectively, these studies suggest that there is a potential role for the use of ¹H-MRS in the differential diagnosis of Parkinsonian related syndromes, and perhaps also for monitoring the effects of treatment in these disorders.[120]

Huntington's disease

Huntington's disease (HD) is an autosomal-dominant inherited neurodegenerative disorder caused by a CAG polyglutamine repeat expansion in exon 1 of the HD gene.[121] Patients affected by this devastating disease suffer early cognitive impairment, motor deficits, and psychiatric disturbances. Symptoms are attributed to cell death in the striatum and disruption of cortical–striatal circuitry.[122] The mechanisms that underlie selective neuronal cell death and dysfunction remain poorly understood, but processes involving mitochondrial abnormalities, excitotoxicity, and abnormal protein degradation have been implicated.[122] The diagnosis may be made by genetic testing.[121] Therapy that slows the progressive neuronal dysfunction or degeneration is unavailable and pharmaceutical therapies are commonly used, with limited benefit, to treat disease symptoms. [121] Several neuroprotective therapies as well as cell replacement strategies such as fetal transplantation have been used with minimal success.[122] In this context, newer neuroimaging technologies may provide surrogate markers of both disease onset and disease progression.[123]

Most of the HD studies using either qualitative or volumetric MRI have focused on the basal ganglia. [123] Several studies have shown evidence of basal ganglia atrophy, even before the onset of motor symptoms.[123,124] Caudate volumes were found to correlate with performance on neuropsychological tests, suggesting that caudate atrophy may play a role in cognitive symptoms.[124] A measurable rate of change in caudate volume over time was reported, which correlated with age at onset and length of trinucleotide repeat, suggesting that striatal volume loss is a potentially important surrogate marker of HD that may be useful in clinical therapeutic trials.[124,125] More recent studies showed that, as suggested by the clinical picture, extrastriatal degeneration also plays an important role in HD with regionally specific degeneration of the cortex.[126,127,128,129] Patients at earlier or even at preclinical disease stages may demonstrate prominent cortical thinning, suggesting that cortical degeneration may play a role in the clinical symptoms. [126,130,131]

^{1}H-MRS

^{1}H-MRS has been essentially used to assess disease mechanisms and, in rare cases, to monitor the effect of treatment in patients with HD.[123,132,133,134]. These studies mainly found reduced NAA and increased Lac in the striatum, occipital cortex, and frontal cortex. [132,133,134,135] In the very first study,[132] Lac levels were increased in the occipital cortex and basal ganglia of symptomatic HD patients and the Lac level correlated with duration of illness. Subsequently, an elevation of Lac signal was found in the striatum and not in the cortex in a few presymptomatic carriers of the HD gene.[135] However, other studies have failed to observe elevated lactate with ^{1}H-MRS in HD patients. Increased glutamine–glutamate level in the striatum also have been reported,[134,136] supporting the theory of glutamate excitotoxicity in HD.[136]

Overall, the ^{1}H-MRS studies support the notion of the altered energy metabolism in HD. Future studies are needed to determine whether these alterations are causative or secondary measures. The use of ^{1}H-MRS in HD for diagnostic purposes is not recommended. Nevertheless, ^{1}H-MRS can definitely provide surrogate markers, especially with respect to the use of potential therapeutic intervention.[120,123,137] In this context, ^{1}H-MRS should be performed in the basal ganglia at either relatively long TE (135 ms) or short TE (30–35 ms) to allow quantification of Lac or glutamine–glutamate, respectively.

Human prion diseases

The human prion diseases (HPD) are a group of fatal, progressive neurodegenerative disorders occurring in inherited, acquired, and sporadic forms.[138] All of them involve modification of the prion protein (PrP). [139] Familial forms (caused by inherited mutations in the PrP encoding gene) are rare and include familial Creutzfeldt–Jakob disease (CJD), Gerstmann–Straussler disease (GSD), and fatal familial insomnia (FFI). [138,140] Acquired prion diseases include variant CJD, iatrogenic CJD, and kuru. The sporadic forms include the sporadic CJD and the more rare sporadic fatal insomnia.[138] The brain damage observed in HPD is generally characterized by spongiform changes, neuronal damage and loss, astrocytosis, and amyloid plaque formation. This causes severe impairment of brain functions in all disorders (e.g. memory and personality changes as well as psychiatric, movement, and sleep disorders). This inexorably worsens over time, leading to death after a period of severe dementia with inability to move and speak.[138]

At present, the definitive diagnosis of HPD is made through brain biopsy (tonsil biopsy is also useful in the variant CJD). However, in the appropriate clinical context, the characteristic periodic EEG (in sporadic CJD) and/or the positive CSF 14–3–3 protein assay can allow a probable diagnosis.[141,142,143] Historically, imaging features are not part of the diagnostic criteria for HPD. There is, however, growing support for MRI to be included in the diagnostic workup.[3] In this regard, MRI plays an extremely important role in early diagnosis, especially with DWI and FLAIR images, which are undoubtedly the most sensitive for the depiction of prion-induced brain lesions. The lesions are characteristically shown as ribbons of cortical hyperintensity, or basal ganglia or thalamic hyperintensity. The cortical and deep lesions may appear alone or together, and although usually bilateral and symmetric, they may be asymmetric or purely unilateral. When these MRI findings are observed in an appropriate clinical context, the diagnosis of HPD is very likely.[3,144]

^{1}H-MRS

This examination is not included in the diagnostic work-up of HPD and its use in these diseases is presently limited to research studies. However, the use of ^{1}H-MRS could be very useful at early disease stages to follow-up, through the assessment of brain NAA levels, the progression of neuronal loss. Several studies have found significant increases in mI and decreases in NAA in the gray matter of HPD patients,[3,145] related to the pronounced gliosis and neuronal damage.[146] In basal ganglia, these changes are pronounced from early stages, making ^{1}H-MRS very useful for longitudinal assessment of disease evolution or potential treatments.[147] Since the gray matter quantification of both mI and NAA is important, short TE (30–35 ms) acquisitions are preferred in the basal ganglia and neocortical regions.

Amyotrophic lateral sclerosis

Amyotrophic lateral sclerosis (ALS) or "Lou Gehrig's disease" (after the famous American baseball player who was diagnosed with the disorder) is a rare degenerative disorder of motor neurons of the cerebral cortex, brain stem, and spinal cord that results in progressive wasting and paralysis of voluntary muscles.[148] The age of onset is in middle adult life. It progresses rapidly and most of the affected individuals die within 3–5 years from onset of symptoms.[149] The cause of the disease is still elusive, except in familial cases (about 10%) where the most common cause is linked to mutations in the gene encoding cytosolic copper–zinc superoxide dismutase (SOD1, an enzyme responsible for scavenging free radicals). [148] Essential features of ALS are progressive signs and symptoms of lower motor neuron dysfunction. These include focal and multifocal weakness, atrophy, cramps, and fasciculations associated with corticospinal tract signs (spasticity, enhanced, and pathological reflexes) in the absence of sensory findings. Usually, the symptoms initially affect a limb, but in some cases there might be a bulbar onset with difficulties in speaking and swallowing. Regardless of the part of the body first affected by the disease, muscle weakness and atrophy spread to other parts of the body as the disease progresses. No treatment prevents, halts, or reverses the disease, although marginal delay in mortality has been noted recently with the drug riluzole.[150]

ALS is a difficult disease to diagnose as there are many other, more treatable diseases which mimic it. Thus, the diagnosis is primarily based on the symptoms and signs observed by the physician and a series of tests to rule out other diseases. To be diagnosed with ALS, patients must have signs and symptoms of both upper and lower motor neuron damage that cannot be attributed to other causes. Electromyography (EMG) provides objective evidence of lower motor neuron involvement. When the lower motor neuron involvement is severe, the upper motor neuron signs may be masked and its involvement can be missed.[151] In this context, MR methods can be very useful to detect early involvement of upper motor neuron involvement, potentially shortening the time to diagnosis. [152] Recent MRI studies have shown the diagnostic utility of hyperintensity of the corticospinal tract on FLAIR sequences in ALS.[153,154] Other studies, using diffusion images[166] or quantitative morphometry, have provided evidence of abnormalities of extramotor areas, supporting the view that ALS is a multisystem degenerative disease.[155,156,157,158,159]

^{1}H-MRS

Among the recent MR technologies used to improve detection of upper motor neuron involvement, ^{1}H-MRS can provide insights into the metabolic integrity of these pathways. Not surprisingly, in the brain of ALS patients, NAA is decreased in a spatially dependent manner that reflects its pathologic distribution and mI is often increased in the motor cortex.[160,161,162] This has led to the proposal of using the NAA/mI ratio, which would enhance the ability of MRS to distinguish patients with ALS from control subjects. In the future, this might become a useful biomarker in ALS.[152,162] In recent pilot studies, ^{1}H-MRS has been used to monitor whether riluzole does have a mild disease-altering effect.[163,164] The finding of an increase in NAA/Cr ratio in the motor cortex of ALS patients after a brief treatment with riluzole

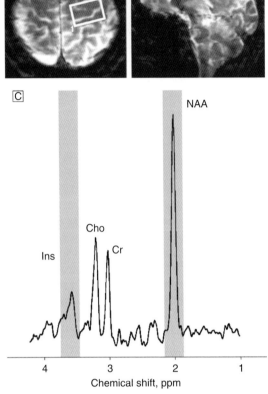

Figure 9.5. Axial (A) and sagittal (B) gradient-echo images demonstrating placement of the VOI in the left motor cortex of a healthy subject. (C) A typical proton magnetic resonance spectrum acquired at 3.0 T from this region is shown on the right. The volume of interest measured 2 × 3 × 2 cm and was centered on and placed parallel to the precentral gyrus to maximize affected tissue content. A stimulated echo acquisition mode sequence was used for single-voxel spectroscopy (TR, 3 s, TE 160 ms, mixing time (TM) 40 ms, 256 averages). These parameters were shown by numerical simulation to be able to detect mI with minimized contributions from glutamate, glutamine, taurine, and macromolecules. [from Kalra et al., *Arch Neurol* 2006; **63**: 1144–8, figure 1, with permission.]

should be confirmed in controlled studies involving larger patient populations.

Since the main goal is to study the regional distribution of metabolites in the corticospinal tract, both single-voxel ^{1}H-MRS and ^{1}H-MRSI can be used (Figure 9.5). The assessment of both mI and NAA should be pursued, therefore short TE (30–35 ms) acquisitions are preferable.

Summary

Presently, ^{1}H-MRS is not used on a routine basis to characterize the metabolic pattern of neurodegenerative diseases, as the diagnostic criteria of these disorders are mainly clinical. However, in conjunction with other quantitative MR techniques, ^{1}H-MRS could allow an early appreciation of the neurological involvement in patients with these disorders and help establish differential diagnoses. Moreover, given the increasing numbers of potential neuroprotective agents available for testing in clinical trials, there is the need for reliable surrogate markers of disease progression and treatment response using neuroimaging technologies. MRS, in combination with other MR modalities, has the potential to contribute to that role.

References

[1] Hsu YY, Du AT, Schuff N, Weiner MW. Magnetic resonance imaging and magnetic resonance spectroscopy in dementias. *J Geriatr Psychiatry Neurol* 2001; **14**: 145–66.

[2] Martin WR. Magnetic resonance imaging and spectroscopy in Parkinson's disease. *Adv Neurol* 2001; **86**: 197–203.

[3] MacFarlane RG, Wroe SJ, Collinge J, Yousry TA, Jager HR. Neuroimaging findings in human prion disease. *J Neurol Neurosurg Psychiatry* 2007; **78**: 664–70.

[4] Kalra S, Arnold D. Neuroimaging in amyotrophic lateral sclerosis. *Amyotroph Lateral Scler Other Motor Neuron Disord* 2003; **4**: 243–8.

[5] Martin WR. MR spectroscopy in neurodegenerative disease. *Mol Imaging Biol* 2007; **9**: 196–203.

[6] Wallin A. Current definition and classification of dementia diseases. *Acta Neurol Scand* 1996; **168**(suppl): 39–44.

[7] Lopez OL, Litvan I, Catt KE, Stowe R, Klunk W, Kaufer DI, et al. Accuracy of four clinical diagnostic criteria for the diagnosis of neurodegenerative dementias. *Neurology* 1999; **53**: 1292–9.

[8] Massoud F, Devi G, Stern Y, Lawton A, Goldman JE, Liu Y, et al. A clinicopathological comparison of community-based and clinic-based cohorts of patients with dementia. *Arch Neurol* 1999; **56**: 1368–73.

[9] Kantarci K. 1H magnetic resonance spectroscopy in dementia. *Br J Radiol* 2007: **80**(Spec No 2): S146–52.

[10] Waldemar G, Dubois B, Emre M, Georges J, McKeith IG, Rossor M, et al. Recommendations for the diagnosis and management of Alzheimer's disease and other disorders associated with dementia: EFNS guideline. *Eur J Neurol* 2007; **14**: e1–26.

[11] Small BJ, Gagnon E, Robinson B. Early identification of cognitive deficits: Preclinical Alzheimer's disease and mild cognitive impairment. *Geriatrics* 2007; **62**: 19–23.

[12] Saxton J, Lopez OL, Ratcliff G, Dulberg C, Fried LP, Carlson MC, et al. Preclinical Alzheimer disease: Neuropsychological test performance 1.5 to 8 years prior to onset. *Neurology* 2004; **63**: 2341–7.

[13] Perneczky R, Pohl C, Sorg C, Hartmann J, Komossa K, Alexopoulos P, et al. Complex activities of daily living in mild cognitive impairment: Conceptual and diagnostic issues. *Age Ageing* 2006; **35**: 240–5.

[14] Forstl H, Kurz A. Clinical features of Alzheimer's disease. *Eur Arch Psychiatry Clin Neurosci* 1999; **249**: 288–90.

[15] Geula C. Abnormalities of neural circuitry in Alzheimer's disease: Hippocampus and cortical cholinergic innervation. *Neurology* 1998; **51**: S18–S29.

[16] Schmitt FA, Davis DG, Wekstein DR, Smith CD, Ashford JW, Markesbery WR. "Preclinical" AD revisited: Neuropathology of cognitively normal older adults. *Neurology* 2000; **55**: 370–6.

[17] Kordower JH, Chu Y, Stebbins GT, DeKosky ST, Cochran EJ, Bennett D, et al. Loss and atrophy of layer II entorhinal cortex neurons in elderly people with mild cognitive impairment. *Ann Neurol* 2001; **49**: 202–13.

[18] Kantarci K, Jack CR, Jr. Quantitative magnetic resonance techniques as surrogate markers of Alzheimer's disease. *Neurorx* 2004; **1**: 196–205.

[19] Heckemann RA, Hammers A, Rueckert D, Aviv RI, Harvey CJ, Hajnal JV. Automatic volumetry on MR brain images can support diagnostic decision making. *BMC Med Imaging* 2008; **8**: 9.

[20] Kloppel S, Stonnington CM, Chu C, Draganski B, Scahill RI, Rohrer JD, et al. Automatic classification of MR scans in Alzheimer's disease. *Brain* 2008; **131**: 681–9.

[21] Fox NC, Crum WR, Scahill RI, Stevens JM, Janssen JC, Rossor MN. Imaging of onset and progression of Alzheimer's disease with voxel-compression mapping of serial magnetic resonance images. *Lancet* 2001; **358**: 201–05.

[22] Pennanen C, Testa C, Laakso MP, Hallikainen M, Helkala EL, Hanninen T, et al. A voxel based morphometry study on mild cognitive impairment. *J Neurol Neurosurg Psychiatry* 2005; **76**: 11–14.

[23] Ashburner J, Csernansky JG, Davatzikos C, Fox NC, Frisoni GB, Thompson PM. Computer-assisted imaging to assess brain structure in healthy and diseased brains. *Lancet Neurol* 2003; **2**: 79–88.

[24] Shonk TK, Moats RA, Gifford P, Michaelis T, Mandigo JC, Izumi J, et al. Probable Alzheimer disease: Diagnosis with proton MR spectroscopy. *Radiology* 1995; **195**: 65–72.

[25] Tedeschi G, Bertolino A, Lundbom N, Bonavita S, Patronas NJ, Duyn JH, et al. Cortical and subcortical chemical pathology in Alzheimer's disease as assessed by multislice proton magnetic resonance spectroscopic imaging. *Neurology* 1996: **47**: 696–704.

[26] Doraiswamy PM, Charles HC, Krishnan KR. Prediction of cognitive decline in early Alzheimer's disease [Letter]. *Lancet* 1998; **352**: 1678.

[27] Schuff N, Capizzano AA, Du AT, Amend DL, O'Neill J, Norman D, et al. Selective reduction of N-acetylaspartate in medial temporal and parietal lobes in AD. *Neurology* 2002; **58**: 928–35.

[28] Kantarci K, Petersen RC, Boeve BF, Knopman DS, Tang-Wai DF, O'Brien PC, et al. 1H MR spectroscopy in common dementias. *Neurology* 2004; **63**: 1393–8.

[29] Rai GS, McConnell JR, Waldman A, Grant D, Chaudry M. Brain proton spectroscopy in dementia: An aid to clinical diagnosis [Letter]. *Lancet* 1999; **353**: 1063–4.

[30] Soher BJ, Doraiswamy PM, Charles HC. A review of 1H MR spectroscopy findings in Alzheimer's disease. *Neuroimaging Clin N Am* 2005; **15**: 847–52, xi.

[31] Valenzuela MJ, Sachdev P. Magnetic resonance spectroscopy in AD. *Neurology* 2001; **56**: 592–8.

[32] MacKay S, Ezekiel F, Di SV, Meyerhoff DJ, Gerson J, Norman D, et al. Alzheimer disease and subcortical ischemic vascular dementia: Evaluation by combining MR imaging segmentation and H-1 MR spectroscopic imaging. *Radiology* 1996; **198**: 537–45.

[33] Schuff N, Amend D, Ezekiel F, Steinman SK, Tanabe J, Norman D, et al. Changes of hippocampal N-acetyl aspartate and volume in Alzheimer's disease. A proton MR spectroscopic imaging and MRI study. *Neurology* 1997; **49**: 1513–21.

[34] Moffett JR, Ross B, Arun P, Madhavarao CN, Namboodiri AM. N-Acetylaspartate in the CNS: From neurodiagnostics to neurobiology. *Prog Neurobiol* 2007; **81**: 89–131.

[35] Mohanakrishnan P, Fowler AH, Vonsattel JP, Husain MM, Jolles PR, Liem P, et al. An in vitro 1H nuclear magnetic resonance study of the temporoparietal cortex of Alzheimer brains. *Exp Brain Res* 1995; **102**: 503–10.

[36] Sweet RA, Panchalingam K, Pettegrew JW, McClure RJ, Hamilton RL, Lopez OL, et al. Psychosis in Alzheimer disease: Postmortem magnetic resonance spectroscopy evidence of excess neuronal and membrane phospholipid pathology. *Neurobiol Aging* 2002; **23**: 547–53.

[37] Miller BL, Moats RA, Shonk T, Ernst T, Woolley S, Ross BD. Alzheimer disease: Depiction of increased cerebral myo-inositol with proton MR spectroscopy. *Radiology* 1993; **187**: 433–7.

[38] Ross BD, BlumI S, Cowan R, Danielsen E, Farrow N, Tan J. In vivo MR spectroscopy of human dementia. *Neuroimaging Clin N Am* 1998; **8**: 809–22.

[39] Chantal S, Braun CM, Bouchard RW, Labelle M, Boulanger Y. Similar 1H magnetic resonance spectroscopic metabolic pattern in the medial temporal lobes of patients with mild cognitive impairment and Alzheimer disease. *Brain Res* 2004; **1003**: 26–35.

[40] Glanville NT, Byers DM, Cook HW, Spence MW, Palmer FB. Differences in the metabolism of inositol and phosphoinositides by cultured cells of neuronal and glial origin. *Biochim Biophys Acta* 1989; **1004**: 169–79.

[41] Kantarci K, Jack CR, Jr., Xu YC, Campeau NG, O'Brien PC, Smith GE, et al. Regional metabolic patterns in mild cognitive impairment and Alzheimer's disease: A 1H MRS study. *Neurology* 2000; **55**: 210–7.

[42] Godbolt AK, Waldman AD, MacManus DG, Schott JM, Frost C, Cipolotti L, et al. MRS shows abnormalities before symptoms in familial Alzheimer disease. *Neurology* 2006; **66**: 718–22.

[43] den Heijer T, Sijens PE, Prins ND, Hofman A, Koudstaal PJ, Oudkerk M, et al. MR spectroscopy of brain white matter in the prediction of dementia. *Neurology* 2006; **66**: 540–4.

[44] Ross BD, Bluml S, Cowan R, Danielsen E, Farrow N, Gruetter R. In vivo magnetic resonance spectroscopy of human brain: The biophysical basis of dementia. *Biophys Chem* 1997; **68**: 161–72.

[45] Chantal S, Labelle M, Bouchard RW, Braun CM, Boulanger Y. Correlation of regional proton magnetic resonance spectroscopic metabolic changes with cognitive deficits in mild Alzheimer disease. *Arch Neurol* 2002; **59**: 955–62.

[46] Kantarci K, Smith GE, Ivnik RJ, Petersen RC, Boeve BF, Knopman DS, et al. 1H magnetic resonance spectroscopy, cognitive function, and apolipoprotein E genotype in normal aging, mild cognitive impairment and Alzheimer's disease. *J Int Neuropsychol Soc* 2002; **8**: 934–42.

[47] Kantarci K, Knopman DS, Dickson DW, Parisi JE, Whitwell JL, Weigand SD, et al. Alzheimer disease: postmortem neuropathologic correlates of antemortem 1H MR spectroscopy metabolite measurements. *Radiology* 2008; **248**: 210–20.

[48] Kantarci K, Weigand SD, Petersen RC, Boeve BF, Knopman DS, Gunter J, et al. Longitudinal 1H MRS changes in mild cognitive impairment and Alzheimer's disease. *Neurobiol Aging* 2007; **28**: 1330–9.

[49] Modrego PJ, Fayed N, Pina MA. Conversion from mild cognitive impairment to probable Alzheimer's disease predicted by brain magnetic resonance spectroscopy. *Am J Psychiatry* 2005; **162**: 667–75.

[50] Krishnan KR, Charles HC, Doraiswamy PM, Mintzer J, Weisler R, Yu X, et al. Randomized, placebo-controlled trial of the effects of donepezil on neuronal markers and hippocampal volumes in Alzheimer's disease. *Am J Psychiatry* 2003; **160**: 2003–11.

[51] Jessen F, Traeber F, Freymann K, Maier W, Schild HH, Block W. Treatment monitoring and response prediction with proton MR spectroscopy in AD. *Neurology* 2006; **67**: 528–30.

[52] Hetherington HP, Pan JW, Mason GF, Adams D, Vaughn MJ, Twieg DB, et al. Quantitative 1H spectroscopic imaging of human brain at 4.1 T using image segmentation. *Magn Reson Med* 1996; **36**: 21–9.

[53] Schuff N, Ezekiel F, Gamst AC, Amend DL, Capizzano AA, Maudsley AA, et al. Region and tissue differences of metabolites in normally aged brain using multislice 1H magnetic resonance spectroscopic imaging. *Magn Reson Med* 2001; **45**: 899–907.

[54] Knopman DS, DeKosky ST, Cummings JL, Chui H, Corey-Bloom J, Relkin N, et al. Practice parameter: Diagnosis of dementia (an evidence-based review). Report of the Quality Standards Subcommittee of the American Academy of Neurology. *Neurology* 2001; **56**: 1143–53.

[55] Small GW, Bookheimer SY, Thompson PM, Cole GM, Huang SC, Kepe V, et al. Current and future uses of neuroimaging for cognitively impaired patients. *Lancet Neurol* 2008; **7**: 161–72.

[56] Moorhouse P, Rockwood K. Vascular cognitive impairment: Current concepts and clinical developments. *Lancet Neurol* 2008; **7**: 246–55.

[57] Nagata K, Saito H, Ueno T, Sato M, Nakase T, Maeda T, et al. Clinical diagnosis of vascular dementia. *J Neurol Sci* 2007; **257**: 44–8.

[58] Wetterling T, Kanitz RD, Borgis KJ. Comparison of different diagnostic criteria for vascular dementia (ADDTC, DSM-IV, ICD-10, NINDS-AIREN). *Stroke* 1996; **27**: 30–6.

[59] Chui HC, Victoroff JI, Margolin D, Jagust W, Shankle R, Katzman R. Criteria for the diagnosis of ischemic vascular dementia proposed by the State of California Alzheimer's Disease Diagnostic and Treatment Centers. *Neurology* 1992; **42**: 473–80.

[60] Roman GC, Tatemichi TK, Erkinjuntti T, Cummings JL, Masdeu JC, Garcia JH, et al. Vascular dementia: Diagnostic criteria for research studies. Report of the NINDS-AIREN International Workshop. *Neurology* 1993; **43**: 250–60.

[61] Holmes C, Cairns N, Lantos P, Mann A. Validity of current clinical criteria for Alzheimer's disease,

vascular dementia and dementia with Lewy bodies. *Br J Psychiatry* 1999; **174**: 45–50.

[62] Victoroff J, Mack WJ, Lyness SA, Chui HC. Multicenter clinicopathological correlation in dementia. *Am J Psychiatry* 1995; **152**: 1476–84.

[63] Chui HC. Subcortical ischemic vascular dementia. *Neurol Clin* 2007; **25**: 717–40, vi.

[64] van Straaten EC, Scheltens P, Barkhof F. MRI and CT in the diagnosis of vascular dementia. *J Neurol Sci* 2004; **226**: 9–12.

[65] Guermazi A, Miaux Y, Rovira-Canellas A, Suhy J, Pauls J, Lopez R, et al. Neuroradiological findings in vascular dementia. *Neuroradiology* 2007; **49**: 1–22.

[66] Mori E, Ishii K, Hashimoto M, Imamura T, Hirono N, Kitagaki H. Role of functional brain imaging in the evaluation of vascular dementia. *Alzheimer Dis Assoc Disord* 1999; **13**(Suppl 3): S91–101.

[67] Hsu YY, Schuff N, Amend DL, Du AT, Norman D, Chui HC, et al. Quantitative magnetic resonance imaging differences between Alzheimer disease with and without subcortical lacunes. *Alzheimer Dis Assoc Disord* 2002; **16**: 58–64.

[68] Pantoni L, Garcia JH. The significance of cerebral white matter abnormalities 100 years after Binswanger's report. A review. *Stroke* 1995; **26**: 1293–301.

[69] Hunt AL, Orrison WW, Yeo RA, Haaland KY, Rhyne RL, Garry PJ, et al. Clinical significance of MRI white matter lesions in the elderly. *Neurology* 1989; **39**: 1470–4.

[70] Meyer JS, Kawamura J, Terayama Y. White matter lesions in the elderly. [Review]. *J Neurol Sci* 1992; **110**: 1–7.

[71] Schmidt R, Hayn M, Fazekas F, Kapeller P, Esterbauer H. Magnetic resonance imaging white matter hyperintensities in clinically normal elderly individuals. Correlations with plasma concentrations of naturally occurring antioxidants. *Stroke* 1996; **27**: 2043–7.

[72] Malloy P, Correia S, Stebbins G, Laidlaw DH. Neuroimaging of white matter in aging and dementia. *Clin Neuropsychol* 2007; **21**: 73–109.

[73] Urresta FL, Medina DA, Gaviria M. Diffusion MRI studies in vascular cognitive impairment and dementia. *Rev Bras Psiquiatr* 2003; **25**: 188–91.

[74] Hanyu H, Imon Y, Sakurai H, Iwamoto T, Takasaki M, Shindo H, et al. Regional differences in diffusion abnormality in cerebral white matter lesions in patients with vascular dementia of the Binswanger type and Alzheimer's disease. *Eur J Neurol* 1999; **6**: 195–203.

[75] Kattapong VJ, Brooks WM, Wesley MH, Kodituwakku PW, Rosenberg GA. Proton magnetic resonance spectroscopy of vascular- and Alzheimer-type dementia. *Arch Neuro* 1996; **53**: 678–80.

[76] MacKay S, Meyerhoff DJ, Constans JM, Norman D, Fein G, Weiner MW. Regional gray and white matter metabolite differences in subjects with AD, with subcortical ischemic vascular dementia, and elderly controls with 1H magnetic resonance spectroscopic imaging. *Arch Neurol* 1996; **53**: 167–74.

[77] Auer DP, Schirmer T, Heidenreich JO, Herzog J, Putz B, Dichgans M. Altered white and gray matter metabolism in CADASIL: A proton MR spectroscopy and 1H-MRSI study. *Neurology* 2001; **56**: 635–42.

[78] McKhann GM, Albert MS, Grossman M, Miller B, Dickson D, Trojanowski JQ. Clinical and pathological diagnosis of frontotemporal dementia: Report of the Work Group on Frontotemporal Dementia and Pick's Disease. *Arch Neurol* 2001; **58**: 1803–09.

[79] Boxer AL, Miller BL. Clinical features of frontotemporal dementia. *Alzheimer Dis Assoc Disord* 2005; **19**(Suppl 1): S3–6.

[80] Knibb JA, Kipps CM, Hodges JR. Frontotemporal dementia. *Curr Opin Neurol* 2006; **19**: 565–71.

[81] Rosen HJ, Gorno-Tempini MI, Goldman WP, Perry RJ, Schuff N, Weiner M, et al. Patterns of brain atrophy in frontotemporal dementia and semantic dementia. *Neurology* 2002; **58**: 198–208.

[82] Rosen HJ, Kramer JH, Gorno-Tempini MI, Schuff N, Weiner M, Miller BL. Patterns of cerebral atrophy in primary progressive aphasia. *Am J Geriatr Psychiatry* 2002; **10**: 89–97.

[83] Josephs KA, Whitwell JL, Jack CR, Parisi JE, Dickson DW. Frontotemporal lobar degeneration without lobar atrophy. *Arch Neurol* 2006; **63**: 1632–8.

[84] Whitwell JL, Jack CR, Jr., Baker M, Rademakers R, Adamson J, Boeve BF, et al. Voxel-based morphometry in frontotemporal lobar degeneration with ubiquitin-positive inclusions with and without progranulin mutations. *Arch Neurol* 2007; **64**: 371–6.

[85] Rabinovici GD, Seeley WW, Kim EJ, Gorno-Tempini MI, Rascovsky K, Pagliaro TA, et al. Distinct MRI atrophy patterns in autopsy-proven Alzheimer's disease and frontotemporal lobar degeneration. *Am J Alzheimers Dis Other Demen* 2007; **22**: 474–88.

[86] Coulthard E, Firbank M, English P, Welch J, Birchall D, O'Brien J, et al. Proton magnetic resonance spectroscopy in frontotemporal dementia. *J Neurol* 2006; **253**: 861–8.

[87] McKeith IG. Consensus guidelines for the clinical and pathologic diagnosis of dementia with Lewy bodies (DLB): Report of the Consortium on DLB International Workshop. *J Alzheimers Dis* 2006; **9**: 417–23.

[88] Gomez-Tortosa E, Irizarry MC, Gomez-Isla T, Hyman BT. Clinical and neuropathological correlates of dementia with Lewy bodies. *Ann NY Acad Sci* 2000; **920**: 9–15.

[89] Gomez-Isla T, Growdon WB, McNamara M, Newell K, Gomez-Tortosa E, Hedley-Whyte ET, et al. Clinicopathologic correlates in temporal cortex in dementia with Lewy bodies. *Neurology* 1999; **53**: 2003–09.

[90] McKeith I, Mintzer J, Aarsland D, Burn D, Chiu H, Cohen-Mansfield J, et al. Dementia with Lewy bodies. *Lancet Neurol* 2004; **3**: 19–28.

[91] Geser F, Wenning GK, Poewe W, McKeith I. How to diagnose dementia with Lewy bodies: State of the art. *Mov Disord* 2005; **20**(Suppl 12): S11–20.

[92] Whitwell JL, Weigand SD, Shiung MM, Boeve BF, Ferman TJ, Smith GE, et al. Focal atrophy in dementia with Lewy bodies on MRI: A distinct pattern from Alzheimer's disease. *Brain* 2007; **130**: 708–19.

[93] Burton EJ, Karas G, Paling SM, Barber R, Williams ED, Ballard CG, et al. Patterns of cerebral atrophy in dementia with Lewy bodies using voxel-based morphometry. *Neuroimage* 2002; **17**: 618–30.

[94] Jankovic J. Parkinson's disease: Clinical features and diagnosis. *J Neurol Neurosurg Psychiatry* 2008; **79**: 368–76.

[95] Bjorklund A, Dunnett SB. Dopamine neuron systems in the brain: An update. *Trends Neurosci* 2007; **30**: 194–202.

[96] Masliah E, Rockenstein E, Veinbergs I, Mallory M, Hashimoto M, Takeda A, et al. Dopaminergic loss and inclusion body formation in alpha-synuclein mice: Implications for neurodegenerative disorders. *Science* 2000; **287**: 1265–9.

[97] Gerlach M, Double K, Riederer P, Hirsch E, Jellinger K, Jenner P, et al. Iron in the Parkinsonian substantia nigra. *Mov Disord* 1997; **12**: 258–60.

[98] Hsu LJ, Sagara Y, Arroyo A, Rockenstein E, Sisk A, Mallory M, et al. alpha-synuclein promotes mitochondrial deficit and oxidative stress. *Am J Pathol* 2000; **157**: 401–10.

[99] Jenner P, Olanow CW. The pathogenesis of cell death in Parkinson's disease. *Neurology* 2006; **66**: S24–S36.

[100] Tolosa E, Wenning G, Poewe W. The diagnosis of Parkinson's disease. *Lancet Neurol* 2006; **5**: 75–86.

[101] Tintner R, Jankovic J. Treatment options for Parkinson's disease. *Curr Opin Neurol* 2002; **15**: 467–76.

[102] Limousin P, Martinez-Torres I. Deep brain stimulation for Parkinson's disease. *Neurotherapeutics* 2008; **5**: 309–19.

[103] Rao G, Fisch L, Srinivasan S, D'Amico F, Okada T, Eaton C, et al. Does this patient have Parkinson disease? *J Am Med Assoc* 2003; **289**: 347–53.

[104] Hughes AJ, Daniel SE, Ben-Shlomo Y, Lees AJ. The accuracy of diagnosis of parkinsonian syndromes in a specialist movement disorder service. *Brain* 2002; **125**: 861–70.

[105] Litvan I, Bhatia KP, Burn DJ, Goetz CG, Lang AE, McKeith I, et al. Movement Disorders Society Scientific Issues Committee report: SIC Task Force appraisal of clinical diagnostic criteria for Parkinsonian disorders. *Mov Disord* 2003; **18**: 467–86.

[106] Seppi K, Schocke MF. An update on conventional and advanced magnetic resonance imaging techniques in the differential diagnosis of neurodegenerative parkinsonism. *Curr Opin Neurol* 2005; **18**: 370–5.

[107] Savoiardo M. Differential diagnosis of Parkinson's disease and atypical parkinsonian disorders by magnetic resonance imaging. *Neurol Sci* 2003; **24** (Suppl 1): S35–7.

[108] Brenneis C, Seppi K, Schocke MF, Muller J, Luginger E, Bosch S, et al. Voxel-based morphometry detects cortical atrophy in the Parkinson variant of multiple system atrophy. *Mov Disord* 2003; **18**: 1132–8.

[109] Lee EA, Cho HI, Kim SS, Lee WY. Comparison of magnetic resonance imaging in subtypes of multiple system atrophy. *Parkinsonism Relat Disord* 2004; **10**: 363–8.

[110] Seppi K, Schocke MF, Wenning GK, Poewe W. How to diagnose MSA early: The role of magnetic resonance imaging. *J Neural Transm* 2005; **112**: 1625–34.

[111] Quattrone A, Nicoletti G, Messina D, Fera F, Condino F, Pugliese P, et al. MR imaging index for differentiation of progressive supranuclear palsy from Parkinson disease and the Parkinson variant of multiple system atrophy. *Radiology* 2008; **246**: 214–21.

[112] Davie C. The role of spectroscopy in parkinsonism [Editorial; Comment]. *Mov Disord* 1998; **13**: 2–4.

[113] Clarke CE, Lowry M. Systematic review of proton magnetic resonance spectroscopy of the striatum in parkinsonian syndromes. *Eur J Neurol* 2001; **8**: 573–7.

[114] Oz G, Terpstra M, Tkac I, Aia P, Lowary J, Tuite PJ, et al. Proton MRS of the unilateral substantia nigra in the human brain at 4 tesla: Detection of high GABA concentrations. *Magn Reson Med* 2006; **55**: 296–301.

[115] Davie CA, Wenning GK, Barker GJ, Tofts PS, Kendall BE, Quinn N, et al. Differentiation of multiple system atrophy from idiopathic Parkinson's disease using proton magnetic resonance spectroscopy. *Ann Neurol* 1995; **37**: 204–10.

[116] Federico F, Simone IL, Lucivero V, Mezzapesa DM, De MM, Lamberti P, et al. Usefulness of proton magnetic resonance spectroscopy in differentiating parkinsonian syndromes. *Ital J Neurol Sci* 1999; **20**: 223–9.

[117] Clarke CE, Lowry M. Basal ganglia metabolite concentrations in idiopathic Parkinson's disease and multiple system atrophy measured by proton magnetic resonance spectroscopy. *Eur J Neurol* 2000; **7**: 661–5.

[118] Watanabe H, Fukatsu H, Katsuno M, Sugiura M, Hamada K, Okada Y, et al. Multiple regional 1H-MR spectroscopy in multiple system atrophy: NAA/Cr reduction in pontine base as a valuable diagnostic marker. *J Neurol Neurosurg Psychiatry* 2004; **75**: 103–09.

[119] Tedeschi G, Litvan I, Bonavita S, Bertolino A, Lundbom N, Patronas NJ, et al. Proton magnetic resonance spectroscopic imaging in progressive supranuclear palsy, Parkinson's disease and corticobasal degeneration. *Brain* 1997; **120**: 1541–52.

[120] Ross BD, Hoang TQ, Bluml S, Dubowitz D, Kopyov OV, Jacques DB, et al. In vivo magnetic resonance spectroscopy of human fetal neural transplants. *NMR Biomed* 1999; **12**: 221–36.

[121] Walker FO. Huntington's disease. *Lancet* 2007; **20**: 218–28.

[122] Ramaswamy S, Shannon KM, Kordower JH. Huntington's disease: Pathological mechanisms and therapeutic strategies. *Cell Transplant* 2007; **16**: 301–12.

[123] Rosas HD, Feigin AS, Hersch SM. Using advances in neuroimaging to detect, understand, and monitor disease progression in Huntington's disease. *Neurorx* 2004; **1**: 263–72.

[124] Aylward EH. Change in MRI striatal volumes as a biomarker in preclinical Huntington's disease. *Brain Res Bull* 2007; **72**: 152–8.

[125] Aylward EH, Codori AM, Rosenblatt A, Sherr M, Brandt J, Stine OC, et al. Rate of caudate atrophy in presymptomatic and symptomatic stages of Huntington's disease. *Mov Disord* 2000; **15**: 552–60.

[126] Rosas HD, Liu AK, Hersch S, Glessner M, Ferrante RJ, Salat DH, et al. Regional and progressive thinning of the cortical ribbon in Huntington's disease. *Neurology* 2002; **58**: 695–701.

[127] Gomez-Anson B, Alegret M, Munoz E, Monte GC, Alayrach E, Sanchez A, et al. Prefrontal cortex volume reduction on MRI in preclinical Huntington's disease relates to visuomotor performance and CAG number. *Parkinsonism Relat Disord* 2008; **15**: 213–9.

[128] Wolf RC, Vasic N, Schonfeldt-Lecuona C, Ecker D, Landwehrmeyer GB. Cortical dysfunction in patients with Huntington's disease during working memory performance. *Hum Brain Mapp* 2009; **30**: 327–39.

[129] Jech R, Klempir J, Vymazal J, Zidovska J, Klempirova O, Ruzicka E, et al. Variation of selective gray and white matter atrophy in Huntington's disease. *Mov Disord* 2007; **22**: 1783–9.

[130] Rosas HD, Hevelone ND, Zaleta AK, Greve DN, Salat DH, Fischl B. Regional cortical thinning in preclinical Huntington disease and its relationship to cognition. *Neurology* 2005; **65**: 745–7.

[131] Rosas HD, Salat DH, Lee SY, Zaleta AK, Pappu V, Fischl B, et al. Cerebral cortex and the clinical expression of Huntington's disease: Complexity and heterogeneity. *Brain* 2008; **131**: 1057–68.

[132] Jenkins BG, Koroshetz WJ, Beal MF, Rosen BR. Evidence for impairment of energy metabolism in vivo in Huntington's disease using localized 1H NMR spectroscopy. *Neurology* 1993; **43**: 2689–95.

[133] Koroshetz WJ, Jenkins BG, Rosen BR, Beal MF. Energy metabolism defects in Huntington's disease and effects of coenzyme Q10. *Ann Neurol* 1997; **41**: 160–5.

[134] Davie CA, Barker GJ, Quinn N, Tofts PS, Miller DH. Proton MRS in Huntington's disease [Letter; Comment]. *Lancet* 1994; **343**: 1580.

[135] Jenkins BG, Rosas HD, Chen YC, Makabe T, Myers R, MacDonald M, et al. 1H NMR spectroscopy studies of Huntington's disease: Correlations with CAG repeat numbers. *Neurology* 1998; **50**: 1357–65.

[136] Taylor-Robinson SD, Weeks RA, Bryant DJ, Sargentoni J, Marcus CD, Harding AE, et al. Proton magnetic resonance spectroscopy in Huntington's disease: Evidence in favour of the glutamate excitotoxic theory. *Mov Disord* 1996; **11**: 167–73.

[137] Ross BD. Re: Long-term fetal cell transplant in Huntington disease: Stayin' alive. *Neurology* 2008; **70**: 815–6.

[138] McKintosh E, Tabrizi SJ, Collinge J. Prion diseases. *J Neurovirol* 2003; **9**: 183–93.

[139] Prusiner SB. The prion diseases. *Brain Pathol* 1998; **8**: 499–513.

[140] Collinge J. Inherited prion diseases. *Adv Neurol* 1993; **61**: 155–65.

[141] Collins S, Boyd A, Fletcher A, Gonzales MF, McLean CA, Masters CL. Recent advances in the pre-mortem diagnosis of Creutzfeldt–Jakob disease. *J Clin Neurosci* 2000; **7**: 195–202.

[142] Wieser HG, Schindler K, Zumsteg D. EEG in Creutzfeldt–Jakob disease. *Clin Neurophysiol* 2006; **117**: 935–51.

[143] Green AJ. Cerebrospinal fluid brain-derived proteins in the diagnosis of Alzheimer's disease and

Creutzfeldt–Jakob disease. *Neuropathol Appl Neurobiol* 2002; **28**: 427–40.

[144] Wada R, Kucharczyk W. Prion infections of the brain. *Neuroimaging Clin N Am* 2008; **18**: 183–91.

[145] Behar KL, Boucher R, Fritch W, Manuelidis L. Changes in *N*-acetylaspartate and myo-inositol detected in the cerebral cortex of hamsters with Creutzfeldt–Jakob disease. *Magn Reson Imag* 1998; **16**: 963–8.

[146] Waldman AD, Cordery RJ, MacManus DG, Godbolt A, Collinge J, Rossor MN. Regional brain metabolite abnormalities in inherited prion disease and asymptomatic gene carriers demonstrated in vivo by quantitative proton magnetic resonance spectroscopy. *Neuroradiology* 2006; **48**: 428–33.

[147] Stewart LA, Rydzewska LH, Keogh GF, Knight RS. Systematic review of therapeutic interventions in human prion disease. *Neurology* 2008; **70**: 1272–81.

[148] Mitchell JD, Borasio GD. Amyotrophic lateral sclerosis. *Lancet* 2007; **369**: 2031–41.

[149] Lomen-Hoerth C. Amyotrophic lateral sclerosis from bench to bedside. *Semin Neurol* 2008; **28**: 205–11.

[150] Meininger V, Lacomblez L, Salachas F. What has changed with riluzole? *J Neurol* 2000; **247**: 19–22.

[151] Rowland LP. Diagnosis of amyotrophic lateral sclerosis. *J Neurol Sci* 1998; **160**(Suppl 1): S6–24.

[152] Cudkowicz M, Qureshi M, Shefner J. Measures and markers in amyotrophic lateral sclerosis. *Neurorx* 2004; **1**: 273–83.

[153] Abe K. MRI in ALS: Corticospinal tract hyperintensity. *Neurology* 2004; **63**: 596–7.

[154] Zhang L, Ulug AM, Zimmerman RD, Lin MT, Rubin M, Beal MF. The diagnostic utility of FLAIR imaging in clinically verified amyotrophic lateral sclerosis. *J Magn Reson Imag* 2003; **17**: 521–7.

[155] Yin H, Cheng SH, Zhang J, Ma L, Gao Y, Li D, *et al.* Corticospinal tract degeneration in amyotrophic lateral sclerosis: A diffusion tensor imaging and fibre tractography study. *Ann Acad Med Singapore* 2008; **37**: 411–5.

[156] Sage CA, Peeters RR, Gorner A, Robberecht W, Sunaert S. Quantitative diffusion tensor imaging in amyotrophic lateral sclerosis. *Neuroimage* 2007; **34**: 486–99.

[157] Kassubek J, Unrath A, Huppertz HJ, Lule D, Ethofer T, Sperfeld AD, *et al.* Global brain atrophy and corticospinal tract alterations in ALS, as investigated by voxel-based morphometry of 3-D MRI. *Amyotroph Lateral Scler Other Motor Neuron Disord* 2005; **6**: 213–20.

[158] Cabello JP, Riverol M, Masdeu JC. ALS corticospinal degeneration on DWI. *Neurology* 2004; **62**: 1834.

[159] Mezzapesa DM, Ceccarelli A, Dicuonzo F, Carella A, De Caro MF, Lopez M, *et al.* Whole-brain and regional brain atrophy in amyotrophic lateral sclerosis. *Am J Neuroradiol* 2007; **28**: 255–9.

[160] Pioro EP, Antel JP, Cashman NR, Arnold DL. Detection of cortical neuronal loss in motor neuron disease by proton magnetic resonance spectroscopic imaging in vivo. *Neurology* 1994; **44**: 1933–8.

[161] Kalra S, Arnold DL. ALS surrogate markers. MRS. *Amyotroph Lateral Scler Other Motor Neuron Disord* 2004; **5**(Suppl 1): 111–4.

[162] Kalra S, Hanstock CC, Martin WR, Allen PS, Johnston WS. Detection of cerebral degeneration in amyotrophic lateral sclerosis using high-field magnetic resonance spectroscopy. *Arch Neurol* 2006; **63**: 1144–8.

[163] Kalra S, Tai P, Genge A, Arnold DL. Rapid improvement in cortical neuronal integrity in amyotrophic lateral sclerosis detected by proton magnetic resonance spectroscopic imaging. *J Neurol* 2006; **253**: 1060–3.

[164] Kalra S, Cashman NR, Genge A, Arnold DL. Recovery of *N*-acetylaspartate in corticomotor neurons of patients with ALS after riluzole therapy. *Neuroreport* 1998; **9**: 1757–61.

[165] Dubois B, Feldman HH, Jacova C, *et al.* Research criteria for the diagnosis of Alzheimer's disease: Revising the NINCDS-ADRDA criteria. *Lancet Neurol* 2007; **6**: 734–46.

[166] Sach M, Winkler G, Glauche V, Liepert J, Heimbach B, Koch MA, *et al.* Diffusion tensor MRI of early upper motor neuron involvement in amyotrophic lateral sclerosis. *Brain* 2004; **127**: 340–50.

Chapter 10
MRS in traumatic brain injury

Key points

- TBI is a major cause of morbidity in young adults and children.
- Low levels of NAA and, if seen, increased lactate, in the early stage of injury are prognostic of poor outcome.
- Other common metabolic abnormalities in TBI (most of which also correlate with poor outcome) include increased levels of choline, *myo*-inositol, and glutamate plus glutamine.
- Metabolic abnormalities are observed with MRS in regions of the brain with normal appearance in conventional MRI.
- MRI and MRS are difficult to perform in acutely ill TBI patients: MRS may be more feasible in mild TBI patients for the purpose of predicting long-term cognitive deficits.
- The role of MRS in guiding TBI therapy is unknown.
- The comparative value of MRS compared to other advanced imaging modalities remains to be determined.

Introduction

Traumatic brain injury (TBI) is a leading cause of death and lifelong disability among children and young adults across the developed world. TBI is estimated to result in greater than $60 billion in direct and indirect annual costs due to health care and work loss disability.[1] The Centers for Disease Control and Prevention (CDC) estimate that each year approximately 1.4 million Americans survive a TBI, among whom approximately 235,000 are hospitalized.[2] Approximately 80,500 new TBI survivors are left each year with residual deficits consequent to their injury, which lead to long-term disabilities that may or may not be improved through rehabilitation. In 2001, 157,708 people died from acute traumatic injury, which accounted for about 6.5% of all deaths in the United States. Teens, young adults, and people over 75, especially males, were far more likely than others to die of traumatic brain injury (NINDS study[3]). The CDC's National Center for Injury Prevention and Control estimates that 5.3 million US citizens (2% of the population) are living with disability as a result of a traumatic brain injury.[4]

Motor vehicle accidents and falls are among the top reasons for TBI.[4] These type of injuries involve impacts that cause sudden acceleration–deceleration type forces leading to linear, rotational, or angular shearing injuries at the gray–white matter boundaries.[5,6,7] The tissue boundaries are especially susceptible because tissues with different physical properties have different momentum, and this difference causes shearing injury of the nerve fibers and tiny vessels along the borderline, tangential to the plane of shift. Such shear mechanisms may cause strain injury to the crossing white matter fibers, and are also responsible for ruptured veins typically seen between the skull bone and the brain leading to sub- or epidural hematoma. The extent of TBI can be quite variable and may also be influenced by the dural folds and the internal structure of the skull.[8,9,10] For example, the sphenoid ridges may provide significant compression on the temporal pole, which in turn could affect numerous structures including the amygdala, hippocampus, and other limbic structures that could lead to deficits in memory, emotion, and other cognitive functions.[11,12]. The injuries may occur on the same side as the impact or could be in areas of the brain opposite to the impact (countercoup) depending upon whether it was an acceleration or deceleration type of injury.[13] The injuries may range from primary axotomy, where axonal connectivity is severed, to a minor disruption in the axoplasma membrane. The effects of such physical insults can be immediate or manifest over several days following the initial injury. Similarly, the sequelae of the injury can be quite varied and depend on the severity and the location of the injury. Secondary injury to the brain includes brain swelling, ischemia, cerebral hypotension, edema, elevated intracranial pressure, and altered metabolism.[14] It has been reported that the cholinergic system is adversely affected following TBI; there is an elevation of acetylcholine in the brain and CSF among patients following head injury, and this is attributed to the reduced ability of acetylcholine binding at the cholinergic receptors.[15,16] It has also been suggested that glutamate may play a role in the pathogenesis of focal contusion. The resting membrane potentials of the cells are altered immediately following TBI due to glutamate release, which in turn increases the intracellular free calcium, ultimately leading to excitotoxic death.[17,18,19] Focal TBI can lead to a complex set of events including vascular dysregulation, edema, ischemia, inflammation, plasma membrane disruption, excitatory neurotransmitter release, mitochondrial dysfunction, reactive oxygen species production, increase in anaerobic metabolism, and lactic acidosis. If left unchecked, all of the above conditions may lead to necrotic or apoptotic cell death and long-term cognitive impairment.[20,21]

Most of the post-concussive syndromes resolve within the first few months following injury. However, continued impairment is exhibited by a significant portion of those injured with estimates as high as 50–60% with wide-ranging symptoms including fatigue, attention disorder, headache, poor memory, language disorder, visuo-perceptual disorder, depressed executive function, difficulty with socializing, lack of motivation, and depression.[22,23,24,25,26]. The effects of TBI may be more profound among the pediatric population, especially those below the age of 10, who have a significantly higher risk of cognitive dysfunction, poor academic performance, and development disorders.[27] Longitudinal studies on young children injured at kindergarten age have shown that brain plasticity may not benefit when diffuse axonal injury (DAI) is involved.[28] It is widely believed that while the pathophysiological changes following TBI are largely similar to that of the adult brain, incompletely myelinated neurons in the younger brain may be more vulnerable to traumatic axonal injury and therefore are at a higher risk for developing long-term effects from the injury.[29,30,31]

Classification of head injury

Head injury is classified in a number ways, including the level of severity, level of consciousness, mental status following head injury, or location of body injury. Typically, upon presentation, an Abbreviated Injury Scale (AIS) is used to assign a seven-digit code that is based on the anatomical system, but it also scores for severity and is based on the seriousness of the lesion and its effect upon mental state.[32] A significant portion of the traumatic brain injury patients suffer DAI. Shearing of deep subcortical white matter tracts is considered to be one of the common components of DAI, particularly in cases where the sudden acceleration–deceleration forces and/or rotational forces may result in axonal stretching, especially at the white matter and gray matter interface, resulting in disruption and eventual destruction of the nerve fibers. Head trauma can be roughly divided into mild, moderate, and severe. The Glasgow Coma Scale (GCS) is widely used to classify degree and severity of head injury, based on motor responses, verbal responses, and eye opening that are assessed independently following the injury.[33] For example, a GCS score between 13 and 15 would be considered mild injury, 9–12 as moderate injury and 3–8 as severe injury. The GCS was initially described to consistently assess the depth and duration of altered consciousness and coma in individual patients after TBI. The GCS is widely accepted because it is practical to use and can be frequently repeated in a wide variety of clinical scenarios by medical staff without special training. Computed tomography (CT) is a good first-line diagnostic tool that classifies the patients into three categories: (a) those with normal intracranial structures, (b) those with focal intra- and extra-axial hematomas, and (c) those that depict a more diffuse pattern of parenchymal injury.[34,35,36,37] However, the role of CT for patients with minor head injury is controversial, as a very low percentage (<0.1%) of these patients will have lesions that would require surgical intervention.[38] Further, CT is not sensitive to subtle injury among patients that have a low GCS,[39] which may be visible on MRI.

MRI in the diagnosis of TBI

MRI is extremely sensitive for the detection of both hemorrhagic and non-hemorrhagic cortical contusions.[40] MRI is suitable for monitoring edema, midline shift, focal atrophy, and the changing status of a hemorrhage and for evaluating lesions that may underlie post-traumatic epilepsy.[41,42,43] The structural images from MR in the subacute state are also helpful as a correlate to the patient's clinical status.

Although most of the pathological changes occur within the first 6 months following injury, brain volume stabilization to normal volumes can be as long as 3 years to be equal to that of a normally aging brain. [44] It should be noted that in a significant number of cases where the patients are in coma or with a very low GCS score (< 5), it is possible that no abnormalities are seen in the CT or conventional MR during admission. The degenerative changes that could occur from such injuries begin to appear on MR a few days following the injury while they are still CT-occult, suggesting that MR is more sensitive in the subacute state and chronic phases compared to CT. MRI is superior to CT scanning in the detection of DAI.[45] DAI represents about 48% of all primary lesions seen in head trauma and is considered to be one of the most common causes of poor clinical outcomes. Most lesions in DAI are non-hemorrhagic, but up to about 20% may contain a small amount of hemorrhage. Typically lobar white matter, corpus callosum, dorsolateral aspect of the upper brain stem, and internal capsule are frequently involved in DAI.[46,47,48]

A typical MRI protocol for TBI may consist of T_1-weighted images (useful for establishing the presence of focal atrophy), T_2-weighted and FLAIR images for visualizing lesions in parenchyma, CSF, and ventricular hemorrhage, and T_2^*-weighted gradient echo image is often used to detect hemorrhage and hemosiderin deposits. High-resolution 3D susceptibility-weighted imaging (SWI), is particularly sensitive to micro-hemorrhages.[49] Diffusion-weighted imaging or diffusion tensor imaging may also be used due to its sensitivity to detect white matter abnormalities.

While MR is very sensitive in detecting and assessing the extent of the injury its usefulness in predicting outcome is not entirely clear. Typically, 80–90% of patients with TBI will present with mild head trauma, while the remainder split evenly between the moderate and severe injury category as defined by the GCS. While the GCS broadly correlates with the Glasgow Outcome Scale (GOS) and the Disability Rating Scale, the heterogeneous nature of the injury results in a wide variation in eventual disability and functional impairment.[50] Although attempts have been made to predict outcomes, since the brain controls multitude facets of human behavior and due to the complex and varied nature of the types of brain injury, it is very difficult to come to a convincing conclusion on the prognostic value of MR imaging. The problem is further compounded when factors such as age, sex, education, cognitive abilities, presence of secondary injury effects, and lesion locations are taken into consideration. Nevertheless, even after taking all these factors into consideration, there is a general consensus that the greater the structural damage, the greater the potential for neuropsychiatric morbidity.[51] MRI studies performed at early and delayed time points can provide useful information with regard to the severity and clinical outcome of patients following TBI. Typically studies have found that late imaging, and not early imaging, correlates with the patient's clinical outcome.[52]

Magnetic resonance spectroscopy

Metabolic studies using nuclear medicine techniques have shown that, immediately following TBI, the brain undergoes a perturbation in the resting state metabolism.[53,54,55,56,57] The energy requirements are increased and the demand for cerebellar glucose is increased in the acute phase after TBI with a concomitant return to normal levels or even a slightly depressed level of glucose utilization. Other studies have also shown that during the same period the oxygen consumption is reduced while maintaining glucose metabolism and that the level of oxygen utilization relates to the final outcome of the patients. [53,54,55,58] At least one study has shown that shunting of metabolite substrate may happen through alternate pathways such as the pentose phosphate pathway, besides glycolysis and glucose oxidation to account for the energy crisis following TBI.[59] A metabolomic study on an animal model of TBI using lateral fluid percussion showed regional changes in brain metabolites in the hippocampus and the cortex 1 h after injury when compared to controls and sham-control surgery. Oxidative stress as determined by decreases in ascorbate, excitotoxic damage as evidenced by decreases in glutamate levels, membrane disruption due to decreases in phosphocholine and glycerophosphocholine, and neuronal injury due to the decrease in N-acetylaspartate were observed within the first hour after TBI in the injured animals was observed compared to the sham-control animals.[60] Although some evidence exists from other experimental TBI studies regarding the reduction of Cho during the initial stages of TBI,[61] almost all human studies report an increase in Cho in the subacute to chronic stage following TBI.[62,63,64,65]

Pediatric TBI

Metabolic changes and prediction of outcome

One of the first studies of pediatric brain injury demonstrated that short echo-time single voxel MRS (TE 20 ms using STEAM localization) could predict the 6–12 month outcome on a relatively large number of pediatric patients ($n = 83$).[66] Spectra obtained from an 8 cm^3 single voxel placed in the occipital gray matter area showed that elevated lactate and decreased NAA/Cr and NAA/Cho ratios were prevalent among patients with a poor clinical outcome at 6–12 months as determined by the GOS. Accuracy was 91% for neonates (≤ 1 months), and 100% in the case of infants (1–18 months) and children (≥ 18 months) when information from MRS was combined with MR imaging and clinical data. This is in contrast to an accuracy of 83%, 84%, and 93%, respectively, when using only MR imaging and the clinical data. However, one weakness of this study was that the patient group included a wide range of pathologies in addition to those with TBI.

Another study looked at the predictive value of MRS on 53 pediatric patients comprising both infants (1–18 months) and children with closed head injury that included both accidental and non-accidental trauma by comparing the MRS findings in the occipital gray matter at the acute stage with their 6–12 month outcomes.[62] Patients with poor outcomes had lower NAA/Cr and NAA/Cho, and a higher Cho/Cr, ratios. Lactate levels were also elevated in 91% of patients with poor outcome. It should be noted that spectroscopic sampling was not necessarily performed at the site of the injury, suggesting widespread metabolic changes may be present at the acute stage of closed head injury. Since the main predictive value is based on metabolites such as NAA and lactate, short or long echo time sequences have generally been found to be equally effective in predicting outcome in children.[67] However, long echo times were found to be more effective in predicting outcomes for neonates, and were in general more effective in detecting lactate among all age groups. The use of long echo time MRS (e.g. TE 140 or 280 msec, see Chapter 3) is helpful as it avoids interference from short T_2 components (especially lipids) and allows for a more accurate quantification of the lactate peak (Figure 10.1). Elevated lactate levels may serve as an early predictor of clinical outcome in non-accidental TBI, such as in the case of shaken baby syndrome.[68]

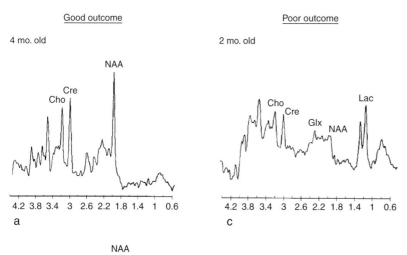

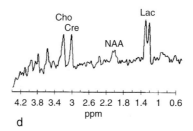

Figure 10.1. Proton MRS in two infants illustrating characteristic patterns seen in short echo time (STEAM; TE 20 ms) and long echo time (PRESS; TE 270 ms) spectra from occipital gray matter in patients with a good outcome (a, b) and a poor outcome (c, d). Both the short TE (a) and long TE (b) spectra are normal for a 4-month-old patient with a mild neurologic outcome after experiencing seizures. The short TE spectrum (c) obtained in a 2-month-old infant with TBI sustained 3 days previously and who subsequently died shows a decreased NAA peak and elevated lactate and possibly increased glutamate/glutamine (Glx) peaks. The long TE spectrum (d) from the same patient also shows a reduced NAA and elevated lactate peak. Both the short and long TE spectra were equally predictive of long-term outcome in children over 1 month of age. (Reproduced with permission from [67].)

Glutamate in TBI

The role of the excitatory neurotransmitter glutamate was examined using short-echo time MRS (TE 20, STEAM) by Ashwal et al.[69] in the subacute stage following TBI among 38 children using a 1.5 Tesla scanner. Although they found an elevation in the glutamate/glutamine (Glx) levels (see Figure 10.2) at an average of 7 days following injury, it did not correlate with the severity of injury or the outcome as determined by the Pediatric Cerebral Performance Category Scale (PCPCS), GCS, or any other metabolite 6–12 months following injury. Other invasive techniques have previously shown an early increase in the levels of glutamate in the ventricular CSF that may lead to edema and eventually neuronal death and the level of glutamate correlated with the patient outcome.[70,71,72,73] It is quite possible that the lack of correlation of glutamate levels with the patient outcomes could be related to the fact that MRS was collected 7 days after injury. Nevertheless, the fact that the glutamate/glutamine (Glx) levels remain elevated 7 days after injury makes it a potential marker for longitudinal follow-up of patients. It is quite likely that the role of glutamate/glutamine will be further examined in the future at the field strengths of 3.0 Tesla and higher, where the quantification and separation of these metabolites should be more accurate.

myo-Inositol in TBI

myo-Inositol (mI) is a form of inositol that is involved in osmoregulatory processes in its free form, and in neurotransmission via its phosphorylated derivates, and is contained in high concentrations in glial cells. In a study of TBI in children, occipital gray matter mI was elevated in all cases 7 days following injury by 22% using short-echo time single-voxel MRS (TE 20 ms, STEAM), and by as much as 35% among children that had a poor 6–12 month neurologic and developmental outcome (Figure 10.3).[74] The authors concluded that elevated mI over a long period is indicative of poor outcome, and may be the result of astrogliosis or disturbance of osmotic function. Although one study did not find a significant increase of mI in the parieto-occipital white matter, others have reported white matter mI increases.[63,75]

Diffuse metabolic abnormalities

As mentioned, several studies have shown that patients with TBI may have metabolic abnormalities in regions of the brain that appear normal on conventional T_1- or T_2-weighted MRI sequences. One study used chemical shift imaging with wide brain coverage to demonstrate that a reduction in the ratio of NAA/Cr and an increase in Cho/Cr in normal appearing white matter predicted 6–12 month neurologic

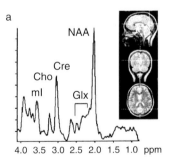

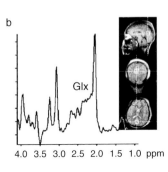

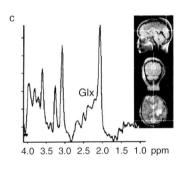

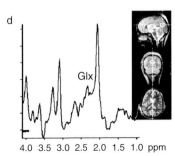

Figure 10.2. MR spectra from a 1.5 Tesla system obtained on a control subject (a); TBI/normal outcome patient (b); TBI/moderate disability outcome patient (c); and TBI/severe disability outcome patient (d). Note that glutamate/glutamine (Glx) (2.1–2.46 ppm) is increased in all TBI groups compared to the control subject seven days after injury. Although the Glx level was elevated in all TBI patients, it did not correlate with injury severity or outcome. Better spectral resolution and quantification of the Glx peaks is expected at the higher field strength of 3.0 Tesla. (Reproduced with permission from [69])

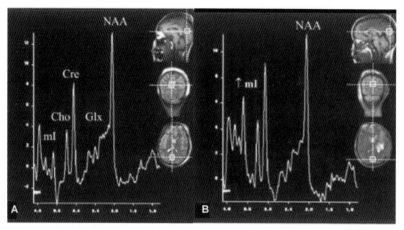

Figure 10.3. (A) Single voxel MRS (STEAM, TE/TR/TM 3000/20/13 ms; 128 averages) spectrum from the mid-occipital gray matter of a 14-year-old male 5 days after he was struck by a motor vehicle while riding on a scooter with an initial GCS of 14. Microhemorrhages were observed in the cortex and subcortical white matter MRI and a large 1-cm diameter hemorrhage with surrounding edema was seen in the periphery of the right parietal lobe. The spectrum shows relatively normal metabolite peaks. At 6–12 month follow-up, he had a normal outcome score. (B) Single-voxel MRS spectrum from the mid-occipital gray matter on an 11-year old female pedestrian with an initial GCS of 4, 2 weeks after she was struck by a motor vehicle. Large hemorrhages were seen in the anterior corpus callosum and left thalamus. Some intraventricular blood and right subdural fluid was also observed. Craniotomy was performed for decompression. The spectrum showed elevated *myo*-inositol (mI) and a moderately reduced NAA peak relative to Cr and Cho. At 6–12 months follow-up she had severe disability. (Adapted with permission from [74].)

outcome on 40 pediatric patients following injury.[76] Reductions of NAA/Cr and increases of Cho/Cr in normal appearing brain at the level of the corpus callosum was 85% more accurate than monitoring the metabolites in the visibly injured area, particularly if the MRI lesion was hemorrhagic. This study suggested that metabolism of normal appearing white matter is an important factor in outcome prediction, and also reinforces the concept that conventional MRI may underestimate the extent of neuroaxonal injury in pediatric TBI.

Cognitive function and metabolism

Along with the long-term outcome measured by the GOS, other measures of cognitive function have been found to correlate with acute measurements of brain metabolism in children with TBI. For instance, intellectual function, expressive language, and written arithmetic were all shown to be lower in pediatric TBI patients followed for 6 weeks to 3 years after TBI compared to controls. These measures were found to correlate with injury severity (as measured by NAA) in frontoparietal areas.[77] A reduction of NAA/Cho ratio between the injured group and the control group with no significant differences in the Cho or Cr levels was observed, implying that the main finding was a reduction in NAA. Although the sample size used in this study was small, these findings support other cognitive and structural imaging studies on TBI patients who performed poorly compared to normal control subjects on information processing tasks such as Trail making tests A & B, Stroop neuropsychological screening test, visuospatial tasks, and tasks that require patients to make inferences about others' mental states.[78,79,80,81]

Another study followed 36 pediatric (6–18 years) TBI patients for 3–26 weeks after TBI using MRS and a battery of cognitive tests (language, motor, visuomotor, and working memory).[82] While no correlations of any of the cognitive measures were found with NAA/Cr and Cho/Cr among the control subjects, significant correlations were found for both these ratios with the cognitive measures among the TBI population. Recovery of the metabolites to their normal levels was also observed by 26 weeks within the anterior region but not the posterior region. Visuomotor and language skills were positively correlated with NAA/Cr and negatively correlated with Cho/Cr. Although there was a slight trend towards a positive correlation of working memory with NAA/Cr this was not statistically significant, and there were no significant correlations between metabolite ratios with the motor skills. Overall these studies provide a justification for the clinical use of MR spectroscopy, both during the acute and chronic state following TBI, as a means of quantifying neuroaxonal damage, and predicting long-term outcome.

Adult TBI

MRS findings in adult TBI are generally similar to those described above for pediatric TBI; however, a significant difference is that the prognostic ability of an elevated lactate signal is much less studied among adults. One of the first MRS reports in adult closed head injury found significant correlations between both the GCS and Rancho Los Amigos Outcome scale and reduced NAA in both the gray and white matter.[83] As in children, diffusely reduced NAA may serve as an early indicator of diffuse neuroaxonal injury. Increased choline levels were also found, and attributed to the breakdown products of myelin following shearing injury in the white matter, and general membrane degradation in the gray matter. Although increased lactate correlated with poor outcome among the 12 children that were examined in this study, this finding was less common among the adult patients. Since this report, many studies have focused on looking at MRS patterns from brain regions far separated from the site of injury using single-voxel techniques. Typically these sites have been in the occipital gray matter, parietal white matter, and areas of the frontal lobe.[63,64,75,84,85,86,87,88,89] A decrease in NAA/Cr is a consistent observation among all these studies during the initial stages following injury in normal appearing gray and white matter. Longitudinal studies have shown that the initial decrease in the NAA/Cr ratio is dependent on the severity of injury and partially normalizes over 6–12 months, reflective of the clinical outcome (Figure 10.4).[64,75,88] It is likely that moderate and severe injury leads to secondary injury such as hypoxia and hypotension, which may adversely affect the reversibility of NAA/Cr

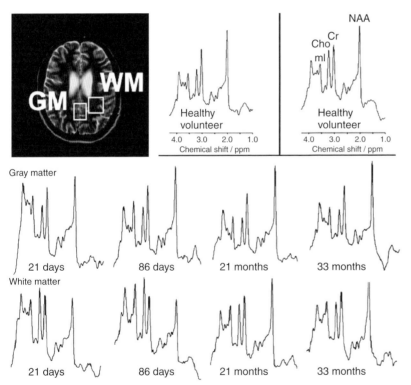

Figure 10.4. Evidence of axonal recovery: spectra from a 28-year-old woman struck by a motor vehicle who suffered severe TBI. She was in a coma for three months (GCS 4–6). One hundred days following that, she made the first purposeful right leg movement. By 21 months, the patient had regained considerable communication skills and was able to walk with assistance. She recovered further by 33 months, where she received and read letters, read short texts, wrote letters using a computer and communicated by telephone and had outcome score of 83 on the functional independence measure 7–8 on the Rancho Los Amigos Scale of cognitive function. MR spectra from occipito-parietal white matter (WM) and mid-occipital gray matter (GM) in the patient 21 and 86 days and 21 and 33 months after the injury, and in a normal healthy adult. The voxel positions are shown on the localizer image at 21 months. MRS at 21 and 86 days post-trauma was in agreement with MRI, showing evidence of diffuse axonal injury, with decreased N-acetylaspartate (NAA) and increased total choline (Cho). By 21 months NAA in white matter was normal, while myo-inositol (ml) was still slightly increased; at 33 months the spectra were similar. Total creatine (Cr) was normal on all studies. (Adapted with permission from [87].)

ratio following injury. Quantitative spectroscopy studies have generally found that Cr levels remain constant over time, so that changes in the NAA/Cr ratio over time reflect changes in NAA (reflective of neuronal loss/dysfunction and recovery) although exceptions do exist (see below).[75,83,85,86,90,91] Nakabayashi et al.[84] observed much higher declines in the NAA/Cr ratio in the peri-contusional area where no apparent abnormalities were detected by conventional MRI. This peri-contusional depression of metabolites may be due to pressure injury from hemorrhage, edema, or a general depression of blood flow near the contused area. Astrocytic swelling and edema in the initial phases after injury may also lead to early reduction of NAA, which may resolve during the later stage, normalizing the NAA concentration. [92,93,94] It should be noted that no lactate was observed in these peri-contusional areas, ruling out anaerobic metabolism. However, in an experimental model, a progressive increase in the lactate concentration was observed up to 7 days following controlled cortical impact injury that resolved after 28 days.[61] This study also found that NAA remained low all through the 28 days of observation, while glutamate was reduced during the initial hours and at 7 days, but showed signs of normalizing at 28 days. There was also an initial reduction in choline and *myo*-inositol at 1 h that normalized over 24 h and then remained significantly elevated after 7 days. The initial reduction of choline was interpreted to be the result of the primary trauma leading to membrane degradation followed by the rapid catabolism of the choline-containing compounds by the activation of Ca-dependent phospholipases. Of particular note in this study is that Cr levels (measured by the LCModel) dropped during the first hour and remained significantly lower by about 40% in comparison to both the contralateral side and the control animals. However, there was a tendency towards normalization of Cr by the seventh day following injury. Working under the assumption that Cr does not change over the course of injury would have underestimated the reduction of NAA by about 40–60% in this case. These experiments suggest that the assumption of a stable Cr peak following injury has to be treated with caution and may depend on the time of injury.

Investigations on other brain structures such as the splenium of the corpus callosum, basal ganglia, hippocampus, and thalamus have often yielded similar results as those seen in perilesional areas. A significantly lowered NAA/Cr in the splenium (and a decreased magnetization transfer ratio (MTR)) was seen among patients with an admission GCS ranging from 3 to 15 (mean 11) that had a 3-month GOS ranging from 1 to 4. Those patients that had a GOS of 5 had a higher NAA/Cr ratio. However, a significant portion of these patients had an MTR abnormality in the normal appearing white matter. Of particular interest would be to assess the utility of MRS changes among those patients that recover well but who have persistent cognitive deficits that negatively impact their ability to return to their pre-injury level of action. The authors caution about the limitation of using the NAA/Cr ratio from a single or few brain regions, and suggest the use of greater brain MRS coverage for better prognostic power of MR spectroscopy.[95]

In one study, NAA/Cho in the basal ganglia correlated with measures of fine motor speed, attention, backward digit span, and Trail making test A when using long echo time PRESS localization (TE 114 ms) on patients with GCS ranging from 3 to 12 that exhibited memory and frontal lobe impairment. NAA/Cho also demonstrated significant correlation with the verbal learning test and a face recognition memory test. These results are expected since the basal ganglia has strong connections to the cortex in the frontal lobe which are involved in cognitive function and the hippocampus is involved in memory-related dysfunctions.[96]

Neuropsychological correlation and vegetative state patients

Vegetative state patients after TBI present a unique clinical challenge. In such patients, advanced imaging techniques including MRS may provide important information both for clinical management and prognosis. In a study of 14 vegetative state patients who underwent bilateral thalamic proton spectroscopy,[97] patients were subdivided into two groups. One group continued to be in the persistent vegetative state after six months, and the other consisted of patients that had regained consciousness. Both groups demonstrated a significant reduction in the NAA/Cr ratio when compared to normal controls, even though the thalamus was normal on conventional MRI. Further, there was a significant decrease in the NAA/Cr ratio among the patients in the persistent vegetative state compared to those who regained consciousness (Figure 10.5). The authors concluded that the reduced NAA/Cr

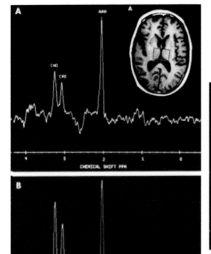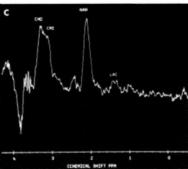

Figure 10.5. Single-voxel spectra from the thalamus demonstrating the usefulness of MRS in the longitudinal evaluation of patients in a persistent vegetative state. MRS spectra obtained (A) from the bilateral thalamus of a healthy volunteer, (B) a patient who regained awareness by the time the study was concluded, and (C) a patient in a persistent vegetative state. Choline remains elevated in the persistent vegetative state with reduced NAA and in a few cases lactate was observed. Inset shows an axial T_1-weighted magnetic resonance image of the site from where MRS of the thalamus was obtained. (Adapted with permission from [97].)

ratio was helpful in differentiating patients who are in a vegetative state and have the potential for some degree of recovery from those who do not. The reduced ratio is presumably reflective of neuronal and axonal loss and secondary transneuronal degeneration. Transneuronal degeneration was observed in neuropathological studies at three months after trauma in 96% of cases after being in the vegetative state for some time.[98,99] The thalamus is involved in processing external stimulus and passing it on to the cerebral cortex, and vice versa. These thalamo-cortico-thalamic circuits are involved in consciousness, regulating arousal, and the level of awareness and may serve as a sensitive location from which to obtain metabolic information to monitor the progress of patients in a vegetative state. In another study, Kirov et al.[100] monitored metabolic changes in the thalamus using multi-voxel spectroscopy in mild TBI patients. Using mixed model regression they found a minimal detectable difference of ± 13% for NAA, ± 13.5% for Cr, and ± 18.8% for Cho among mildly injured patients compared to the normal control subjects. Thus the thalamus appears to be involved metabolically in both severe and mild TBI.

Lactate in adult TBI

There are very few reports of elevated lactate in the mild and moderate TBI adult populations; however, lactate may be present in severe TBI. One case study reported elevation of lactate for weeks following severe TBI in a 43-year-old male patient with repeated assault using a blunt instrument who was admitted with a GCS of 6 T.[101] Spectra using both short and long echo times at the level of the thalamus, midline occipital gray matter, and frontal gray matter were obtained 9 and 23 days following injury. NAA was reduced and Cho was elevated as is typically seen in severe adult TBI, as well as increased lactate. Since intracranial pressure was normal at this time, it was thought that elevation of lactate may not be due to ischemia. Although elevated levels of lactate have been shown to be indicative of poor outcome among a pediatric population, the overall role of lactate remains to be established in adults. [68,102] Other studies have indicated that the presence of lactate may actually serve as a fuel during metabolic recovery.[103]

Microdialysis studies in TBI: Relevance to MRS

In a study on 19 TBI patients with admission GCS ranging from 3 to 15, microdialysis probes were placed within 2 cm of the contusion to monitor extracellular NAA, Lac, pyruvate, glycerol, and glutamate at 12-h intervals.[104] Each of these markers to some degree were correlated with survival and GOS. Eleven patients died, and a significantly lowered NAA was found among these patients compared to the surviving patients. The time course of NAA changes was significantly different between the survivors and non-survivors; NAA decreased among non-survivors 4 days after injury while NAA was stable among the survivors. Similar results were obtained with lactate-to-pyruvate ratio, and glycerol, while no significant differences were found in glutamate levels between the two groups. The investigators suggest that the late fall in extracellular NAA could be due to sustained metabolic dysfunction in the neuronal population leading to impaired mitochondrial biosynthesis, extracellular degradation, reuptake or clearance of the compound. The marked reduction in NAA among the non-survivors is supported by animal studies that have shown similar results where NAA is reduced with a concomitant decrease in the ATP concentration, suggesting that NAA depletion is related to energy impairment which may get worse in the case of secondary injuries.[105] These studies clearly indicate poor prognosis among patients with persistent low level concentrations of NAA over an extended period of time.

MRS studies of concussion and repeated brain injury

Experimental studies have shown the existence of a temporal window of brain vulnerability after mild TBI, which can be significantly affected if a second concussive event were to happen within this window. [106,107,108] From the few human studies that have examined the temporal window, it appears that it falls within the first 3–4 days following injury. A study on athletes observed a decrease in NAA/Cr ratio 3 days after the initial concussion which returned to normal levels by the 30th day, indicating full metabolic recovery although these athletes were declared as having completely resolved their symptoms within only 3 days. Those athletes that had a second concussive injury within 15 days of the first injury showed a continuing decline of NAA/Cr ratio until 30 days, which resolved to near normal levels at 45 days. In the case of these athletes the symptoms were declared to have resolved within 30 days. In both cases, although the symptoms were resolved, there was a continued metabolic imbalance which is potentially vulnerable to repeat exposure to concussions. The existence of this temporal window as demonstrated on experimental models and athletes might have significant consequences on decisions made regarding the return of athletes to their sports, or soldiers to the battlefield.

Magnetic resonance spectroscopic imaging studies in TBI

Although there is evidence that metabolic abnormalities can be seen far from the injury location among TBI patients, it is only recently that the investigators have began to shift their focus from single voxel to MR spectroscopic imaging (MRSI) techniques. Besides a larger coverage MRSI also usually allows the acquisition of spectra from smaller voxel sizes. MRSI may be helpful when conventional MRI is negative in patients who are presenting with clinical symptoms (Figure 10.6). Signoretti et al.[109] demonstrated the usefulness of MRSI in diffuse axonal injury patients (GCS ≤ 8) where they saw a uniform reduction of NAA throughout the brain. NAA remained significantly reduced up to 30 days among patients with poor outcome, whereas recovery to normal levels was observed in patients with good outcome. In a study that sought to find evidence of early ischemic neurochemical changes in patients suffering TBI (n = 18, mean GCS 5.5) and severe subarachnoid hemorrhage (SAH, n = 6, mean GCS 8) using 2D-MRSI at the level of the midbrain, it was found that NAA was lowered in both the TBI and SAH patients in both normal and abnormal appearing areas on T_2-weighted MRI.[110] Patients suffering from SAH but not TBI also exhibited elevation in choline and creatine in both T_2-normal and -abnormal areas. Lactate was not seen in either of the two patient populations, suggesting that ischemia does not play a prominent role in the pathogenesis of either disorder. However, it should be noted that the mean time to image following injury onset was about 9 days, during which time lactate may have resolved if it was initially elevated.

In another 2D-MRSI study of TBI,[111] it was demonstrated that the pooled metabolite ratios (NAA/Cr and Cho/Cr) from the corpus callosum during the early stages of injury (7 ± 4 days) predicted the outcomes of patients with 83% accuracy (Figure 10.7). Follow-up scans 6–12 months after injury in subjects who had a poor outcome demonstrated that the pooled NAA/Cr ratio remained low, and the Cho/Cr remained elevated.[111,112]

Chapter 10: MRS in traumatic brain injury

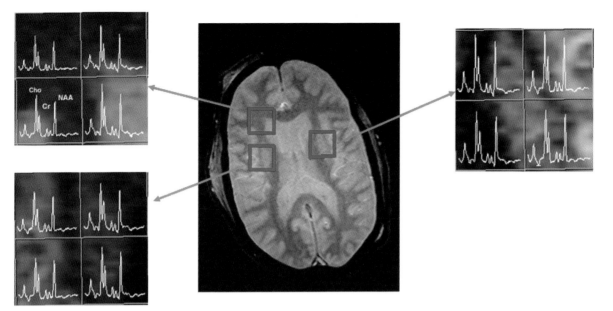

Figure 10.6. Spectra using MRSI (3D-PRESS, TE 135 ms) from a 37-year-old male motor vehicle accident victim with an initial GCS of 5, with no visible abnormality on CT, conventional MRI, or susceptibility-weighted imaging. The genu of the corpus callosum, internal and external capsule, and basal ganglia show a general decrease in NAA and an increase in the Cho/Cr ratio.

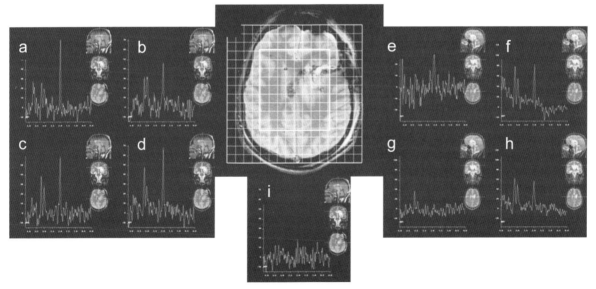

Figure 10.7. Seventeen-year-old male who was involved in a motorcycle accident with an admission GCS of 5. Patient was unconscious for 33 days after injury, and ventilated for 15 days. MRI/MRS was done 4 days after injury (PRESS, TE 144 ms). MRI showed hemorrhage in the left frontal lobe extending into the basal ganglia. MRS showed marked abnormalities with decreased NAA involving the left frontal lobe (b) and anterior and mid-corpus callosum (d) and extending into the left parieto-occipital area (c). Note that the metabolites are normal in the right frontal lobe. Some areas show near absence of metabolites most likely due to presence of hemorrhage (i). Follow-up MRSI was performed 6 months after injury. MRSI shows decreased NAA in multiple areas, including bi-frontal (e), right parietal (f), anterior and mid-corpus callosum (g, h). Glasgow Outcome Scale at 6 months and 1 year for this patient was 4 (moderate disability). [Courtesy of Dr Barbara Holshouser, Loma Linda University School of Medicine.]

Metabolic changes in the early stages of TBI

Early changes in cerebral metabolites following TBI in patients with various degrees of severity (covering the entire GCS scale) have been reported.[113] It was found that NAA (expressed relative to the sum of other metabolites) was significantly decreased, and Cho was increased compared to control subjects (see Figure 10.8). Lactate was elevated in half the patients examined in this study. Further, a strong correlation of NAA and Lac with the GCS upon admission ($r = 0.73$ and -0.62, respectively) and GOS at 3 months ($r = -0.79$ and 0.79, respectively) was found, suggesting that biochemical changes following TBI are clinically relevant and may be important indicators for predicting outcome. In contrast, another study found no significant changes in NAA during the initial period following severe TBI.[114] However, it was observed that a gradual decrease in NAA occurred which reached its lowest point at about 10 days, and then reversed among those patients that had a relatively good outcome. In contrast, those patients who continued to show continued NAA decline after 10 days had a poor outcome. It was also found that there was a good correlation between the late NAA/Cho ratio (beyond 10 days) and the GOS at 6 months (Figure 10.9). Based on this and previous animal studies, it was suggested that there may be a threshold beyond which NAA loss is irreversible in TBI patients with poor outcome. NAA reductions may be due to either neuroaxonal death or mitochondrial dysfunction. When this mitochondrial dysfunction is overcome the metabolic energy is stabilized with a subsequent return of the NAA to its normal levels. However, a prolonged impairment in energetics may cause cellular damage and ultimately lead to neuronal loss.[105] The authors suggest that NAA measurements be made at about 10 days following injury, as measurements

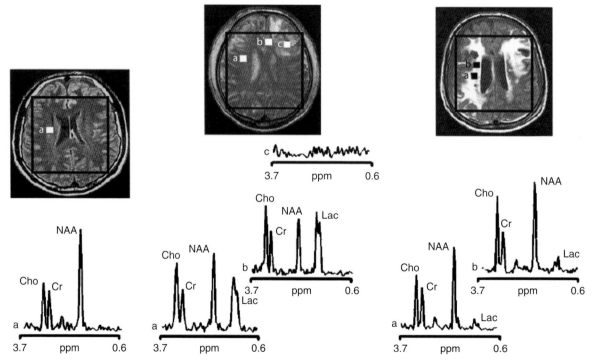

Figure 10.8. Proton MRS and conventional MRI showing the volume of interest for spectroscopic imaging of a normal control (left panel), a 69-year-old male patient (Patient 1) involved in a fall with an initial GCS of 4 with hematoma in the left frontoparietal area on MRI (central panel), and a 77-year-old male patient (Patient 2) who also suffered a fall with initial GCS of 4 and exhibiting diffuse abnormalities on MRI (right panel). Spectra show decreases of N-acetylaspartate (NAA) and increases of choline (Cho) and lactate (Lac) in patients with TBI (a and b in central and right panels) with respect to the normal control (a in left panel). The spectra from Patient 1 in central panel show more pronounced metabolic abnormalities than those of Patient 2 in the right panel, despite the fact that Patient 2 showed markedly more abnormalities on conventional MRI. In the spectra of Patient 1, metabolic abnormalities are clearly evident in the normal appearing brain. Notice the absence of metabolites in the voxel inside the focal hematoma (c in central panel) on Patient 1. (Reproduced with permission from [113].)

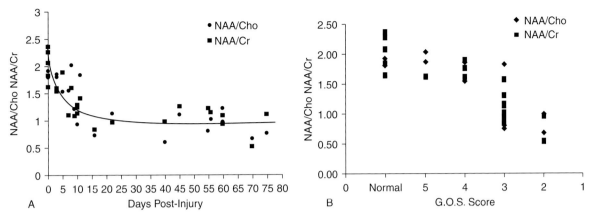

Figure 10.9. (A) Graph showing the time course of NAA/Cho and NAA/Cr decline in patients with head injuries who had poor outcomes. The NAA reduction was gradual and non-linear, reaching its lowest point at about 10 days and showing no recovery up to 60 days. (B) Scatter plot demonstrating the association between NAA reduction and 6-month outcome (Glasgow Outcome Score, GOS). Patients with good outcomes exhibited mean ratios >1.50. Conversely, those with poor outcomes were characterized by ratios <1.50, identifying a possible threshold of irreversible neurochemical damage. Note that the metabolite ratios in patients with GOS scores indicating good recovery or moderate disability are not significantly different from those of healthy volunteers. The metabolite ratios of patients with poor outcomes at 6 months were significantly lower than controls ($p < 0.01$). (Adapted with permission from [114].)

made before that time may underestimate the amount of neuronal loss. Although this study used a fairly large number of patients, it was more focused on the severe brain injury category compared to the study by Marino et al.,[113] which may cause the disparity in results between the two studies. More work is required to study the time course of NAA following TBI and to investigate whether it might reflect a temporal window for therapeutic intervention. The clinical significance of elevated lactate signals in the early time course of TBI also needs to be studied in more detail.

Volumetric and whole-brain spectroscopy studies in TBI

3D echo planar spectroscopic imaging techniques have been used to study mild TBI with near whole-brain coverage.[115] Twenty-five brain regions in 14 subjects were analyzed, including the brain stem, corpus callosum, white matter, and cerebellum. The volumetric scan utilized chemical shift-selective water suppression and inversion recovery for lipid nulling, and an echo-planar readout at an echo time of 70 ms and a TR of 1800 ms for a total acquisition time of 31 min. With the volumetric sequence it was possible to obtain spectra from at least eight 1.5-cm contiguous slices across the brain with linewidths of the spectra ranging from 4 to 8 Hz from each of the 1.15-cm^3 voxels. Significant changes in NAA/Cr, Cho/Cr, and NAA/Cho were found in the temporal lobe, frontal and occipital gray and white matter, but not in the cerebellum or brain stem, in these mildly injured patients. Metabolite ratios were not correlated with admission GCS, but were weakly correlated with the GOS at discharge.

Another technique, which makes no attempt at spatial localization, has been used to estimate whole-brain NAA concentrations in mildly injured patients with GCS 13–15 who had lost consciousness for at least 30 s.[116] Quantification of whole-brain NAA was performed with acquisition parameters of TE/TI/TR of 0/940/10,000 ms, respectively. The concentration of NAA was obtained using a reference 3-l sphere containing a known concentration of NAA. It was found that an average NAA declined 12% among patients compared to controls, with larger decreases in older subjects. It was also found that age-adjusted gray and white matter atrophy and whole-brain NAA decline was no different between patients with and without MRI-visible lesions.[116] Although assessment of the whole-brain NAA may be obtained quickly using this technique, it is of limited value when specific regional information is required. Nevertheless, the technique may have value in evaluating patients in the acute stage as the scan time is short. Future studies could benefit from acquisition schemes that cover a larger area of the brain and circumvent some of the limitations posed from limited coverage techniques such as single-voxel and single-slice 2D-MRSI.

Conclusion

The studies reported to date indicate that proton MRS and MRSI are sensitive for the detection of microstructural changes in TBI, even when conventional brain MRI is normal. However, very little work has been done on MRS during the very early stages of TBI where administration of neuroprotective treatment could be most critical. Decreases in NAA and increases in Cho are consistently documented among patients with poor outcome in the subacute and chronic stages of TBI. A continued increase in Cho levels beyond four days to a week following TBI has been demonstrated to lead to a poor outcome, and there is evidence that the presence of lactate may indicate poor prognosis, at least in the case of the pediatric population. Abnormal metabolite levels in the midbrain region including the thalamus, basal ganglia, and the brain stem over prolonged periods following TBI are indicative of poor outcome. Elevation of glutamate and glutamine and *myo*-inositol levels may be indicative of cellular dysfunction that may ultimately lead to compromised cognitive function.

Despite the literature evidence supporting the sensitivity of MRS in the assessment of TBI, no clear guidelines have emerged regarding the optimum MRS technique to use and its timing following TBI. Part of the reason may be due to the lack of concordance with the specific type of pulse sequence used, the echo times employed, and the brain locations used to obtain quantitative metabolite information. Standardized acquisition and processing techniques are likely to provide more objective results that will lead to accurate interpretation of MRS in the clinical setting. Whole-brain MRSI may overcome the issue of which brain location to obtain spectra from. Given that the literature has shown a clear indication that both short and long echo time spectra are useful in the diagnosis of TBI patients, if sufficient time is available it is recommended both short and long TE spectra be acquired using MRSI if possible. As described in Chapter 3, short echo time spectroscopy enables the measurement of metabolites such as *myo*-inositol, glutamate, and glutamine (in addition to Cho, Cr, and NAA), whereas long echo spectra provide fewer compounds, but have generally flatter baselines and are suitable for detecting lactate and with less lipid contamination than at short TE. Outer volume suppression, water suppression, and good shimming are crucial for obtaining quality spectra.

MRS is able to detect metabolic dysfunction in normal appearing brain regions removed from the foci of injury, or in the absence of any focal injury. This suggests that MRS may be useful among subjects whose clinical condition cannot be explained through CT or conventional MRI. Given the diffuse nature of TBI, MRS with wide spatial coverage may be valuable both at an early stage, as well as prior to discharge for an objective measure of diffuse axonal injury. Studies suggest that metabolic deficits may be useful correlates of cognitive status, and also provide prognostic information on long-term outcome. However, obtaining MRS during the acute stage of the injury is often extremely difficult for logistical issues. The more severely injured patients will have to be stabilized first and depending on the instrumentation applied on the patient during the stabilization process, the patient may or may not be eligible for MR studies. Patients with mild TBI are more likely to be eligible for MR studies, and given that this group generally forms the greater proportion of admitted TBI patients, an MRS investigation could be beneficial prior to discharge as a predictor of long-term cognitive deficit.

Finally, it should be mentioned that other advanced imaging modalities including diffusion tensor imaging, susceptibility-weighted imaging, arterial spin labeling, and functional MRI (fMRI) are being developed and which may also be very sensitive for the detection of TBI.[49,117,118,119,120] For instance, one study explored the impact of brain atrophy, diffusion-weighted imaging, MRS, and fMRI on patients that were in more than 3 years of long-lasting persistent vegetative state over 24 months. Although no specific conclusions were possible due to the variable nature of the initial injury among the patients studied, a progressive decrease in NAA/(Cho+Cr) ratio that corresponded with the progressive tissue injury as depicted by increased diffusivity in various regions and decreased excitability of primary cortices including the motor, auditory, and the visual system.[121] Although future multimodal imaging TBI studies are required to establish the relative value of MRS compared to these other techniques, it appears likely that MRS will play a significant role in the assessment of TBI in the future.

References

[1] Finkelstein E, Corso P, Miller T. *The Incidence and Economic Burden of Injuries in the United States*. New York (NY): Oxford University Press, 2006.

[2] Langlois JA, Rutland-Brown W, Thomas KE. *Traumatic Brain Injury in the United States: Emergency Department*

Visits, Hospitalizations, and Deaths. Atlanta, GA: Centers for Disease Control and Prevention, National Center for Injury Prevention and Control, 2004.

[3] National Institute of Neurological Disorders and Stroke. *Traumatic Brain Injury: Hope Through Research*. Bethesda, MD: National Institutes of Health; Feb. NIH Publication No.: 02-158, 2002.

[4] Thurman D, Alverson C, Dunn K, Guerrero J, Sniezek J. Traumatic brain injury in the United States: A public health perspective. *J Head Trauma Rehabil* 1999; **14**: 602–15.

[5] Ommaya AK, Gennarelli TA. Cerebral concussion and traumatic unconsciousness. Correlation of experimental and clinical observations of blunt head injuries. *Brain* 1974; **97**: 633–54.

[6] Ommaya AK, Goldsmith W, Thibault L. Biomechanics and neuropathology of adult and pediatric head injury. *Br J Neurosurg* 2002; **16**: 220–42.

[7] Povlishock J, Becker DP, Cheng CL, Vaughan GW. Axonal change in minor head injury. *J Neuropathol Exp Neurol* 1983; **42**: 225–42.

[8] Elson LM, Ward CC. Mechanisms and pathophysiology of mild head injury. *Semin Neurol* 1994; **14**: 8–18.

[9] McCrory P, Johnston KM, Mohtadi NG, Meeuwisse W. Evidence-based review of sport-related concussion: Basic science. *Clin J Sport Med* 2001; **11**: 160–5.

[10] Johnston KM, McCrory P, Mohtadi NG, Meeuwisse W. Evidence-based review of sport-related concussion: Clinical science. *Clin J Sport Med* 2001; **11**: 150–9.

[11] Graham DI. Neuropathology of head injury. In *Neurotrauma*, Narayan RK, Wilberger JE, Povlishock JT, Eds. Philadelphia: W. B. Saunders; 1996: 43–60.

[12] Graham DI, Gennarelli TA, McIntosh TK. Trauma. In *Greenfield's Neuropathology*, 7th edn, Graham DJ, Lantos PL, Eds. London: Arnold Press, 2002.

[13] Gentry LR. Imaging of closed head injury. *Radiology* 1994; **191**: 1–17.

[14] DeKosky ST, Kochanek PM, Clark RS, Ciallella JR, Dixon CE. Secondary injury after head trauma: Subacute and long-term mechanisms. *Semin Clin Neuropsychiatry* 1998; **3**: 176–85.

[15] Gorman LK, Fu K, Hovda DA, Murray M, Traystman RJ. Effects of traumatic brain injury on the cholinergic system in the rat. *J Neurotrauma* 1996; **13**: 457–63.

[16] Pike BR, Hamm RJ. Activating the post-traumatic cholinergic system for the treatment of cognitive impairment following traumatic brain injury. *Pharmacol Biochem Behav* 1997; **57**: 785–91.

[17] Globus MY, Alonso O, Dietrich WD, Busto R, Ginsberg MD. Glutamate release and free radical production following brain injury: Effects of posttraumatic hypothermia. *J Neurochem* 1995; **65**: 1704–11.

[18] LaPlaca MC, Lee VM, Thibault LE. An in vitro model of traumatic neuronal injury: Loading rate-dependent changes in acute cytosolic calcium and lactate dehydrogenase release. *J Neurotrauma* 1997; **14**: 355–68.

[19] Farooqui AA, Horrocks LA. Lipid peroxides in the free radical pathophysiology of brain diseases. *Cell Mol Neurobiol* 1998; **18**: 599–608.

[20] Nemetz PN, Leibson C, Naessens JM, Beard M, Kokmen E, Annegers JF, et al. Traumatic brain injury and time to onset of Alzheimer's disease: A population-based study. *Am J Epidemiol* 1999; **149**: 32–40.

[21] Lye TC, Shores EA. Traumatic brain injury as a risk factor for Alzheimer's disease: A review. *Neuropsychol Rev* 2000; **10**: 115–29.

[22] Chaumet G, Quera-Salva MA, Macleod A, Hartley S, Taillard J, Sagaspe P, et al. Is there a link between alertness and fatigue in patients with traumatic brain injury? *Neurology* 2008; **71**: 1609–13.

[23] Makley MJ, English JB, Drubach DA, Kreuz AJ, Celnik PA, Tarwater PM. Prevalence of sleep disturbance in closed head injury patients in a rehabilitation unit. *Neurorehabil Neural Repair* 2008; **22**: 341–7.

[24] Nampiaparampil DE. Prevalence of chronic pain after traumatic brain injury: A systematic review. *J Am Med Assoc* 2008; **300**: 711–9.

[25] Immonen RJ, Kharatishvili I, Gröhn H, Pitkänen A, Gröhn OH. Quantitative MRI predicts long-term structural and functional outcome after experimental traumatic brain injury. *NeuroImage* 2008 [Epub ahead of print].

[26] Menzel JC. Depression in the elderly after traumatic brain injury: A systematic review. *Brain Inj* 2008; **22**: 375–80.

[27] Muscara F, Catroppa C, Anderson V. The impact of injury severity on executive function 7–10 years following pediatric traumatic brain injury. *Dev Neuropsychol* 2008; **33**: 623–36.

[28] Nybo T, Koskiniemi M. Cognitive indicators of vocational outcome after severe traumatic brain injury (TBI) in childhood. *Brain Inj* 1999; **13**: 759–66.

[29] Adelson PD, Kochanek PN. Head injury in children. *Child Neurol* 1998; **13**: 3–15.

[30] Kochanek PM, Clark RS, Ruppel RA, Adelson PD, Bell MJ, Whalen MJ, et al. Biochemical, cellular, and molecular mechanisms in the evolution of secondary damage after severe traumatic brain injury in infants

and children. Lessons learned from the bedside. *Pediatr Crit Care Med* 2000; **1**: 4–19.

[31] Levin HS, Aldrich EF, Saydjari C, Eisenberg HM, Foulkes MA, Bellefleur M, *et al*. Severe head injury in children: Experience of the Traumatic Coma Data Bank. *Neurosurgery* 1992; **31**: 435–44.

[32] Garthe E, States JD, Mango NK. Abbreviated injury scale unification: The case for a unified injury system for global use. *J Trauma* 1999; **47**: 309–23.

[33] Teasdale G, Jennett B. Assessment of coma and impaired consciousness: A practical scale. *Lancet* 1974; **13**: 81–3.

[34] Zimmerman RA. Craniocerebral trauma. In *Cranial MRI and CT*, 4th edn, Lee SH, Rao KCVG, Zimmerman RA, Eds. New York: McGraw-Hill; 1999: 413–52.

[35] Schunk JE, Rodgerson JD, Woodward GA. The utility of head computed tomographic scanning in pediatric patients with normal neurologic examination in the emergency department. *Pediatr Emerg Care* 1996; **12**: 160–5.

[36] Cihangiroglu M, Ramsey RG, Drohmann GJ. Brain injury: Analysis of imaging modalities. *Neurol Res* 2002; **24**: 7–18.

[37] Zee CS, Go J. CT of head trauma. *Neuroimaging Clin N Am* 1998; **8**: 541–8.

[38] Borzcuk P. Mild head trauma. *Emerg Med Clin North Am* 1997; **15**: 563.

[39] Sojka P, Stalnacke BM. Normal findings by computer tomography do not exclude CNS injury. *Lakartidningen* 1999; **96**: 616.

[40] Gentry LR. *Head Trauma. Magnetic Resonance Imaging of the Brain and Spine* (3rd edn), SW Atlas Ed. Philadelphia: Lippincott Williams & Wilkins, 2002; 1059–98 (Chapter 20).

[41] Bigler ED. The lesion(s) in traumatic brain injury: Implications for clinical neuropsychology. *Arch Clin Neuropsychol* 2001; **16**: 95–131.

[42] Bigler ED. Structural and functional neuroimaging of traumatic brain injury. In *State of the Art Reviews in Physical Medicine and Rehabilitation: Traumatic Brain Injury*, McDaeavitt JT, Ed.. Philadelphia: Hanley and Belfus; 2001: 349–61.

[43] Bigler ED, Tate DF. Brain volume, intracranial volume, and dementia. *Invest Radiol* 2001; **36**: 539–46.

[44] Blatter DD, Bigler ED, Gale SD, Johnson SC, Anderson CV, Burnett BM, *et al*. MR-based brain and cerebrospinal fluid measurement after traumatic brain injury: Correlation with neuropsychological outcome. *Am J Neuroradiol* 1997; **18**: 1–10.

[45] Adams JH, Graham DI, Gennarelli TA, Maxwell WL. Pathology of non-missile head injury. *J Neurol Neurosurg Psychiatry* 1991; **54**: 481–3.

[46] Parizel PM, Ozsarlak, Van Goethem JW, van den Hauwe L, Dillen C, Verlooy J, *et al*. Imaging findings in diffuse axonal injury after closed head trauma. *Eur Radiol* 1998; **8**: 960–5.

[47] Diaz-Marchan PJ, Haymam LA, Carrier DA, Feldman DJ. Computed tomography of closed head injury. In *Neurotrauma*, Narayan RK, Wilberger JE, Povlishock JT, Eds. New York: McGraw-Hill; 1996: 137–50.

[48] Shibata Y, Matsumura A, Meguro K, Narushima K. Differentiation of mechanism and prognosis of traumatic brain stem lesions detected by magnetic resonance imaging in the acute stage. *Clin Neurol Neurosurg* 2000; **102**: 124–8.

[49] Reichenbach JR, Venkatesan R, Schillinger DJ, Kido DK, Haacke EM. Small vessels in the human brain: MR venography with deoxyhemoglobin as an intrinsic contrast agent. *Radiology* 1997; **204**: 272–7.

[50] Thornhill S, Teasdale GM, Murray GD, McEwen J, Roy CW, Penny KI. Disability in young people and adults one year after head injury: Prospective cohort study. *Br Med J* 2000: **320**: 1631–5.

[51] Bigler ED. Quantitative magnetic resonance imaging in traumatic brain injury. *J Head Trauma Rehabil* 2001; **16**: 117–34.

[52] Levin HS, Williams DH, Eisenberg HM, High WM Jr, Guinto FC Jr. Serial MRI and neurobehavioural findings after mild to moderate closed head injury. *J Neurol Neurosurg Psychiatry* 1992; **55**: 255–62.

[53] Bergsneider M, Hovda DA, Shalmon E, Kelly DF, Vespa PM, Martin NA, *et al*. Cerebral hyperglycolysis following severe traumatic brain injury in humans: A positron emission tomography study. *J Neurosurg* 1997; **86**: 241–51.

[54] Bergsneider M, Hovda DA, Lee SM, Kelly DF, McArthur DL, Vespa PM, *et al*. Dissociation of cerebral glucose metabolism and level of consciousness during the period of metabolic depression following human traumatic brain injury. *J Neurotrauma* 2000; **17**: 389–401.

[55] Glenn TC, Kelly DF, Boscardin WJ, McArthur DL, Vespa P, Oertel M, *et al*. Energy dysfunction as a predictor of outcome after moderate or severe head injury: Indices of oxygen, glucose, and lactate metabolism. *J Cereb Blood Flow Metab* 2003; **23**: 1239–50.

[56] Hovda DA, Villablanca JR, Chugani HT, Phelps ME. Cerebral metabolism following neonatal or adult hemineodecortication in cats: I. Effects on glucose metabolism using [14C]2-deoxy-D-glucose autoradiography. *J Cereb Blood Flow Metab* 1996; **16**: 134–46.

[57] Hovda DA, Becker DP, Katayama Y. Secondary injury and acidosis. *J Neurotrauma* 1992; **9**(Suppl 1): S47–60.

[58] Jaggi JL, Obrist WD, Gennarelli TA, Langfitt TW. Relationship of early cerebral blood flow and metabolism to outcome in acute head injury. *J Neurosurg* 1990; **72**: 176–82.

[59] Dusick JR, Glenn TC, Lee WNP, Vespa PM, Kelly DF, Lee SM, *et al.* Increased pentose phosphate pathway flux after clinical traumatic brain injury: A [1,2-13C2] glucose labeling study in humans. *J Cereb Blood Flow Metab* 2007; **27**: 1593–602.

[60] Viant MR, Lyeth BG, Miller MG, Berman RF. An NMR metabolomic investigation of early metabolic disturbances following traumatic brain injury in a mammalian model. *NMR Biomed* 2005; **18**: 507–16.

[61] Schuhmann MU, Stiller D, Skardelly M, Bernarding J, Klinge PM, Samii A, *et al.* Metabolic changes in the vicinity of brain contusions: A proton magnetic resonance spectroscopy and histology study. *J Neurotrauma* 2003; **20**: 725–43.

[62] Ashwal S, Holshouser BA, Shu SK, Simmons PL, Perkin RM, Tomasi LG, *et al.* Predictive value of proton magnetic resonance spectroscopy in pediatric closed head injury. *Pediatr Neurol* 2000; **23**: 114–25.

[63] Garnett MR, Blamire AM, Rajagopalan B, Styles P, Cadoux-Hudson TAD. Evidence for cellular damage in normal-appearing white matter correlates with injury severity in patients following traumatic brain injury: A magnetic resonance spectroscopy study. *Brain* 2000; **123**: 1403–09.

[64] Garnett MR, Blamire AM, Corkill RG, Cadoux-Hudson TAD, Rajagopalan B, Styles P. Early proton magnetic resonance spectroscopy in normal-appearing brain correlates with outcome in patients following traumatic brain injury. *Brain* 2000; **123**: 2046–54.

[65] Condon B, Oluoch-Olunya D, Hadley D, Teasdale G, Wagstaff A. Early 1H magnetic resonance spectroscopy of acute head injury: Four cases. *J Neurotrauma* 1998; **15**: 563–71.

[66] Holshouser BA, Ashwal S, Luh GY, Shu S, Kahlon S, Auld KL, *et al.* 1H-MR spectroscopy after acute CNS injury: Outcome prediction in neonates, infants and children. *Radiology* 1997; **202**: 487–96.

[67] Holshouser BA, Ashwal S, Shu S, Hinshaw DB. Proton MR spectroscopy in children with acute brain injury: Comparison of short and long echo time acquisitions. *J Magn Reson Imaging* 2000; **11**: 9–19.

[68] Makoroff KL, Cecil KM, Care M, Bass Jr WS. Elevated lactate as an early marker of brain injury in inflicted traumatic brain injury. *Pediatr Radiol* 2005; **35**: 668–76.

[69] Ashwal S, Holshouser B, Tong K, Serna T, Osterdock R, Gross M, *et al.* Proton MR spectroscopy detected glutamate/glutamine is increased in children with traumatic brain injury. *J Neurotrauma* 2004; **21**: 1539–52.

[70] Baker AJ, Moulton RJ, Macmillan VH, Shedden PM. Excitatory amino acids in cerebrospinal fluid following traumatic brain injury in humans. *J Neurosurg* 1993; **79**: 369–72.

[71] Bullock R, Zauner A, Woodward JJ, Myseros J, Choi SC, Ward JD, *et al.* Factors affecting excitatory amino acid release following severe human head injury. *J Neurosurg* 1998; **89**: 507–18.

[72] Gopinath SP, Valadka AB, Goodman JC, Robertson CS. Extracellular glutamate and aspartate in head injured patients. *Acta Neurochir* 2000; **76**(suppl): 437–8.

[73] Ruppel RA, Clark RS, Bayir H, Satchell MA, Kochanek PM. Critical mechanisms of secondary damage after inflicted head injury in infants and children. *Neurosurg Clin N Am* 2002; **13**: 169–82.

[74] Ashwal S, Holshouser B, Tong K, Serna T, Osterdock R, Gross M, *et al.* Proton spectroscopy detected myoinisitol in children with Traumatic Brain Injury. *Pediatr Res* 2004; **56**: 630–8.

[75] Brooks WM, Stidley CA, Petropoulos H, Jung RE, Weers DC, Friedman SD, *et al.* Metabolic and cognitive response to human traumatic brain injury: A quantitative proton magnetic resonance study. *J Neurotrauma* 2000; **17**: 629–40.

[76] Holshouser BA, Tong KA, Ashwal S. Proton MR spectroscopic imaging depicts diffuse axonal injury in children with traumatic brain injury. *Am J Neuroradiol* 2005; **26**: 1276–85.

[77] Hunter JV, Thornton J, Wang ZJ, Levin HS, Roberson G, Brooks WM, *et al.* Late proton MR spectroscopy in children after traumatic brain injury: Correlation with cognitive outcomes. *Am Neuroradiol* 2005; **26**: 482–8.

[78] Flemingham KL, Baguley IJ, Green AM. Effects of diffuse axonal injury on speed of information processing following severe traumatic brain injury. *Neuropsychology* 2004; **18**: 564–71.

[79] Bibby H, McDonald S. Theory of mind after traumatic brain injury. *Neuropsychologia* 2005; **43**: 99–114.

[80] Azouvi P. Neuroimaging correlates of cognitive and functional outcome after traumatic brain injury. *Curr Opin Neurol* 2000; **13**: 665–9.

[81] Verger K, Junque C, Levin HS, Jurado MA, Perez-Gomez M, Bartres-Faz D, *et al.* Correlation of atrophy measures on MRI with neuropsychological sequelae in children and adolescents with traumatic brain injury. *Brain Injury* 2001; **15**: 211–21.

[82] Yeo RA, Phillips JP, Jung RE, Brown AJ, Campbell RC, Brooks WM. Magnetic resonance spectroscopy detects brain injury and predicts cognitive functioning in children with brain injuries. *J Neurotrauma* 2006; **23**: 1427–35.

[83] Ross BD, Ernst T, Kreis R, Hasseler LJ, Bayer S, Danielsen E, et al. 1H MRS in acute traumatic brain injury. *J Magn Reson Imaging* 1998; **8**: 829–40.

[84] Nakabayashi M, Suaki S, Tomita MD. Neural injury and recovery near cortical contusions: A clinical magnetic resonance spectroscopy study. *J Neurosurg* 2007; **106**: 270–7.

[85] Friedman SD, Brook WM, Jung RE, Hart BL, Yeo RA. Proton MR spectroscopic findings correspond to neuropsychological function in traumatic brain injury. *Am J Neuroradiol* 1998; **19**: 1879–85.

[86] Friedman SD, Brooks WM, Jung RE, Chiulli SJ, Sloan JH, Montoya BT, et al. Quantitative proton MRS predicts outcome after traumatic injury. *Neurology* 1999; **52**: 1384–91.

[87] Danielsen ER, Christensen PB, Arlien-Soborg P, Thomsen C. Axonal recovery after severe traumatic brain injury demonstrated in vivo by 1H MR spectroscopy. *Neuroradiology* 2003; **45**: 722–4.

[88] Shutter L, Tong KA, Holshouser BA. Proton MRS in acute traumatic brain injury: Role for glutamate/glutamine and choline for outcome prediction. *J Neurotrauma* 2004; **21**: 1693–705.

[89] Yoon SJ, Lee JH, Kim ST, Chun MH. Evaluation of traumatic brain injured patients in correlation with functional status by localized 1H-MR spectroscopy. *Clin Rehabil* 2005; **19**: 209–15.

[90] Garnett MR, Corkill RG, Blamire AM, Rajagopalan B, Manners DN, Young JD, et al. Altered cellular metabolism following traumatic brain injury: A magnetic resonance spectroscopy study. *J Neurotrauma* 2001; **18**: 241–6.

[91] Zampolini M, Tarducci R, Gobbi G, Franceschini M, Todeschini E, Presciutti O. Localized in vivo 1H-MRS of traumatic brain injury. *Eur J Neurol* 1997; **4**: 246–54.

[92] Bullock R, Maxwell WL, Graham DI, Teasdale GM, Adams JH. Glial swelling following human cerebral contusion: An ultrastructural study. *J Neurol Neurosurg Psychiatry* 1991; **54**: 427–4.

[93] Liang D, Bhatta S, Gerzanich V, Simard JM. Cytotoxic edema: Mechanisms of pathological cell swelling. *Neurosurg Focus* 2007; **22**: E4.

[94] Simard JM, Kent TA, Chen M, Tarasov KV, Gerzanich V. Brain oedema in focal ischaemia: Molecular pathophysiology and theoretical implications. *Lancet Neurol* 2007; **6**: 258–68.

[95] Sinson G, Bagley LJ, Cecil KM, Torchia M, McGowan JC, Lenkinski RE, et al. Magnetization transfer imaging and proton MR spectroscopy in the evaluation of axonal injury: Correlation with clinical outcome after traumatic brain injury. *Am J Neuroradiol* 2001; **22**: 143–51.

[96] Ariza M, Junque C, Mataro M, Paca MA, Bargallo N, Olondo M, et al. Neuropsychological correlates of basal ganglia and medial temporal lobe NAA/Cho reductions in Traumatic Brain Injury. *Arch Neurol* 2004; **61**: 541–4.

[97] Uzan M, Albayram S. Dashti SGR, Aydin S, Hanci M, Kuday C. Thalamic proton magnetic resonance spectroscopy in vegetative state induced by traumatic brain injury. *J Neurol Neurosurg Psychiatry* 2003; **74**: 33–8.

[98] Jennett B, Adams JH, Murray LS, Graham DI. Neuropathology in vegetative and severely disabled patients after head injury. *Neurology* 2001; **56**: 486–90.

[99] Graham DI, Maxwell WL, Adams JH, Jennett B. Novel aspects of the neuropathology of the vegetative state after blunt head injury. *Prog Brain Res* 2005; **150**: 445–55.

[100] Kirov I, Fleysher L, Babb JS, Silver JM, Grossman RI, Gonen O. Characterizing 'mild' in traumatic brain injury with proton MR spectroscopy in the thalamus: Initial findings. *Brain Injury* 2007; **21**: 1147–54.

[101] Hillary FG, Liu WC, Genova HM, Maniker AH, Kepler K, Greenwald BD, et al. Examining lactate in severe TBI using proton magnetic resonance spectroscopy. *Brain Injury* 2007; **21**: 981–91.

[102] Unterberg AW, Stover J, Kress B, Kiening KL. Edema and brain trauma. *Neuroscience* 2004; **129**: 1021–9.

[103] Vespa M, McArthur D, O'Phelan K, Glenn T, Etchepare, Kelly D, et al. Persistently low extracellular glucose correlates with poor outcome 6 months after human traumatic brain injury despite a lack of increased lactate: A microdialysis study. *J Cereb Blood Flow Metab* 2003; **23**: 865–77.

[104] Bello A, Sen J, Petzold A, Russo S, Kitchen N, Smith M, et al. Extracellular N-acetylaspartate depletion in traumatic brain injury. *J Neurochim* 2006; **96**: 861–79.

[105] Signoretti S, Marmarou A, Tavazzi B, Lazzarino G, Beaumont A, Vagnozzi R. N-acetylaspartate reduction as a measure of injury severity and mitochondrial dysfunction following diffuse traumatic brain injury. *J Neurotrauma* 2001; **18**: 977–91.

[106] Vagnozzi R, Tavazzi B, Stefano S, Amorini AM, Belli A, Cimatti M, et al. Temporal window of metabolic brain vulnerability to concussions: Mitochondrial-related impairment – Part 1. *Neurosurgery* 2007; **61**: 379–89.

[107] Tavazzi B, Vagnozzi R, Signoretti S, Amorini AM, Belli A, Cimatti M, *et al*. Temporal window of metabolic brain vulnerability to concussions: Oxidative and nitrosaive stresses – Part II. *Neurosurgery* 2007; **61**: 390–5.

[108] Vagnozzi R, Signoretti S, Tavazzi B, Floris R, Ludovici A, Marziali S, *et al*. Temporal window of metabolic brain vulnerability to concussion: A pilot 1H-magnetic resonance spectroscopic study in concussed athletes – Part III. *Neurosurgery* 2008; **62**: 1286–96.

[109] Signoretti S, Marmarou A, Fatouros P, Hoyle R, Beaumont A, Sawauchi S, *et al*. Application of chemical shift imaging for measurement of NAA in head injured patients. *Acta Neurochir* 2002; **81**(suppl): 373–5.

[110] Macmillan CSA, Wild JM, Wardlaw JM, Andrews PJD, Marshall I, Easton VJ. Traumatic brain injury and subarachnoid hemorrhage: In vivo occult pathology demonstrated by magnetic resonance spectroscopy may not be "ischaemic". A primary study and review of the literature. *Acta Neurochir* 2002; **144**: 853–62.

[111] Holshouser BA, Tong KA, Ashwal S, Oyoyo U, Ghamsary M, Saunders D, *et al*. Prospective longitudinal proton magnetic resonance spectroscopic imaging in adult traumatic brain injury. *J Magn Reson Imaging* 2006; **24**: 33–40.

[112] Shutter L, Tong KA, Lee A, Holshouser BA. Prognostic role of proton magnetic resonance spectroscopy in acute traumatic brain injury. *J Head Trauma Rehabil* 2006; **21**: 334–49.

[113] Marino S, Zei E, Battaglini M, Vittori C, Buscalferri A, Bramanti P, *et al*. Acute metabolic brain changes following traumatic brain injury and their relevance to clinical severity and outcome. *J Neurol Neurosurg Psychiatry* 2007; **78**: 501–07.

[114] Signoretti S, Marmarou A, Aygok GA, Fatouros PP, Portella G, Bullock RM. Assessment of mitochondrial impairment in traumatic brain injury using high-resolution proton magnetic resonance spectroscopy. *J Neurosurg* 2008; **108**: 42–52.

[115] Govindaraju V, Gauger GE, Manley GT, Andreas E, Meeker M, Maudsley AA. Volumetric proton spectroscopic imaging of mild traumatic brain injury. *Am J Neuroradiol* 2004; **25**: 730–7.

[116] Cohen BA, Inglese M, Rusinek H, Babb JS, Grossman RO, Gonen O. Proton MR spectroscopy and MRI-volumetry in mild traumatic brain injury. *Am J Neuroradiol* 2007; **28**: 907–13.

[117] Babikian T, Freier MC, Tong KA, Nickerson JP, Wall CJ, Holshouser BA, *et al*. Susceptibility-weighted imaging: Neuropsychological outcome and pediatric head injury. *Pediatr Neurol* 2005; **33**: 179–89.

[118] Ashwal S, Babikian T, Gardner-Nichols J, Freier M-C, Tong KA, Holshouser BA. Susceptibility-weighted imaging and proton magnetic resonance spectroscopy in assessment of outcome after pediatric traumatic brain injury. *Arch Phys Med Rehabil* 2008; **82**: S50–7.

[119] Beaumont A, Fatouros P, Gennarelli T, Corwin F, Marmarou A. Bolus tracer delivery measured by MRI confirms edema without blood–brain barrier permeability in diffuse traumatic brain injury. *Acta Neurochir* 2006; **96**(suppl): 171–4.

[120] Hendrich KS, Kochanek PM, Williams DS, Schiding JK, Marion DW, Ho C. Early perfusion after controlled cortical impact in rats: Quantification by arterial spin-labeled MRI and the influence of spin-lattice relaxation time heterogeneity. *Magn Reson Med* 1999; **42**: 673–81.

[121] Rousseau MC, Confort-Gouny S, Catala A, Graperon J, Blaya J, Soulier E, *et al*. A MRS-MRI-fMRI exploration of the brain. Impact of long-lasting persistent vegetative state. *Brain Injury* 2008; **22**: 123–34.

Chapter 11
MRS in cerebral metabolic disorders

Key points

- MR spectroscopy is a valuable tool to direct biochemistry work-up of patients with inborn errors of metabolism.
- Multivoxel MR spectroscopic imaging is the best method to study the heterogeneous anatomic distribution of metabolic diseases.
- The interpretation of MR spectra and MR images together increases diagnostic accuracy.
- Abnormal MR spectral peaks are diagnostic of a few hereditary metabolic disorders.
- Lactate is elevated in about half of patients with mitochondrial disorders, in most patients with leukoencephalopathies with demyelination or rarefaction of white matter, and in few with organic acidopathies targeting the subcortical gray matter nuclei.
- In patients with leukoencephalopathy, H-MRSI is a valuable tool for identifying one of the following three underlying tissue pathophysiologies: hypomyelination, demyelination, and rarefaction of white matter.
- MRS may be useful to monitor response to therapy when available.

Introduction

The advent of magnetic resonance (MR) imaging has changed the clinical approach to the evaluation of metabolic disorders. MR imaging is highly sensitive and plays a prominent role in the diagnostic evaluation of patients with metabolic disorders of the central nervous system (CNS). However, the structural and signal abnormalities detected on *conventional* MR imaging are often not specific enough to suggest a definite diagnosis in many of these complex disorders.[1]

With advances in MR technology, proton MR spectroscopy (^{1}H-MRS) has become more widely available, and now it can be performed with *conventional* MR imaging in the same study session. Nowadays, a complete imaging exam lasts no longer than 30 min at 1.5 Tesla or higher magnetic fields. In the last 15 years, in vivo single-voxel ^{1}H-MRS and multivoxel MR spectroscopic imaging (^{1}H-MRSI) have been applied to an increasing number of metabolic disorders. ^{1}H-MRS has been shown to be particularly useful as it can simultaneously provide chemical–pathological correlates of changes occurring within and outside MR-visible lesions. ^{1}H-MRS techniques increase diagnostic accuracy and better understanding of pathology in many of these disorders. It must also be emphasized, however, that ^{1}H-MRS is usually complementary to MR imaging, except in a handful of disorders where disease-specific spectral abnormalities have been identified.

The aim of this chapter is to give an overview of the main spectral abnormalities associated with inborn errors of metabolism and how to use spectroscopy to improve diagnostic imaging accuracy. Clinical and MR imaging pertinent findings will be presented for the most relevant diseases.

Classification of metabolic disorders

Metabolic disorders include a very numerous and heterogeneous group of diseases due to an acquired or genetic (inborn) defect that is responsible for malfunctioning of one or more cell types in the nervous tissue. Acquired metabolic disorders may be due to six main categories: noninfectious-inflammatory, infectious-inflammatory, toxic, nutritional, hypoxic–ischemic, and traumatic. They are discussed in Chapters 6, 7 and 10.

The classification of inborn error of metabolism is complex due to the multitude of genes and proteins involved. Many attempts have been made to classify inborn errors of metabolism; two of the most useful classification schemes rely on the biochemical pathway involved or the cellular organelle involved. The former classification scheme includes organic acidopathies, amino acidopathies, disorders of carbohydrate and metal metabolism. The latter classification includes *lysosomal, peroxisomal,* and *mitochondrial disorders.* Lysosomes are the cell's cleaning station and contain multiple enzymes, whose primary role is to break down macromolecules. A genetic defect in one of these lysosomal enzymes will cause accumulation of non-degradable macromolecules and different subtypes of lysosomal storage disease: mucopolysaccharidoses, sphingolipidoses, oligosaccharidoses, mucolipidoses. One of the many functions of peroxisomes is the breakdown of very long chain fatty acids. Mitochondria are the only organelles that contain their own DNA. They are the energy room of the cell: defects in any of the many mitochondrial pathways can cause mitochondrial disorders.

Metabolic disorders may also be classified according to the brain tissue predominantly involved: disorders affecting gray matter go under the name of *polioencephalopathies,* those affecting white matter *leukoencephalopathies.* The neutral word *polioencephalopathy* is preferable to poliodystrophy because it emphasizes the preferential anatomic location of the malfunction without implying that the disease is sparing the white matter. The distinction between the anatomic structure involved is very useful in neuroradiology because it facilitates pattern recognition analysis.

In the last two decades, MR imaging and genetic methods have dictated the tempo of new disease discovery. MR imaging has taken the place of neuropathology in the identification of clusters of patients sharing homogeneous imaging and clinical findings. New disease entities like leukoencephalopathy with vanishing white matter (VWM), megalencephalic leukoencephalopathy with subcortical cysts (MLC), leukoencephalopathy with calcifications and cysts, leukoencephalopathy with involvement of brain stem and spinal cord and elevated white matter lactate (LBSL), hypomyelination with atrophy of the basal ganglia and cerebellum have all been identified with the critical aid of MR, ^{1}H-MRS, and other imaging methods.

Ultimately, any type of classification scheme should be regarded as provisional, and there is no doubt that it will have to be adapted in the future to include hereditary metabolic diseases whose putative defects are yet to be elucidated. In this chapter uses a practical clinically based approach is adopted. First, disorders with specific ^{1}H-MRS features are reviewed separately, to give more emphasis to these diseases that can be diagnosed directly with ^{1}H-MRS. Then the polioencephalopathies, which show prevalent involvement of the gray matter (mitochondrial disorders, aminoacid and organic acidopathies) are discussed. Finally, the leukoencephalopathies with hypomyelination, demyelination, and rarefaction of the white matter are addressed.

Conventional MR imaging

The advent of MR imaging has provoked a dramatic acceleration in the recognition and diagnosis of metabolic diseases involving the brain. MR imaging is probably the most important test in the initial diagnostic work-up of children with a suspected metabolic disorder. A defect of metabolism will ultimately be diagnosed and treated according to its genetic mutation; nevertheless, brain imaging will continue to increase its pivotal role in identification and characterization of tissue pathologies. The initial assessment of the *conventional* MR images should determine whether the disease involves predominantly the cortical gray matter, the white matter, the subcortical gray nuclei, or whether it is a diffuse process.

When signal abnormalities in the subcortical gray nuclei are the most important feature, the location of the affected structures becomes very important for the differential diagnosis. Hallervorden–Spatz, methylmalonic acidemia and L-2-hydroxyglutaric aciduria (L-2-OHGA) are among the diseases targeting the globus pallidus. Leigh syndrome, MELAS, a few organic acidemias, Wilson disease, and Huntington disease target the striatum.

In the differential diagnosis of leukoencephalopathies the following features are very useful. The presence of megalencephaly suggests diagnosis of either Alexander, Canavan, L-2-OHGA, or MLC. When early in the disease the subcortical white matter (U fibers) is spared the diagnosis of metachromatic leukodystrophy (MLD), globoid cell leukodystrophy (GCL), or adrenoleukodystrophy (ALD) are more likely. Canavan and Kearn–Sayre disease (a mitochondriomyopathy) target the subcortical white matter instead. It is of interest that metabolic diseases rarely show blood–brain barrier damage and enhance following intravenous gadolinium administration: Alexander disease and ALD are the exceptions to this rule. Other important features are initial distribution of signal abnormalities in the white matter and the occurrence of atrophy.

Tips for performing ^{1}H-MRS exams in metabolic disorders

For a full discussion of MRS methods, please see Chapters 1 and 2. Single-voxel ^{1}H-MRS is an easy and robust method that is widely available on 1.5 Tesla and higher field strength MR systems without any hardware modifications. A high-quality spectrum is acquired from a selected volume of brain within a reasonable short acquisition time typically in the range of 2–5 min. The voxel is positioned in the brain where the pathology is assumed to be, and acquiring a second spectrum from an analogous brain location spared by the pathological process is recommended for comparison whenever possible. The typical voxel size is around 8 ml. The choice of the echo-time (TE) is very flexible (see below).

Multi-voxel 2D ^{1}H-MRSI is also robust and acquires multiple spectra from one or more slices. The spatial resolution is usually better than with the single-voxel method. For instance, at 1.5 T it is possible to use a nominal resolution of 0.54 ml (FOV = $200 \times 200 \times 15$ mm^3; 24×24 matrix) with a total scan time of 14 min. Metabolic maps are produced for the main peaks: choline, creatine, NAA, lipids, and lactate. The volume of interest (VOI) is usually positioned at the level of the slice with signal abnormalities on T_2-weighted images. In metabolic disorders in particular, the centrum ovale and the basal ganglia are the most frequently chosen VOI. Intermediate or long TE (TE 140 or 280 msec) have been used for many years because they produce clean spectra with flat baselines and relatively few overlapping peaks. Recently, multi-voxel sequences with very short echo times (e.g. TE = 15–35 ms), high-quality spectra, and spatial resolution of ~1 cm^3 have been introduced.[2] Multi-slice sequences or 3D MRSI sequences (if available – currently not all manufacturers support multi-slice MRSI) are valuable because they offer more coverage of the brain. The use of out-of-volume lipid suppression is highly recommended, since strong peri-cranial lipid signals may otherwise interfere with the signal of other metabolites. Reconstruction of the MRSI data is more time-consuming than single-voxel, and often in the past required dedicated custom software. However, currently all major commercial vendors provide MRSI reconstruction with their spectroscopy packages of varying degrees of sophistication. The metabolic maps and a grid with multiple spectra must first be inspected for quality and presence of artifacts (specifying types of artifacts), then interpretation of the spectroscopic findings is usually accomplished by comparison of areas with pathology with those spared by the disease and presumed normal.

The signal intensity of each metabolite depends on its T_2 relaxation time; therefore spectra with TE<30 ms should be acquired when the detection of a compound with a short T_2 is the object of the study. Alternatively, TE = 140 ms (intermediate) may be chosen when detection of lactate is one of the priorities of the study, and short T_2 compounds (e.g. glutamate, myo-inositol, lipids) are not required. Very long TE values (TE = 270–280 ms) have been used extensively in the past, since they offer a very clean baseline and an in-phase lactate signal. However, these long TE values also suffer from variable T_2 losses, and are not recommended for high field (e.g. $\geq$ 3 T), as T_2 relaxation times decrease at higher fields.

Before prescribing an MR exam, it is also important to review current and past MR imaging studies. A brief discussion about the patient's symptoms and the clinical differential diagnoses with the referring physician is often very informative. A better understanding of the metabolic pathways that may be affected by the disease will improve the neuroradiologist's skills in choosing the optimal MRS sequence and interpreting the spectra.

Diagnosis-specific ^{1}H-MRS findings

As stated before, in most metabolic disorders the ^{1}H-MRS changes detected are not disease-specific. However, in some disorders ^{1}H-MRS can disclose pathognomonic abnormalities in the spectrum that suggest a unique diagnosis.

Specific ^{1}H-MRS findings can be classified in three groups: (a) absence of a peak that is present in the normal brain spectrum; (b) appearance of a new abnormal peak; and (c) abnormal elevation of one or more peaks already detectable in the normal spectrum. Recently, two new disorders have been discovered with MR spectroscopy, and they belong to the first group.

Creatine deficiency

Creatine (Cr) deficiency syndromes result in an early-onset progressive encephalopathy with severe language impairment and mental retardation, extrapyramidal

Chapter 11: MRS in cerebral metabolic disorders

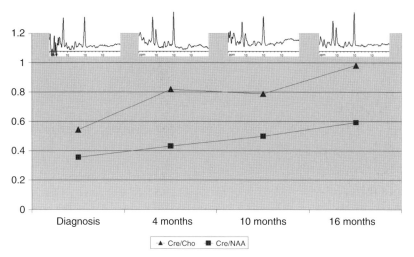

Figure 11.1. Longitudinal MRSI (PRESS: TR/TE = 1500/136 ms; 16 × 16 matrix; FOV = 160 × 160 × 20 mm^3) follow-up study in a 2-year-old boy diagnosed with *creatine GAMT deficiency*. At diagnosis the spectrum acquired in the right centrum semiovale showed a complete absence of the Cr peak at 3.0 ppm. The plot shows progressive recovery of the Cr/Cho and Cr/NAA ratios following oral diet supplementation with Cr monohydrate (700 mg/day) at 4, 10, and 16 months after diagnosis. In this particular case, MRSI was useful to confirm the pediatrician's suspicion that the parents had stopped supplementing Cr to the infant's diet between 8 and 10 months. Cr reached normal levels at 16 months after resuming the diet.

motor signs, and drug-resistant epilepsy. Symptoms are related to depletion of Cr in the brain.[3]

Cr synthesis occurs in the liver, pancreas, and kidneys, and two enzymes are involved. Cr is transported through the blood to tissues void of Cr synthesis. In the brain, Cr is transported into the neuronal and glial cells through the Cr transporter 1. The majority of patients have a defect of the second enzyme involved in Cr synthesis: guanidinoacetate methyltransferase (GAMT);[4,5] others have a defect in arginine–glycine amidinotransferase.[6] The two enzymatic defects are treatable with oral Cr supplementation (~400 mg kg^{-1} day^{-1}). ^{1}H-MRS is valuable both to suggest the initial diagnosis (based on absence of Cr in the spectrum) and to document restoration of the Cr pool in the brain after treatment (Figure 11.1). If ^{1}H-MRS shows that oral Cr supplementation is totally ineffective, even at high doses, with an absence of cerebral Cr in the post-treatment spectrum, the possibility of a defective Cr transporter must be considered. This Cr deficiency syndrome due to a defect of the transporter is X-linked and it is due to molecular defects in the CrT1 gene *SLC6A8*.[7]

Global NAA deficiency

Another condition with a very specific ^{1}H-MRS spectrum is the unique case of a child with severe developmental delay and absence of cerebral NAA, in whom the most prominent peak of ^{1}H-MRS was undetectable.[8] The authors hypothesized a defect in the biosynthesis of acetyl-CoA-L-aspartate-N-acetyltransferase (ANAT). NAA is predominantly located in neurons and their prolongings. Decreased NAA is associated with neuronal loss in a wide range of diseases. A few reports of reversible NAA signal loss in multiple sclerosis, acute disseminated encephalomyelitis, and mitochondrial encephalopathies suggest that reduction of the NAA signal may also be due to temporary neuronal dysfunction. The report of a 3-year-old boy with microencephaly and severe developmental delay without language acquisition, but normal MR imaging and EEG, challenged the concept of using NAA as a marker of neuroaxonal integrity. Notwithstanding, the importance of NAA as a surrogate marker of neuronal health is supported by the extreme rarity of this defect, and the severity of cognitive impairment in this child.

The absence of abnormalities on conventional MR images in Cr and NAA deficiency syndromes confirms the potential of ^{1}H-MRS in revealing metabolic abnormalities in children with severe developmental disorders but normal-appearing tissue on MR imaging.

The second group of diseases we discuss include diseases involves accumulation of molecules that are normally absent (or present in concentrations below ≈ 0.5 mM g^{-1} brain tissue[9]), and are thus below the usual threshold of detectability of ^{1}H-MRS at 1.5 Tesla.

Phenylketonuria

In phenylketonuria, an aminoacidopathy due to impairment of the phenylalanine hydroxylating pathway, ^{1}H-MRS shows an abnormal peak in the spectrum at 7.36 ppm (to the left ("low-field") side of the water resonance) due to accumulation of phenylalanine. Early diagnosis is usually made thanks to worldwide neonatal screening programs that test phenylalanine levels in the blood. ^{1}H-MRS can be used to follow the influx of phenylalanine from the blood into brain tissue as well as to monitor the response to dietary therapy.[10,11]

Leukoencephalopathy and neuropathy of unknown origin

Two abnormal peaks between 3.5 and 3.8 ppm at short and long TE were observed in the MR spectrum of a 14-year-old boy with leukoencephalopathy and neuropathy of unknown origin.[12] The abnormal resonances were assigned to arabitol and ribitol, which are involved in the metabolism of polyhydric alcohols (polyols). The abnormal peaks were associated with signal loss of the main metabolites in both white and gray matter regions. MR imaging showed diffuse T_2-weighted signal abnormalities in the cerebral white matter, with relative sparing of the periventricular areas, corpus callosum, and internal capsule.

Polyols are particularly abundant in the CNS and they derive from reduction of sugars. Elevated polyols have also been found in diabetes mellitus and galactose intoxication, which are associated with a peripheral neuropathy with myelin vacuolization. Abnormally elevated concentrations of polyols have been observed in metabolic (hypergalactosemia and hepatic encephalopathy) and neurodegenerative diseases (Alzheimer and Down syndrome) as well.

Succinate-dehydrogenase deficiency

The presence of an abnormal peak at 2.42 ppm in the cerebral and cerebellar white matter is distinctive of succinate-dehydrogenase deficiency (SDH),[1,13,14,15,16,17] a rare cause of mitochondrial leukoencephalopathy associated with reduction of SDH and respiratory chain complex II activities. The spectroscopic singlet originates from the two equivalent methylene groups of succinate. So far only six children from three families (Italian, Turkish and Norwegian) with this disease have been reported. In the three children from the Italian pedigree, spectroscopic and imaging findings were very similar and consisted of succinate accumulation in the white matter, also associated with mild elevation of lactate and severe depletion of Cho, Cr, and NAA (Figure 11.2). MR imaging showed symmetric extensive leukoencephalopathy with cavitations, best demonstrated on FLAIR. The disease seems to spare the gray matter, where spectra and MR imaging appear normal.[1] Three additional children have been described by Hanefeld *et al.*[15,16]

SDH consists of only four subunits encoded in the nucleus and shares two subunits (Ip and Fp) with complex II. Mutations in the Fp subunit have been reported in three patients with leukoencephalopathy. It is interesting to note that the appearance of succinate occurs in only a minority of patients with decreased complex II activity, while lactate is elevated in most patients. Recently, a mutation on chromosome 19 associated with SDH deficiency has been identified in the Italian family.[17] Although few cases of SDH have been identified to date, as more children with leukoencephalopathy become evaluated with ^{1}H-MRS, it may become more commonly diagnosed. No satisfactory treatment is currently available for complex II deficiency. Stabilization of clinical symptoms was documented in three children after treatment with riboflavin.[18]

Maple syrup urine disease

Maple syrup urine disease (MSUD) is an autosomal recessive amino acidopathy with many heterogeneous phenotypes. The disease is caused by a deficiency in activity of a mitochondrial multi-enzyme complex catalyzing the oxidative decarboxylation of branched-chain α-keto acids (BCKA). The concentration of the intermediate BCKAs (leucine, valine, and isoleucine) increases in blood, urine, and CSF as the result of the defect. The most common and most severe phenotype is the neonatal classical phenotype. Infants are normal at birth, but a few days later become lethargic with poor feeding, hypotonia alternating to hypertonia. An odor of maple syrup is frequently noted. Early diagnosis and treatment with dietary restriction of essential branched-chain amino acids may reverse the neurological symptoms, but sequelae are common in adequately treated patients.

The acute phase of the neonatal phenotype is associated with diffuse brain edema and diffuse signal hyperintensity on T_2-weighted MR images. Diffusion-weighted imaging is very useful for demonstrating a characteristic pattern of white matter involvement with decreased ADC in myelinated tracts, whereas ADC is increased in the lobar unmyelinated white matter.[19] Accumulation of branched-chain α-keto and amino acids are believed to trigger myelin instability in the acute phase of MSUD. In myelinated

Chapter 11: MRS in cerebral metabolic disorders

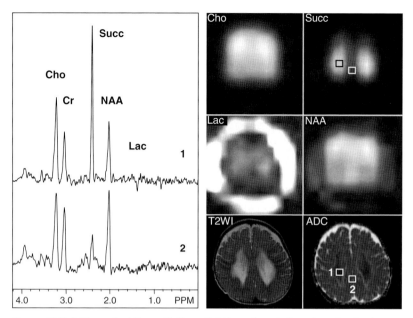

Figure 11.2. A 12-month-old boy with 3 month history of arrest of psychomotor development and tetraparesis. He was diagnosed with mitochondrial encephalopathy due to **succinate dehydrogenase deficiency** (SDH).
MRSI (PRESS: TR/TE = 1200/135 ms; 24 × 24 matrix; FOV = 200 × 200 × 15 mm^3) was acquired at the level of the centrum semiovale. Cho, succinate (Succ), lactate (Lac) and NAA maps with two selected spectra are illustrated.
The white matter spectrum from the right centrum semiovale (1) showed an abnormally elevated succinate peak at 2.42 ppm, associated with moderate NAA and mild Cr signal losses. Lactate was mildly elevated in white matter voxels. The gray matter spectrum from the adjacent parasagittal parietal cortex showed only minimal succinate accumulation and mild NAA signal loss. The Succ map elegantly demonstrates accumulation selectively in white matter, confirming that the metabolic defect targets white matter and spares gray matter. In the Lac map the bright signal around the brain arises from the lipid signal in the scalp.
On the T_2-weighted MR images at the level of the centrum semiovale, note the signal hyperintensity in the deep white matter with relative sparing of the subcortical white matter. On the ADC map diffusivity is reduced in the deep periventricular white matter; it was reduced also in the corpus callosum and corticospinal tracts (images not shown).

white matter water molecules accumulate, splitting the lamellae and forming intramyelinic vacuoles. The condition is known as intramyelinic edema. The rate of water diffusion decreases within vacuoles if their diameter is small, while in poorly myelinated tracts, the ADC is increased due to blood–brain barrier damage and diffuse vasogenic edema in the interstitium.

During a metabolic crisis, multiple peaks are detectable around 0.9 ppm in the brain spectrum. This region of the spectrum is usually attributed to lipids, but in MSUD the peak is believed to be formed by resonances of methyl protons from branched-chain α-keto and amino acids that accumulate as a result of defective oxydative decarboxylation of leucine, isoleucine, and valine.[20] A similar abnormal peak in the spectrum at 0.9 ppm has been reported also in patients with Niemann–Pick type C disease, probably due to a defective metabolism of cholesterol with ceramide accumulation.[21,22] In both MSUD and Niemann–Pick type C diseases, the abnormal peaks detectable at 0.9 ppm seem to disappear after appropriate therapy.[19,20] Occasionally, in the acute phase of MSUD, lactate may be elevated and NAA decreased.

Canavan disease

Canavan disease (CD) is a rare organic acidopathy leading to severe spongiform leukoencephalopathy. It is due to a defect of the enzyme aspartoacylase (ASPA) and affects infants and children. ASPA breaks down NAA to free acetate and aspartate, and its deficiency is responsible for the abnormally high levels of NAA found in the brain and urine of sick children.[23,24] CD has an autosomal recessive mode of inheritance with high prevalence among Ashkenazi Jews.[25] The infantile form is the most common; the child is apparently normal at birth and symptoms appear within the first six months of life. In 1990, ^{1}H-MRS showed that NAA is indeed elevated (particularly ratios of NAA/Cr) in the brain of patients with CD spongiform

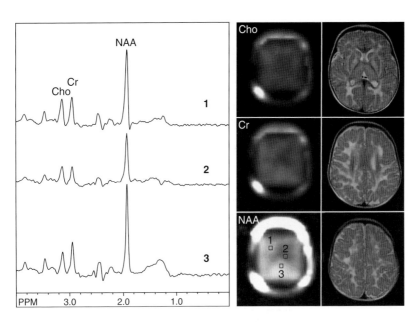

Figure 11.3. Ten-month-old girl presenting with hypotonia with poor head control, mild spasticity, and macrocephaly, diagnosed with **Canavan disease**.
 MRSI (PRESS: TR/TE = 1200/135 ms; 24 × 24 matrix; FOV = 200 × 200 × 15 mm^3) was acquired at the level of the lateral ventricles. Cho, Cr, and NAA maps with three selected spectra are illustrated.
 Note abnormal marked elevation of the NAA peak in the right centrum semiovale (spectrum 1) and in the parasagittal parietal lobe (spectrum 3). The elevation of NAA is milder in the left corona radiata where less signal hyperintensity is seen on the T_2-weighted MR images (right column).

leukoencephalopathy for the first time.[26–28] MR imaging is also characteristic and shows mild swollen subcortical white matter with T_2 prolongation effects, bilateral involvement of the globi pallidi and thalami, brainstem tracts, and cerebellum. Sparing of the neostriatum results in an MR image that is very typical of CD. Diffusion weighted MR images (DWI) show decreased water diffusivity in the subcortical white matter and thalami early in the disease, probably due to intramyelin edema, whereas increased apparent diffusion coefficient (ADC) is found later in the course of the disease probably due to further degeneration of the white matter infrastructure. The DWI findings are similar to those observed in MSUD.[29] Multi-voxel ^{1}H-MRSI best illustrates the heterogeneous distribution of the metabolic abnormalities (Figure 11.3): NAA is most elevated in the subcortical white matter and thalami which are affected earlier by the characteristic spongy degeneration of myelin described by van Bogaert and Bertrand in 1949. Cho is often slightly decreased throughout the white matter while Cr may be elevated, especially in the subcortical white matter. With the advent of ^{1}H-MRS and MR imaging, more patients with CD have been diagnosed, suggesting that the disease may be more common than previously thought. It should also be noted that an elevated "NAA" signal can also be found in Salla disease and severe infantile sialic acid storage diseases. However, in these conditions the high NAA signal reflects an accumulation of N-acetyl-neuraminic acid (NANA) rather than accumulation of NAA itself. Moreover, mild NAA elevation has been reported in the white matter of few patients with Pelizaeus–Merzbacher disease (PMD). This apparent increase in NAA may reflect increased axonal density due to thinner myelin sheaths around the axons.[30–32] Elevation of NAA must therefore be confirmed in urine and the deficit of ASPA demonstrated in cultured fibroblasts for a definitive diagnosis of CD. However, the assessment of ASPA activity is difficult and not always reliable. Ultimately, the diagnosis is confirmed by demonstration of mutations in the ASPA gene. This step is particulary important in the diagnosis of the milder juvenile phenotype.[33] At present there is no specific therapy for Canavan patients.

Free sialic acid storage disorders

As mentioned above, sialic acid storage disorders are characterized by accumulation of free sialic acid (N-acetyl-neuraminic acid, NANA) in lysosomes due to defective transport across their membranes.[34] Sialic acid storage diseases are caused by autosomal recessive mutations of a lysosomal sialic acid transporter called sialin.[35] They include a severe form and a milder variant (also known as Salla disease [34]). Sialin mutants found in the lethal infantile type are inactive, whereas residual transport activity is present in the milder clinical form. The most severe neonatal phenotype presents with coarse facial

Chapter 11: MRS in cerebral metabolic disorders

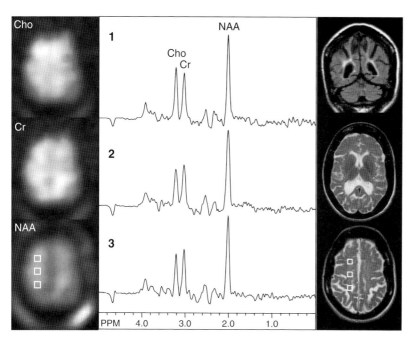

Figure 11.4. 37-year-old woman with mild motor developmental delay, mild cognitive impairment. Deterioration of her motor and cognitive symptoms occurred in the last four years. She was diagnosed with **Salla disease**.

MRSI (PRESS: TR/TE = 1200/135 ms; 24 × 24 matrix; FOV = 200 × 200 × 15 mm^3) was acquired at the level of the centrum semiovale. Cho, Cr, and NAA maps with three selected spectra are illustrated.

Note mild elevation of the NAA peak in all three spectra (numbered 1, 2, 3 from anterior to posterior) in the right centrum semiovale. Average ratio measured in the centrum semiovale were as follows: Cho/NAA = 0.45; Cho/Cr = 0.76; NAA/Cr = 1.67. NAA was elevated despite diffuse white matter signal hyperintensity and discrete brain volume loss as demonstrated on the coronal FLAIR and axial T_2-weighted MR images (right column). The diagnosis was confirmed by detection of increased sialic acids in the urine.

features and generalized hypotonia, then followed by failure to thrive and severe delayed development. Spastic tetraparesis develops with hypertonia and hyperreflexia. Most patients with the severe form commonly die within 5 years of age. Over 100 patients have been diagnosed with Salla disease in Finland, where the disorder occurs with a relatively high frequency. After an uneventful perinatal period, the first symptoms appear at 6–9 months with generalized hypotonia, ataxia, and nystagmus. Patients then develop spasticity with signs of athetosis. The majority of patients become able to walk and speak few words with short sentences.[36] Mental development is delayed and those reaching adulthood are mentally retarded. Life expectancy may be only slightly decreased in patients who do not present with the severe childhood form. As mentioned above, an apparent increase of the "NAA" resonance has been reported in Salla disease.[34] The signal elevation at 2.02 ppm may be due to accumulation of free NANA (which also has an N-acetyl moiety, like NAA, and hence resonates at the same spectral frequency) that offsets possible losses of NAA inside the axons (Figure 11.4). In the severe form, neuropathology studies have shown paucity of white matter with a very thin corpus callosum. Also the cerebellum, brain stem, and spinal cord are severely reduced in size. In the white matter, a marked reduction of axons is associated with pronounced astrogliosis. MR imaging may show some atrophy of the brain, a thin corpus callosum with marked dilatation of the lateral ventricles. The reduction in volume of the white matter is remarkable. In Salla disease, the reduction of white matter volume is less severe; however, a thin corpus callosum is often observed. The diffuse subtle hyperintensity on T_2-weighted MR images due to scarce myelination has a similar appearance to that seen in PMD patients, who may also show mild elevation of NAA on brain MRS. In white matter, Cho may be slightly decreased, while Cr may be slightly increased. Histological studies have shown vacuoles in neurons and astrocytes throughout the brain with storage of large amounts of lipofuscin in the cerebral cortex, basal ganglia, thalamus, brain stem nuclei, cerebellar cortex, and spinal cord. Accumulation of sialic acid in neurons and astrocytes may lead to dysfunction in myelin formation and maturation, that is the result of a close interaction of these cells with oligodendrocytes.

Nonketotic hyperglycinemia

Nonketotic hyperglycinemia is an autosomal recessive amino acidopathy caused by a defect in the glycine cleavage system which presents with neonatal, infantile, and later-onset phenotypes. The neonatal form is the most frequent. After a brief normal postnatal interval, lethargy, hypotonia, myoclonic jerks, and seizures appear with progressive severity. The majority of patients will die within a few weeks.

MR imaging may show hyperintense signal abnormalities on T_2-weighted sequences in the posterior limb of the internal capsules, dorsal midbrain, and pons. Diffusion imaging shows low ADC values in these regions, probably related to intramyelin vacuolation. The corpus callosum may be extremely thin. These imaging findings are similar, but less prominent than those found in MSUD. The corticospinal tracts, cerebellar peduncles, optic nerves, and tracts are the most severely affected in the neonatal form, while tracts that have completed myelination before birth (anterior and posterior spinal roots) or that are unmyelinated are less affected. Widespread spongiosis of actively myelinating white matter is observed on histopathology.

^{1}H-MRS shows an abnormally elevated signal resonating at 3.55 ppm. The assignment to glycine can be confirmed at long echo time (e.g. 280 ms),[37] since glycine has a longer T_2 value than *myo*-inositol, which co-resonates with glycine at 3.56 ppm. Diagnosis should be confirmed with demonstration of glycine elevation in CSF, plasma, and urine in the absence of ketoacidosis. In earlier studies,[38,39] the reduction of glycine resonance with treatment corresponded more reliably with clinical findings than levels in plasma and cerebrospinal fluid, suggesting that ^{1}H-MRS can be a valuable tool in the diagnosis and monitoring of treatment effects in patients with this rare disorder.

Mitochondrial disorders

Mitochondrial disorders are genotypically and phenotypically very heterogeneous and the classification is complex. In 2003, Di Mauro proposed a revised classification in three main groups of mitochondrial disorders:[40] (a) respiratory chain disorders due to defects of mitDNA; (b) respiratory chain disorders due to defects of nuclearDNA; and (c) disorders with indirect involvement of the respiratory chain.

From a biochemical viewpoint, it is useful to distinguish mitochondrial disorders into three types according to the dysfunctional energy production pathway: (a) pyruvate metabolism; (b) respiratory chain; and (c) fatty acid oxidation.

There is much overlap in terms of MR imaging and spectroscopic abnormalities between different genetic and clinical phenotypes.[41] In this section, a brief description of the clinical syndromes will be followed by a discussion of their imaging and spectroscopic findings.

Disorders of pyruvate metabolism

Pyruvate metabolism is an essential step in energy production and gluconeogenesis in all organs with high energy demand. The pyruvate dehydrogenase complex transforms pyruvate into acetyl-CoA that enters the tricarboxylic acid cycle. The two most common abnormalities of pyruvate metabolism are *deficiency of pyruvate dehydrogenase complex* and *pyruvate carboxylase*. The former is the most common cause of lactic acidosis. Pyruvate is then converted into lactate, which produces significantly less ATP per glucose molecule compared to metabolism through the TCA cycle.

Pyruvate carboxylase deficiency is a rare autosomal recessive disorder with three clinical phenotypes: neonatal, infantile, and a milder form in patients who may survive until the third decade. Involvement of the brain is present in all subtypes. The disorder may present with a severe progressive encephalopathy with signs of respiratory instability, feeding problems, hypotonia, pyramidal signs, and ocular abnormalities. On MRI, the cerebral white matter is mainly affected, with poor myelination and cystic degeneration.

Defects of the respiratory chain

Mitochondrial disorders due to defects of the respiratory chain are the most common metabolic disorder with an estimated incidence of 1 per 10,000 live births. They are extremely heterogeneous with variable age of onset, progression, and severity. Similar mutations may present with different clinical phenotypes. The same clinical syndrome may be associated with multiple genotypes. Diseases associated with defects of the mitDNA are Kearns–Sayre syndrome (KSS), mitochondrial encephalopathy with lactic acidosis (MELAS), Leber hereditary optic neuropathy (LHON), myoclonus epilepsy and ragged-red fibers (MERRF), neuropathy, ataxia and retinitis pigmentosa (NARP), and progressive external ophthalmoplegia (PEO). In recent years, interest has shifted toward respiratory chain disorders due to mutations in the nuclear DNA. Mutations in structural components of the respiratory chain have thus far been found only in complex I and II deficiency, and have been associated with severe neurologic disorders of childhood, such as Leigh syndrome and a few leukoencephalopathies.

Fatty acid oxidation disorders

Disease entities due to a defect in the fatty acid β-oxidation pathway are difficult to diagnose but they are probably more common than previously thought. The β-oxidation of fatty acids leads to formation of ketone bodies, which are a particularly important source of energy metabolism during fasting when glucose availability is limited. Glucose and ketone bodies, followed by lactate, are the most important energy sources in the brain, whereas fatty acids and ketones are most important in muscles. The oxidation of fatty acids is a complex process with defects at multiple step levels: carnitine cycle defects, β-oxidation disorders, electron transfer flavoprotein defects and ketone body synthesis and degradation. They are characterized by multiorgan failure with signs of acute or chronic encephalopathy. The aim of treatment is correction of carnetine deficiency and reduction of lipolysis to prevent fatty acid overload in liver. An example of a defect in ketone body degradation is 3-hydroxy-3-methylglutaryl coenzyme A lyase, which will be discussed later in the organic acidopathies section. Glutaric aciduria type 2 is a severe mitochondrial disorder caused by deficiency of the electron transfer flavoprotein, failing to transport electrons by dehydrogenase enzymes. MR imaging abnormalities include enlarged sylvian fissures due to underdeveloped frontal and temporal lobes, hypoplasia of the corpus callosum, and agenesis of the cerebellar vermis.

MR imaging characteristic findings

The most characteristic MR imaging diagnostic clues are the following: (a) symmetric signal abnormalities in specific gray nuclei: basal ganglia, thalami, subthalamic nuclei; (b) involvement of pontine tegmentum and midbrain tectum; (c) diffuse atrophy; (d) cortical territories without a vascular distribution (i.e. stroke-like lesions in MELAS); (e) basal ganglia calcifications; and (f) diffuse cerebral and cerebellar white matter.

Symmetric T_2-signal abnormalities in the subthalamic nuclei and/or in the gray nuclei of the brain stem are suggestive of mitochondrial encephalopathy in infants and children with Leigh disease.[42] Cerebral or cerebellar atrophy without any signal abnormality is another possible and relatively frequent imaging finding associated with mitochondrial encephalopathy. Signal abnormalities in the cortex and subcortical white matter that do not match a vascular distribution are a characteristic feature of MELAS. Diffuse white matter signal changes are the predominant abnormality in about 10% of patients: in a retrospective study on 110 children with mitochondrial disorders, 8 patients had leukoencephalopathy without basal ganglia or cortical signal abnormalities.[43] Cavitations in the white matter may develop in a few of these patients.

MR spectroscopy

^{1}H-MRS is a valuable tool for the investigation of suspected mitochondrial disorders; the most common finding is elevation of lactate in the brain and/or in CSF.[44,45] Although lactate may be elevated in mitochondrial disorders,[46] it should be emphasized that lactate levels are not elevated in all patients and in all anatomic structures.[13,14,47] Lactate elevation beyond areas with MR signal abnormality may be indicative of widespread energy failure associated with mitochondrial dysfunction.[48,49]

In a ^{1}H-MRSI study, lactate was elevated in 17 of 25 children with mitochondrial encephalopathy at least in one of three compartments: brain, CSF, and plasma.[13] The incidence of lactate elevation in the brain was about 40%. The same study found that elevation of lactate is more frequent in patients with leukoencephalopathy than in patients with atrophy or symmetric subcortical nuclei signal abnormalities on conventional MR imaging. Lin et al. evaluated sensitivity and specificity of lactate detection in 29 children with or without genetic or biochemically confirmed diagnosis of a mitochondrial disorder.[47] They found that when lactate is detected, the probability that the patient has a mitochondrial encephalopathy was high; however, the diagnosis cannot be excluded when lactate is not elevated. In general, NAA is often markedly decreased in mitochondrial disorders (especially in MELAS and leukoencephalopathies), while Cho and Cr are usually mildly decreased.

Lactate may be elevated in the acute phase of Leigh disease with abnormal signal hyperintensities on T_2-weighted MR images in the basal ganglia (Figure 11.5). However, in the chronic phase, spectral changes are often not specific with moderate NAA signal decrease and no evidence of lactate accumulation. Lactate is frequently elevated in the white matter of patients with features of leukoencephalopathies with or without cavitations (Figure 11.6). Lactate accumulation in the white matter may be detected even in the chronic stage of the disease for years. In areas of white matter cavitation, Cho, Cr, and NAA are decreased and the differential diagnosis includes vanishing white matter (VWM). Mitochondrial encephalopathies with predominant signal changes in the white matter are likely

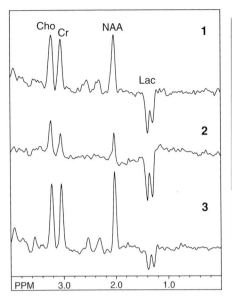 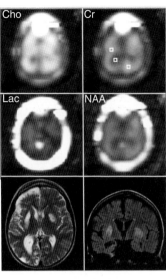

Figure 11.5. Seven-year-old girl with developmental delay since the first year of age, with ataxia and dysarthria; at 5 years onset of seizures. She was diagnosed with a **mitochondrial encephalopathy due to complex I respiratory chain defect**.

MRSI (PRESS: TR/TE = 1200/135 ms; 24 × 24 matrix; FOV = 200 × 200 × 15 mm^3) was acquired at the level of the basal ganglia. Cho, Cr, and NAA maps with three selected spectra (numbered 1, 2, 3 from anterior to posterior), axial T_2-weighted and coronal FLAIR MR images are illustrated.

Note abnormal moderate elevation of the Lac peak in the right putamen (1) and in the right ventricle (2). Elevation of Lac is also detected to a lesser degree in the left occipital lobe (3), an area without T_2-signal abnormality. NAA is only mildly decreased in the right putamen, whereas Cho and Cr are within the normal range.

Note the remarkable difference in imaging abnormalities of this case with the patient presented in Figure 11.6, despite the similar biochemical defect. On the contrary, MR spectra are very similar in these two patients with mitochondrial encephalopathy.

due to isolated defects of respiratory chain complexes I, II, or IV. The association of elevated succinate and lactate is pathognomonic of isolated complex II deficiency due to SDH reduced activity (Figure 11.2), so far the only mitochondrial disease with specific diagnostic spectroscopic findings.[14,15]

In one of the earliest reports, it was shown that ^{1}H-MRS was a valuable tool in assessing response to therapy in a 2-year-old boy with Leigh disease, progressive spasticity, dysarthria, and optic atrophy, and symmetric basal ganglia T_2-signal abnormalities on MR imaging due to a pyruvate dehydrogenase complex defect. After a ketonemic diet, the abnormal lactate peak was no longer detectable and clinical symptoms became stable.[50] However, in other case reports, lactate in the white matter remained elevated for years, despite improvement or stabilization of clinical symptoms.[18]

In other very rare metabolic disorders such as *ethylmalonic encephalopathy*[51] and a recently defined type of *leukoencephalopathy with brainstem and spinal cord involvement and lactate elevation* (LBSL),[52] the findings of diffuse brain mitochondrial impairment have strongly contributed to the interpretation of the complex pathogenetic mechanisms of these disorders. In both cases, lactate elevation detected by a multi-voxel ^{1}H-MRSI study may suggest a possible primary mitochondrial disorder. [53,54] Furthermore, in a very rare metabolic disorder such as cerebrotendinous xanthomatosis, the diffuse ^{1}H-MRS increase in brain lactate detected with a single-voxel ^{1}H-MRS study added to morphological and biochemical evidence of mitochondrial dysfunction,[55] probably secondary to the toxic effect of high cholestanol and/or bile alcohol levels.[56,57]

Leukoencephalopathy with brainstem and spinal cord involvement and lactate elevation

In view of lactate elevation in the white matter, it is no surprise that LBSL was eventually found to be a mitochondrial disease. LBSL is characterized by slowly progressive pyramidal, cerebellar, and often dorsal column dysfunction, sometimes with a mild cognitive deficit or decline. Mutations in gene *DARS2* of chromosome 1 cause a deficit in mitochondrial aspartyl-tRNA synthetase. Thus it has been hypothesized that LBSL might be a particular type of mitochondrial encephalopathy. So far, LBSL is the only known disease involving a mitochondrial aminoacyl-tRNA synthetase. The high expression of mitochondrial tRNAs in the brain may explain cerebral vulnerability in this disease. MR imaging has a distinct pattern with variable extent of cerebral and cerebellar white matter

Chapter 11: MRS in cerebral metabolic disorders

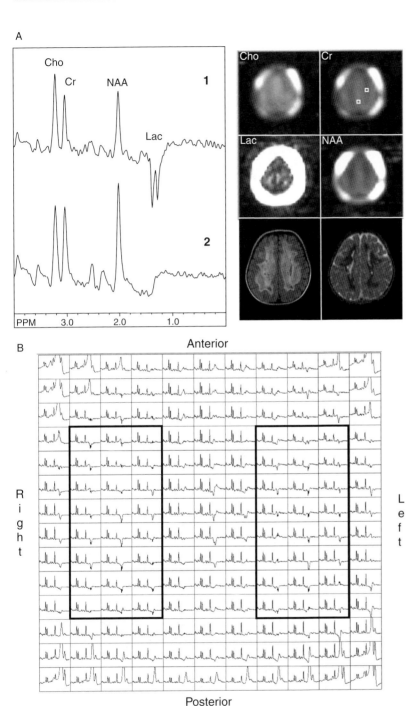

Figure 11.6. 16-month-old girl who presented with ataxia and severe developmental delay. She was diagnosed with **mitochondrial encephalopathy due to complex I respiratory chain defect**:

(A) MRSI (PRESS: TR/TE = 1200/135 ms; 24 × 24 matrix; FOV = 200 × 200 × 15 mm^3) was acquired at the level of the centrum semiovale. Cho, Cr, and NAA maps with two selected spectra (anterior 1, posterior 2), axial FLAIR and ADC maps at the same level are illustrated.

Note abnormal moderate elevation of Lac, associated with moderate NAA signal loss in the right centrum semiovale. A normal spectrum from the parasagittal parietal gray matter is displayed for comparison. Diffuse signal hyperintensity on axial FLAIR MR image are associated with moderate decrease of the ADC on the maps. In the deeper part of the centrum semiovale the ADC is elevated probably due to cavity formation.

(B) Evaluation of the multivoxel display from the same data set shows extensive elevation of lactate in the white matter of the centrum semiovale (outlined by the two rectangles), whereas in the gray matter there is no Lac accumulation and Cho, Cr, and NAA have normal signal intensities.

involvement, with sparing of the arcuate fibers.[52] MR signal abnormalities may progress over time. The corticospinal tracts are preferentially affected at the level of the corona radiata, posterior limb of the internal capsule, brain stem, and lateral columns of the spinal cord. The intra-axial segment of the trigeminal nerves, and the spinothalamic tracts over their entire length in the posterior columns of the spinal cord, and in medial lemniscus, are also affected. The cerebellar peduncles are also selectively involved: T_2-signal abnormalities occur early in the inferior and superior cerebellar peduncles, later in the middle peduncles.

¹H-MRS shows elevated lactate in the majority but not in all patients. The elevation of lactate may vary from very high to minimal lactate accumulation in a few white matter voxels. A significant decrease in NAA with near normal Cho, Cr, and *myo*-inositol elevation are common findings in the cerebral white matter of LBSL patients; the spectra in gray matter are normal.

Amino acidopathies

Amino acid metabolism takes part in many complex biochemical pathways. Neurological signs occur when the interrupted biochemical pathway plays a role in brain metabolism, structure, or neurotransmitter function. Amino acidopathies may cause extensive white matter demyelination and spongiform degeneration. There is a close interrelationship between amino acidopathies and disorders linked to accumulation of their organic acid intermediates. Several amino acidopathies have specific features on the MR spectrum and have been described in the appropriate section. Relevant amino acidopathies causing brain disorders and associated with relevant spectroscopic findings are reported in Table 11.1.

Urea cycle defects

The urea cycle is involved in elimination of toxic nitrogen in the form of ammonia, producing urea and arginine. Impairment of the urea cycle has several severe consequences with hyperammonemia and impaired metabolism of arginine, alanine, glutamine, and citrulline. Clinical signs of metabolic dysfunction can present at any time from early infancy to adulthood, and may be associated with diet changes or infections. Enzymes involved in the urea cycle are located within the cytosol and partly within the mitochondria. A defect in a urea cycle enzyme will have two main consequences: arginine becomes an essential amino acid, and nitrogen accumulates in a variety of molecules, particularly ammonia. Ammonia is highly toxic and diffuses easily into the brain. Increased levels of ammonia will influence the glutamine–glutamate–GABA balance. The disturbance of this balance between glutamate (the most important excitatory neurotransmitter) and GABA (the most important inhibitory neurotransmitter) may contribute to brain dysfunction due to accumulation of glutamine within astrocytes, diffuse brain edema, and swelling probably due to osmosis. On light microscopy, astrocyte

Table 11.1. Classification of hereditary metabolic disorders with the characteristic metabolic MRS findings.

	Cho	Cr	NAA	Lac	Others	Indication
Defects in gene encoding myelin protein						
Pelizaeus–Merzbacher disease	N	N	N	0	–	Recommended
Connexin 47	N	N	N	0	–	Recommended
Mitochondrial encephalomyopathies						
MELAS	−1	−1	−1	+2	–	Valuable
Leigh syndrome	−1	−1	−1	+2	–	Valuable
Menkes disease	N	−1	−2	+2	–	Reasonable
Respiratory chain complex	−1	−2	−3	+2	–	Valuable
SDH-Complex II deficiency	−2	−2	−3	+2	Suc	Valuable
Nuclear DNA repair defects						
Cockayne syndrome	N	N	N	N	–	Reasonable
Amino acid and organic acidurias						
Phenylketonuria	N	N	N	0	Phe	Recommended
Nonketotic hyperglycinemia	N	N	N	N	Gly	Recommended
Maple syrup urine disease	N	N	−2	+1	BCAA, BCKA	Valuable
Canavan disease	−1	+1	+3	0	–	Valuable
Glutaric aciduria type I	N	N	−2	+1	–	Reasonable

Table 11.1. (cont.)

	Cho	Cr	NAA	Lac	Others	Indication
Propionic acidemia	N	N	−2	+2	Glx	Recommended
L-2 Hydroxyglutaric aciduria	+1	+1	−2	0	−	Reasonable
Hyperhomocysteinemias	N	−1	−1	−1	Met	Valuable
Urea cycle defects	−1	−1	−1	+1	Gln	Valuable
Lysosomal storage disease						
Metachromatic leukodystrophy	+2	−1	−3	+1	−	Valuable
Globoid cell leukodystrophy	+2	−1	−3	+1	−	Valuable
Fucosidosis	N	N	−1	+1	−	Reasonable
Sialic acid strorage disease	N	+1	+2	0	−	Recommended
Neuronal ceroid lipofuscinoses	N	−1	−2	0	−	Reasonable
Peroxisomal disorders						
X-linked adrenoleukodystrophy	+3	−2	−3	+1	−	Valuable
Adrenomyeloneuropathy	N	N	N	0	−	Reasonable
Leukoencephalopathies						
Alexander disease	+3	−2	−3	+2	−	Valuable
Megalencephalic leukoencephalopathy with subcortical cysts	−2	−2	−2	+1	−	Recommended
Congenital muscular dystrophies	N	N	N	0	−	Reasonable
Leukoencephalopathy with vanishing white matter	−2	−2	−3	+1	−	Recommended
Leukoencephalopathy with involvement of brain stem and spinal cord and elevated white matter lactate	+1	N	−2	+2	−	Recommended
Hypomyelination with atrophy of the basal ganglia and cerebellum	N	N	N	0	−	Reasonable
Cerebral autosomal dominant arteriopathy with subcortical infarcts and leukoencephalopathy (CADASIL)	N	N	−2	+1	−	Reasonable
Cerebral autosomal recessive arteriopathy with subcortical infarcts and leukoencephalopathy (CARASIL)	N	N	−2	+1	−	Reasonable
Other genetic defects						
Creatine deficiency	N	−3	N	0	−	Valuable
NAA deficiency	N	N	−3	0	−	Valuable

Indication to request an MRS study is classified in three groups of diminishing importance: valuable, recommended, reasonable. Scale of metabolic single changes are indicated: 0=none,1=mild, 2=moderate, 3=marked
Abbreviations:
BCAA, BCKA= branched chain amino acid, branched-chain alfa-keto acid
Gln=glutamime
Glx=glutamate plus glutamine
Gly=glycine
met=methionine
Phe=phenylalanine
Suc=succinate

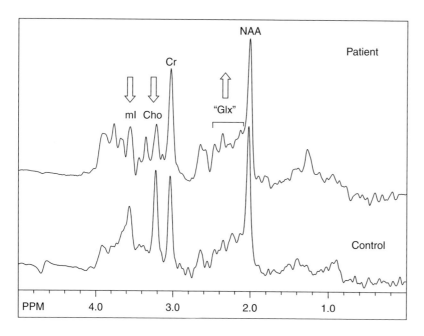

Figure 11.7. Hepatic encephalopathy. Example of single-voxel PRESS spectra from parietal white matter (short echo time, TE 35 msec) from a 2-year-child with liver failure and an age-matched control subject. Note the increased signal from Glx (due to increased glutamine) and decreased Cho in the patient. There is also a small decrease in *myo*-inositol.

swelling with Alzheimer type II changes can be observed. Ammonia appears to be the only cause of the acute encephalopathy in urea cycle defects, with the exception of hyperargininemia.

MR imaging abnormalities change according to age at presentation and the stage of the disease. Lesions are often focal and asymmetric. In neonates, severe brain swelling due to vasogenic edema in the white matter is the prominent finding during the acute stage. Myelinated white matter tracts are often more severely involved than unmyelinated tracts. This observation facilitates the differential diagnosis with MSUD. In the most severe cases there is also involvement of basal ganglia and cortex. MR spectroscopy at short TE (e.g. 20–35 ms) may show elevation of glutamine (Gln) at 2.2 ppm.[58] The major metabolites (NAA, Cho, and Cr) are frequently decreased, whereas lactate may be elevated. If the patient survives the acute crisis then severe atrophy with encephalomalacia will follow.

It should also be noted that elevated brain glutamine can occur in patients with hepatic encephalopathy (HE), as a result of increased glutamine synthesis secondary to high blood ammonia levels due to poor liver function. Such changes may be seen in mild or even preclinical cases of HE; a triad of metabolic changes are usually found, with decreased choline and *myo*-inositol as well as increased Gln. The reasons for this are unclear, but it may be due to osmotic changes, or decreased Cho delivery to the brain (Cho is synthesized in the liver). Figure 11.7 shows examples of short echo time spectra (TE 35 ms) in a child with HE, and a normal control subject for comparison.

Hyperhomocysteinemias

Hyperhomocysteinemias are associated with multiple medical conditions that may also involve the nervous system, and can cause mental retardation, cerebral infarction, and degeneration of the spinal cord. A heterogeneous group of biochemical defects may result in elevated concentrations of homocysteine in the blood. Methylation of methionine produces homocysteine, an important branch point of the sulfur amino acid metabolism. The metabolism of homocysteine requires folate and cobalamine; therefore a defect in either folate or cobalamin (see also methylmalonic aciduria) metabolism may also cause hyperhomocysteinemia. High levels of homocysteine in the blood stream are a risk factor for arterial disease in the heart, kidney, and brain. Homocysteine may also be toxic to the CNS with combined subacute degeneration of brain and spine, with spongiform white matter degeneration and demyelination. Lesions in the brain are usually less symptomatic than those in the dorsal and lateral columns of the spinal cord: they are usually small, ill-defined, and perivascular. In a minority of

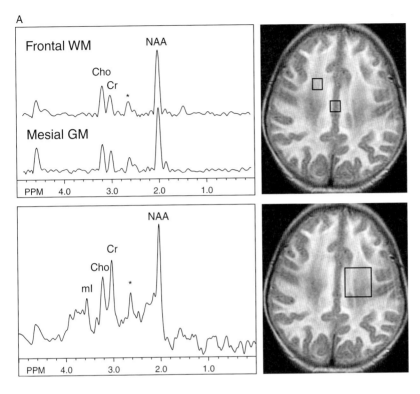

Figure 11.8A. Ten-year-old female with **Cystathionine B-Synthase (CBS) deficiency**, hyperhomocysteinemia, and elevated plasma methionine levels. T_2-MRI at presentation shows global elevation of white matter signal intensity, while selected spectra from gray and white matter (MRSI, TR/TE 2300/270 ms, top row) show near-normal metabolite levels, suggesting vasogenic edema rather than demyelination or axonal loss. Short TE white matter spectrum (bottom row) also shows near-normal metabolite levels. (*) indicates unknown resonance at approximately 2.6 ppm, possibly due to homocysteine.

cases, however, diffuse white matter signal abnormalities have been reported. In a 10-year-old girl with B6 unresponsive cystathione B-synthase deficiency with high methionine and homocysteine plasma concentrations, MR showed diffuse T_2-signal hyperintensity in the white matter, midbrain, and pons.[59] In this girl, ^{1}H-MRSI showed an abnormal peak at 2.6 ppm detected at TE = 30 and TE = 280 ms associated with minimal signal loss of the main metabolites (Figure 11.8A). The abnormal peak was tentatively assigned to methionine. Ex vivo MR spectra from a few amino acids are illustrated in Figure 11.8B. Homocysteine would resonate at 2.02 and 2.65 ppm; however, it has never been demonstrated in the brain with in vivo MRS. Hyperhomocysteinemia has been associated with low concentrations of NAA and Cr, but not Cho, in the white matter of centrum semiovale in a single-voxel ^{1}H-MRS study on 113 consecutive patients with coronary artery disease.[60]

Organic acidopathies

Organic acids are intermediates in carbohydrate, lipid, steroid, and biogenic amine metabolism. Primary defects in organic acid metabolism are difficult to recognize and diagnose. Most organic acids are concentrated in the urine. Organic acid concentrations in plasma or urine may be modified by dietary change, environmental contamination, and drug metabolism.

The majority of organic acidopathies are devastating diseases of the neonatal and infantile period that typically involve the basal ganglia. White matter involvement is usually less prominent and frequent than gray matter involvement, with the exception of L-2-hydroxyglutaric aciduria and a rare phenotype of methylmalonic aciduria (MMA). Relevant organic acidopathies causing brain disorders and associated with relevant spectroscopic findings are reported in Table 11.1. Canavan disease has characteristic MRS features. Accumulation of lactate is often elevated in other organic acidopathies.

Propionic (PA) and MMA acidurias are the most common in humans. Both disorders are autosomal recessive and typically present in the neonatal period with recurrent episodes of ketoacidotic metabolic crisis. Fewer patients will present in the first year of life with failure to thrive and recurrent episodes of vomiting and acidosis. These metabolic crises may mimic mitochondrial disorders, since lactate and ketone accumulation may inhibit succinate dehydrogenase activity. The globus pallidus is particularly sensitive to mitochondrial dysfunction.

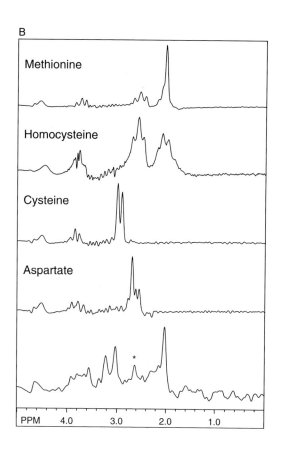

Figure 11.8B. Spectra from individual metabolites, showing that the 2.6 ppm peak observed in vivo could be consistent with either aspartate or homocysteine.

Propionic aciduria

PA is a complex disorder due to primary deficiency of propionyl coenzyme A carboxylase. It catalyzes conversion of propionyl coenzyme A into methylmalonyl coenzyme A as part of the catabolic pathway of amino acids isoleucine, methionine, and valine. Typical imaging findings are symmetrical T_2-signal abnormalities in the basal ganglia with or without dentate nucleus involvement. Diffuse brain swelling with edema can be seen during an acute metabolic crisis. Diffusion images usually show low ADC in the involved gray matter that eventually will reverse to high ADC in the chronic stage. Impairment of energy metabolism leads to accumulation of lactate. Increase of glutamate/glutamine has been described on short TE spectra, in association with decrease of NAA and *myo*-inositol.[61]

Methylmalonic aciduria

MMA is a complex and biochemically heterogeneous disorder due to primary deficiency of methylmalonyl coenzyme A mutase, the next step on the catabolic pathway. There are two main clinical phenotypes. Vitamin B_{12} levels are low in the therapy-responsive phenotype that will improve with cobalamin (B_{12}) supplemental diet.

In the unresponsive phenotype, the most striking MR findings are symmetrical T_2-hyperintensities associated with decreased ADC on diffusion imaging in the globus pallidus. ^{1}H-MRS has shown lactate elevation and NAA signal loss in the basal ganglia.[62] Methylmalonyl acid resonates with a doublet at 1.24 ppm; however, its concentration in the involved globi pallidi must be below the detectable threshold, since it has never been documented in vivo. Elevated methylmalonyl acid has been detected in the CSF of patients with vitamin B_{12} deficiency with in vitro ^{1}H-MRS at 7 Tesla.[63] Patients with symmetrical globus pallidus lesions present with a extrapyramidal syndrome.

The cobalamin-responsive phenotypes are related to intracellular utilization of vitamin B_{12}, which is a cofactor of methylmalonyl coenzyme A mutase. Most patients present during the immediate postnatal period or early infancy with microcephaly, poor feeding, failure to thrive, hypotonia, and seizures. Rarely, the disorder may manifest in adolescence with spastic quadriplegia and confusion. In this less common presentation MR imaging may show diffuse white matter T_2-signal hyperintensities with brain swelling and elevated choline as signs of demyelination on ^{1}H-MRS. Throughout the course of the disease, there is progressive loss of white matter volume with decreasing choline signal and NAA depletion. The characteristic imaging abnormalities in the globus pallidus may develop also in this phenotype.

Ethylmalonic aciduria

Ethylmalonic aciduria is another organic acidopathy that may be related to a defect in mitochondrial fatty acid oxidation, respiratory chain, or isoleucine metabolism. Patients present with metabolic crises with elevated lactate acidosis and mild hypoglycemia without ketoacidosis. Increased urinary levels of ethylmalonic acid and methylsuccinic acid, as well as hyperlactic acidemia, are usually found in these patients. MR imaging shows patchy and heterogeneous T_2-hyperintensities in the caudate nucleus and putamen, which may help to differentiate the disease from other organic acidopathies. The globus pallidus and dentate nucleus may be involved. On ^{1}H-MRS, lactate may be very elevated in the basal ganglia, while NAA is decreased with variable choline elevation.[51]

3-Hydroxy-3-methylglutaryl coenzyme A (HMG CoA) lyase

3-Hydroxy-3-methylglutaryl coenzyme A (HMG CoA) lyase is a mitochondrial and peroxisomal enzyme that is active in leucine degradation and ketone body synthesis. A defect in HMG CoA lyase usually presents in the neonatal or infantile period with acute metabolic crises with hypoglycemia and acidosis but without ketosis, especially in cases of insufficient alimentary supply due to fasting or vomiting or high glucose utilization, such as with intercurrent infections. The MR imaging findings may vary according to the different clinical phenotype and the age of the patient. The most frequent findings are symmetrical signal abnormalities in the basal ganglia and especially in older patients in the white matter with relative sparing of the arcuate fibers. ^{1}H-MRS may show an abnormal peak resonating at 2.42 ppm on long TE spectra that has been tentatively assigned to increased levels of 3-hydroxyisovalerate and/or 3-hydroxy-3-methylglutarate,[64] in association with elevated lactate, increased choline, and loss of NAA.[65] Definite assignment of the abnormal peaks has been done with two-dimensional chemical shift correlated spectroscopy in the urine specimens of two patients.[66]

Glutaric aciduria type I

Glutaric aciduria type I is another autosomal recessive disorder caused by deficiency of glutaryl-CoA dehydrogenase, a flavoprotein inside the mitochondria, involved in the degradation of lysine, hydroxylysine, and tryptophan. Many different mutations have been described for the gene of glutaryl-CoA dehydrogenase that is located on chromosome 19p13.2 with relatively different phenotypes. All mutations are characterized by increased levels of glutaric acid. MR imaging shows signal abnormalities in the basal ganglia, midbrain tegmentum and white matter. The signal abnormalities in the white matter are typically patchy and scattered, and seen in the subcortical arcuate fibers. Another clue may be the presence of atrophy in the basal ganglia and in the frontal and temporal lobes with dilatation of the sylvian fissures, in association with macrocephaly. On MRS, elevated lactate has been reported in a few cases.

L-2-hydroxyglutaric aciduria

Another organic aciduria with macrocephaly and subcortical white matter T_2-signal abnormality is L-2-hydroxyglutaric aciduria (L-2-OH-GA). The biochemical hallmark of this autosomal recessive disorder is elevated urinary excretion of L-2-hydroxyglutaric acid in association with elevated lysine in blood and CSF. Unlike other organic acidurias, L-2-OH-GA has a relatively late onset with a slowly progressive course, and diagnosis may occur in late childhood or early adolescence. Patients usually survive into adulthood. L-2-OH-GA is caused by a deficiency of a FAD-linked, membrane-bound mitochondrial dehydrogenase enzyme, encoded by a gene called DURANIN, located on chromosome 14q22.1. MR imaging shows T_2-signal abnormalities in the subcortical white matter with relative sparing of the lobar and periventricular white matter, in association with basal ganglia and cerebellar dentate nuclei involvement. The higher vulnerability of the arcuate fibers may be due to spongy degeneration occurring in the bundles which myelinate later in life.[67] Abnormal signal hypointensity may occur on T_2-weighted images in the thalami, possibly due to accumulation of paramagnetic iron. Despite spongiform changes on histopathology, there are no signs of decreased ADC on diffusion imaging in the few cases reported. MRS shows very mild, non-specific signal changes of the main metabolites. Unlike other organic acidurias, lactate is not elevated in L-2-OH-GA, but (like most other organic acidurias) NAA may be slightly decreased, choline and/or creatine may be only minimally increased. In at least four cases, an association with development of brain tumors has been reported. This association suggests a possible role of L-2-OH-GA in predisposing to brain tumorigenesis.[68] Thus MRSI may have an additional important role to monitor choline signal levels in patients at risk of developing a brain tumor.

Hereditary leukoencephalopathies

Hereditary leukoencephalopathies are a large group of inherited disorders that predominantly involve the white matter. Enzymatic defects leading to immature myelin deposition, dysfunction or breakdown of myelin have been identified in an increasing number of leukoencephalopathies with onset in infancy, childhood, adolescence, or adulthood. Nevertheless, approximately 30% of leukoencephalopathies remain of unknown etiology, despite extensive clinical, imaging, biochemical, and genetic investigation.[1]

Three main pathophysiologic mechanisms may affect myelin and cause a leukoencephalopathy. "Hypomyelination" indicates a deficit in the formation and maturation of myelin, whereas "demyelination" indicates destruction of previously formed myelin. A third mechanism leading to "rarefaction" of white matter tissue due to intramyelinic cavitation or cyst formation has been recently identified with MR imaging.[1,69,70] In a quantitative study on 70 children with leukoencephalopathy, it has been recently shown that multivoxel ^{1}H-MRSI alone can predict the underlying pathophysiology with a relatively good accuracy of 75%.[1] In another MR study on 41 children, single-voxel ^{1}H-MRS was combined with diffusion tensor and magnetization transfer MR imaging: the combination of 7 MR parameters showed an accuracy of 95%.[70] Both studies demonstrated that MR spectroscopy is a very valuable tool in the classification of leukoencephalopathies. Qualitative assessment of the spectra is usually sufficient to differentiate the three pathophysiological forms, based on the white matter (intravoxel) ratios Cho/NAA, Cho/Cr, and NAA/Cr (Figure 11.9). Normalized ratios, expressed

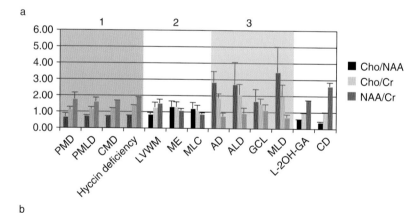

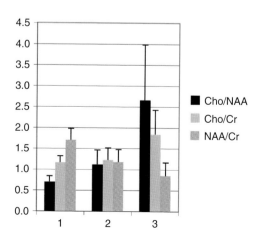

Figure 11.9. The bar graph shows white matter **intra-voxel ratios of Cho/NAA, Cho/Cr, and NAA/Cr** measured in the centrum semiovale with multivoxel MRSI of 47 children with hereditary leukoencephalopathies. The ratios are reported by diagnosis (A) and by pathophysiology group (B). Groups 1, 2, and 3 correspond to hypomyelination, rarefaction, and demyelination, respectively.

Note that the profiles of the intra-voxel ratios are different among the three groups by pathophysiology. In hypomyelinating diseases, ratios are in the normal range (Cho/NAA~0.6; Cho/Cr~1.0; NAA/Cr~1.5). In demyelinating leukoencephalopathies Cho/NAA and Cho/Cr are elevated, whereas NAA/Cr is reduced. In leukoencephalopathies with rarefaction, the three ratios are around one. (From [1].)

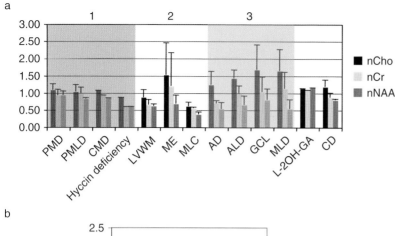

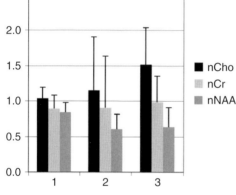

Figure 11.10. Bar graph showing the **normalized ratios of nCho, nCr, and nNAA** in the white matter (normalized to parasagittal gray matter) measured with multivoxel MRSI in the white matter of 47 children with leukoencephalopathies. The ratios are reported by diagnosis (A) and by pathophysiology group (B). Groups 1, 2, and 3 correspond to hypomyelination, rarefaction, and demyelination, respectively.

nCho is elevated in diseases with demyelination, near-normal in hypomyelinating diseases, and decreased in diseases leading to rarefaction. The study showed that profile of intravoxel ratios (Figure 11.8) were more practical than normalized ratios to differentiate the three groups by pathophysiology. (From [1].)

relative to the same metabolite in the contralateral hemisphere (nCho, nCr, and nNAA) (Figure 11.10) also contribute to differentiating these three pathophysiology groups.

Hypomyelinating diseases

The process of myelination is complex: it begins in utero, continues during infancy, and is completed at the end of the second decade of life. However, from the MR viewpoint, the most dramatic changes occur during the first two years of life, after which most imaging parameters (T_1, T_2, spectroscopy, and diffusion fractional anisotropy) have matured into the adult-like patterns. Hypomyelinating leukoencephalopathies are a heterogeneous group of genetic disorders involving significant, permanent deficits in myelin deposition and maturation. Several recently discovered hypomyelinating disorders include sialic acid storage and Salla diseases, 18-q syndrome, hypomyelination, and atrophy of basal ganglia and cerebellum (HABC), Cockaine syndrome, and trichothiodystrophy with photosensitivity (Tay syndrome).

In all hypomyelinating disorders, MR imaging shows subtle diffuse signal hyperintensity on T_2-weighted images. The presence on CT scan of bilateral calcifications in the basal ganglia may favor the diagnosis of Cockaine or Tay syndrome. Both disorders are associated with a defect of DNA repair.

Near-normal Cho, Cr, and NAA without accumulation of lactate is the H-MRS hallmark of hypomyelinating disorders.

Pelizaeus–Merzbacher disease

Pelizaeus–Merzbacher disease (PMD) is one of the most commonly encountered hypomyelinating diseases, and is caused by mutations in the proteolipid protein (*PLP*) gene located in the X chromosome. Individuals with PMD fail to form and maintain compacted myelin and present with hypotonia, nystagmus, motor delay, cerebellar ataxia, and spasticity. Normal or only very mild metabolic MRS signal abnormalities have been reported in PMD. Most studies have shown no significant spectroscopic changes in the white and gray matter, despite diffuse signal hyperintensity on T_2-weighted MR images.

Chapter 11: MRS in cerebral metabolic disorders

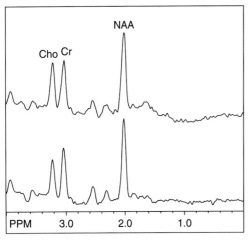
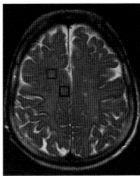

Figure 11.11. 39-year-old man who presented with an 8-year history of slowly progressive walking difficulties, leg stiffness, slurred speech, unsteady gait, dysarthria, cerebellar ataxia, and nystagmus. Since childhood he had minimal writing difficulties. The current neurological exam showed spastic gait, bilateral Babinski signs, mild dysarthria, and mild saccadic eye movements. He was diagnosed with a mild phenotype of **Pelizaeus–Merzbacher-like disorder (PMLD)** due to a deficit in gap junction connexin 47.
MRSI (PRESS: TR/TE = 1200/135 ms; 24 × 24 matrix; FOV = 200 × 200 × 15 mm^3) was acquired at the level of the centrum semiovale. Selected white and gray matter spectra and axial T_2-weighted MR image are illustrated.
Note the normal profiles of Cho, Cr, and NAA in the spectra from the right centrum semiovale and parasagittal gray matter, despite subtle diffuse hyperintensity in the white matter of the centrum semiovale.

A normal MRS scan is therefore suggestive of a hypomyelinating leukoencephalopathy in patients presenting with white matter disease of unknown etiology.[1]

Pelizaeus–Merzbacher-like disorder

Patients of both genders who do not have *PLP* mutations may nevertheless exhibit a slowly progressive clinical phenotype with MR imaging features indistinguishable from PMD. Mutations responsible for few of these patients with a PMD-like disorder (PMLD) have recently been identified in the autosomal *GJA12* gene that encodes the gap junction connexin 47 (C×47) protein, which is highly expressed in oligodendroglia. [71–73]

MR spectroscopy is useful in establishing a diagnosis of PMLD.[71,73] In the large majority of patients, MRS shows near normal spectra in both white and gray matter, despite diffuse signal abnormalities in the white matter (Figure 11.11). In a few patients with advanced disease and significant atrophy, MRS may show mild NAA loss. In others, a mild decrease in Cho has been reported.

Hypomyelination and atrophy of basal ganglia and cerebellum

HABC is a recently identified disorder.[74] Onset is in infancy or early childhood, with a variable degree of severity. Extensive metabolic work-up is unrevealing. MR imaging shows a paucity of myelin with signal abnormalities suggestive of hypomyelination. MR spectroscopy shows near-normal spectra that support the diagnosis of a hypomyelinating disorder. However, more striking associated atrophic changes are diagnostic: marked atrophy in the putamen, moderate in the caudate and cerebellum. The globus pallidus and the thalamus are not involved.

Demyelinating and dysmyelinating diseases

The term *demyelination* indicates destruction of previously, completely formed myelin, with selective primary involvement of myelin sheets or oligodendroglia. The term "dysmyelination", introduced by Poser in 1957, indicates disorders in which myelin is not formed properly. Classic examples of demyelinating leukoencephalopathies are metachromatic leukoencephalopathy (MLD), globoid cell leukoencephalopathy (GCL), adrenoleukoencephalopathy (ALD), and Alexander disease (AD). Cho and lactate elevation in association with Cr and NAA loss are the common MRS changes in demyelinating disorders.

Metachromatic leukodystrophy

MLD involves sphingolipidosis due to deficiency of the lysosomal enzyme arylsulfatase A, that catalyzes the hydrolysis of sulfatide, the sulfate ester of cerebroside, a myelin component. The *ARSA* gene is located on chromosome 22q13.3. There are three clinical phenotypes with onset in late infancy, adolescence, and adulthood, probably related to different levels of residual enzyme activity. One possible pathogenetic mechanism is that myelin becomes increasingly unstable. Demyelination starts when myelin stability reaches a critical point. Another possible mechanism is sulfatide accumulation in lysosomes of oligodendroglia and Schwann cells with cellular dysfunction and death.

MR imaging shows symmetrical signal abnormalities in the periventricular and lobar white matter, including the corpus callosum and the corticospinal tracts. The arcuate fibers are relatively spared, at least in the early phase. There are no signs of signal abnormality or atrophy in the cortex and basal ganglia until the advanced stages of the disease. MRS shows the characteristic profile of a demyelinating disease: Cho elevation with NAA signal loss and mild lactate accumulation in the deep white matter. Mild Cr elevation is often seen.

Globoid cell leukoencephalopathy

GCL involves a sphingolipidosis due to deficiency of the lysosomal enzyme galactosylceramidase that catalyzes cerebroside degradation into galactose and ceramide. It is also known as Krabbe disease. The most common subtype has onset in early infancy; other subtypes have connatal, late-infantile, juvenile, and adult onset. The *GALC* gene is located on chromosome 14q31. Prior to myelination, cerebroside is virtually absent. As soon as myelination starts, and myelin turnover begins, cerebroside accumulates inside phagocytic cells that undergo a globoid transformation. Loss of myelin sheets is then associated with apoptosis that is selectively induced in oligodendrocytes.

MR imaging in early-onset GCL is very similar to MLD with symmetric periventricular white matter signal abnormalities, relative sparing of the arcuate subcortical fibers, and involvement of the corpus callosum, internal capsule, corticospinal tracts, and cerebellar white matter. In the neonatal subtype the myelinated corticospinal tracts and cerebellar peduncles are the first structures to be involved. In the adult subtype, MR imaging shows a much milder pattern with selected signal abnormalities in the corticospinal tracts that may also be asymmetric.[75,76]

MR spectroscopy shows elevated Cho and loss of NAA in white matter with T_2-signal abnormalities. Lactate is often mildly elevated; occasionally Cr is also elevated. Cho may be elevated also in normally appearing white matter (Figure 11.12).

Adrenoleukodystrophy

ALD involves lipidosis caused by impairment of saturated very long chain fatty acids (VLCFA) β-oxidation

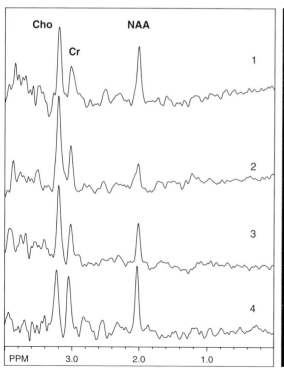
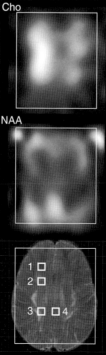

Figure 11.12. Five-year-old boy who presented with unsteady gait, dysarthria, cerebellar ataxia, and nystagmus. He was diagnosed with **globoid cell leukoencephalopathy (Krabbe disease).**
MRSI (PRESS: TR/TE = 1500/135 ms; 32 × 32 matrix; FOV = 160 × 160 × 20 mm^3) was acquired at the level of the centrum semiovale. Cho and NAA maps, four selected spectra are illustrated. The location of the PRESS-VOI and of the four selected spectra is overlaid on the T_2-weighted MR image.
Note increased Cho and decreased NAA in the spectra from the right centrum semiovale (1, 2, 3). Cho is increased also in the anterior third where the T_2-signal abnormalities are very subtle. The gray matter spectrum (4) shows mild NAA signal loss.

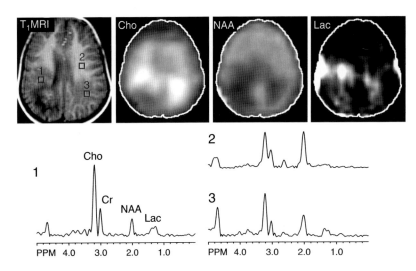

Figure 11.13. Adrenoleukodystrophy (ALD). Multi-voxel MRSI study showing T_1-localizer image, Cho, NAA, and Lac metabolic images, and selected spectra from a 21-year-old patient with onset, hemiplegia and dysarthria. Symptom onset was at age 11. The posterior regions of chronic demyelination have low levels of all metabolites, whereas spectra from the rim of the lesion (active demyelination) show high levels of Cho and elevated lactate bilaterally (voxels 1 and 3). Note that normal-appearing white matter (voxel 2) is also metabolically abnormal, with an NAA/Cho ratio lower than normal.

in peroxisomes, and perhaps in mitochondria. ALD is caused by mutations in the *ABCD1* gene, located on chromosome Xq28, which encodes ALDP, a peroxisomal ATP-binding cassette transmembrane transported protein. The relation between the ALD protein and the accumulation of VLCFA is unknown. ALD involves the CNS and the adrenal cortex. The clinical phenotypes share the same gene defect and multiple subtypes may occur in the same family. The x-ALD subtype presents in childhood with a rapidly progressive course; the adrenomyeloneuropathy (AMN) subtype presents later with a slower clinical progression; the Addison subtype involves only the adrenal cortex and accounts for about 50% of childhood cases.

Diagnosis is made with demonstration of abnormally high levels of saturated VLCFA in plasma, fibroblasts, or tissue. Accumulation of VLCFA may cause rapidly progressive demyelination with a marked inflammatory reaction in childhood x-ALD. Peptide and lipid antigens may play a key role in the pathogenesis of inflammatory demyelination. Reactive astrocytes, macrophages, and CD8[+] T-lymphocytes are found in white matter lesions. There is no inflammatory response in the milder AMN subtype.

MR imaging usually shows very characteristic signal abnormalities: the splenium of the corpus callosum is involved first, then the abnormalities spread out into the posterior periventricular white matter, sparing the arcuate fibers, before spreading anteriorly. Three zones can be identified on histopathology, MR imaging, and spectroscopic imaging.[77,78] Typically, the inner (posterior and older) zone appears burned-out and completely demyelinated; MR spectra show severe NAA loss in association with moderate Cho and lactate elevation. The outer (anterior and advancing) zone shows signs of active demyelination without inflammation: choline is elevated, whereas NAA is only mildly decreased; lactate accumulation is low and Cr is within normal ranges (Figure 11.13). In the intermediate zone, enhancement on T_1-weighted images after intravenous injection of gadolinium is the most striking imaging feature. The spectra show intermediate changes between the inner and outer zones. Significant NAA signal loss has been found in the normal appearing white matter not only in patients with x-ALD, but also with the milder AMN subtype.[79]

MR and MRS signal abnormalities are frequently symmetric and advancing forward; however, in about 5% of cases the lesion arises in one lobe and spreads first to the ipsilateral hemisphere and then to the contralateral cerebral white matter. The long corticospinal tracts are commonly involved with more severe abnormalities in the occipito-parieto-temporo-pontine and pyramidal tracts, with relative sparing of the fronto-pontine tracts. The lateral lemniscus is also often abnormal on MR imaging.

MR imaging and spectroscopy also have an important role in monitoring progression of the disease, determining which patients may benefit from a particular therapy, and in evaluating response to treatment.[80] Currently, dietary therapy and bone marrow transplantation are the two main treatment options.

Alexander disease

Alexander disease (AD) is a rare and sporadic autosomal dominant disorder due to de novo mutations in the *GFAP* gene located on the chromosome 17q21.[81,82] Laboratory tests are not helpful for the diagnosis. In the past, a definite diagnosis was made only with a brain biopsy or autopsy. Three clinical subtypes can be identified: infantile, juvenile, and adult. Familial cases are seen only in adult AD.

The characteristic astrocytic involvement was first described by Alexander in 1949.[83] Neuropathology shows hypertrophic reactive astrocytes with countless cytoplasmic aggregates containing glial fibrillary acidic protein (GFAP) and heat shock protein 27 (HSP27), known as Rosenthal fibers. Cellularity is increased in the white matter with a multitude of abnormal astrocytes; oligodendrocytes appear normal, despite the lack of myelin sheets demonstrated on electron microscopy; there are no signs of inflammatory reaction. In infantile AD, disturbed myelination may prevail in the frontal lobes, while demyelination occurs in areas with mature myelin. GFAP is an intermediate filament protein that is specific to astrocytes and it is involved in the structure and function of the cytoskeleton. GFAP is thought to contribute to the growth of astrocyte processes and the maintenance of their mechanical strength and shape. The relationship between GFAP and AD is not completely understood. It has been proposed that mutations in the GFAP gene result in a gain of function with formation of Rosenthal fibers. These aggregates may reduce astrocytes' interaction with oligodendrocytes and integrity of the blood–brain barrier.

The most frequent infantile phenotype presents as a progressive leukoencephalopathy at about 6 months of age with macrocephaly, failure to thrive, feeding problems, spasticity, ataxia, and rapid signs of neurological deterioration. The course is severe and infants usually die in a few years. In juvenile AD, the onset of symptoms is more subtle with progressive bulbar and pseudobulbar signs, delayed speech, and swallowing problems. Macrocephaly is less common. In adult AD, the onset of symptoms may occur at any age with bulbar and pseudobulbar signs, cerebellar ataxia, spastic paraparesis, nystagmus, palatal myoclonus, and occasionally dementia.[84] The course may be episodic or chronic progressive.

Specific MR imaging criteria for the diagnosis of infantile AD have been established.[85] Neuroradiological diagnosis of infantile AD can be made when four of the following five MR imaging criteria are met: (i) signal abnormalities in cerebral white matter with frontal predominance, (ii) signal abnormalities in basal ganglia and thalami, (iii) signal abnormalities in the brain stem, (iv) signal abnormalities in the subependymal region, and (v) contrast enhancement in one of the above structures. On sagittal views a periventricular rim with hyperintensity on T_1-weighted, hypointensity on T_2-weighted images and enhancement after gadolinium administration is a very suggestive finding of infantile AD (Figure 11.14). The density of Rosenthal

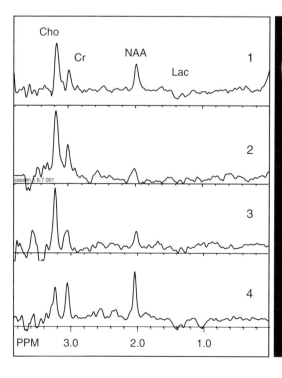
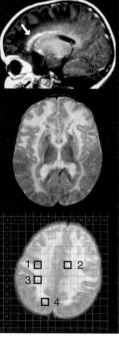

Figure 11.14. Two-year-old girl with severe developmental delay and macrocephaly since the first year of life, muscle hypotonia, dysphagia, and cerebellar ataxia. She was diagnosed with **Alexander disease**.

MRSI (PRESS: TR/TE = 1500/135 ms; 32 × 32 matrix; FOV = 160 × 160 × 20 mm^3) was acquired at the level of the centrum semiovale. The location of four selected spectra is overlaid on the T_2-weighted MR image.

Note increased Cho with decreased NAA and Cr in the spectra from the right and left centrum semiovale (1, 2, 3). Minimal lactate elevation is also detected (1), while the gray matter spectrum (4) is normal. Swelling with diffuse signal hyperintensity throughout the white matter, in the basal ganglia, and subependymal regions is illustrated in the axial T_2-weighted MR images. Note "enhancement" of the periventricular rim (arrow) after i.v. gadolinium administration in the sagittal T_1-weighted MR image, which is a very suggestive finding of infantile AD.

fibers is very high in the subependymal, perivascular and subpial regions. In the acute stage, extensive usually symmetric signal abnormalities in frontal white matter are associated with swelling, whereas cystic degeneration and atrophy occur in the chronic stage.

The imaging features are very different, but also highly suggestive in adult onset AD. Atrophy and T_2-signal hyperintensity are seen in the medulla oblongata and upper cervical spinal cord. In a few patients these changes are associated with moderate symmetric T_2-hyperintensity in the posterior periventricular white matter.

MRS has been shown to be useful in order to confirm a suspected diagnosis of infantile AD:[1] in four infants with AD, H-MRSI showed elevated Cho, decreased Cr and NAA, and mild lactate accumulation in the white matter (Figure 11.14). Elevation of Cho and mI in the basal ganglia and white matter have been demonstrated with short TE spectra,[16,85,86] probably due to a combination of active demyelination and increased cellularity (astrocyte proliferation). However, Cho and mI elevation are a less consistent finding in the milder juvenile and adult subtypes.[84,86–88] This discrepancy of MRS changes among subtypes, despite similar neuropathological features, may in part be explained by the milder disease expression in the latter two subtypes.

Diseases leading to white matter rarefaction

Rarefaction may be the result of intramyelin vacuolation in a subtype of mitochondrial disorders due to respiratory chain deficiency,[43] and in MLC,[89] or the result of cystic degeneration in leukoencephalopathy with VWM.[90–93]

A significant signal drop of Cho, Cr, and NAA resonances in white matter compared with normal-appearing gray matter is characteristic of disorders with white matter rarefaction. Variable degree of lactate accumulation occurs in these disorders: elevation is mild in MLC and in the early stages of VWM; it is marked in mitochondrial leukoencephalopathy.

Leukoencephalopathy with vanishing white matter

Leukoencephalopathy with VWM has an autosomal recessive inheritance and it is due to a mutation in any of the genes encoding one of the five subunits of the eukariotic translation initiation factor eIF2B.[94,95] Translation of mRNA into polypeptides is highly regulated, and it has been highly conserved throughout evolution. Protein synthesis is inhibited under a variety of stress conditions, known as the heat shock response. Cellular stress conditions may lead to misfolding and denaturation of proteins, contributing to cell dysfunction and death. Inhibition of RNA translation during a stress condition, such as febrile infections and other forms of cellular stress, is thought to enhance cell survival and recovery. VWM is one of the recently identified leukoencephalopathies: this was possible with the fundamental contribution of MR imaging and spectroscopy.[69,91–93,96] Despite its recent description, VWM is one of the most prelevant leukoencephalopathies in children, with an incidence similar to that of MLD. VWM is exacerbated by episodes of infection and minor trauma. The childhood subtype is the most common with onset between 2 and 6 years. The child shows signs of progressive neurological deterioration with predominant cerebellar ataxia, associated with mild spasticity and occasional optic atrophy, but only mildly impaired higher cognitive functions. Characteristically, additional episodes of major and rapid neurological deterioration are associated with febrile infections. A neonatal subtype has been described, but is relatively rare. In the adult subtype, motor deterioration, seizure or psychiatric symptoms may be the first signs. Later onset implies a milder subtype, a slower progression, and more protracted disease course. Laboratory tests are not helpful in VWM. Glycine is consistently elevated in CSF, but is not necessarily a specific finding.

Neuropathology examinations have shown rarefied white matter with preservation of gray matter structures. The cerebellum and brain stem are much less affected, with the exception of the central tegmental tract in the pons that is often involved. The periventricular deep white matter is more commonly involved with relative sparing of the corpus callosum, anterior commissure, anterior limb of the internal capsules, arcuate fibers, and temporal lobes.[97] On microscopic examination, vacuolation and cystic changes, thin myelin sheets without overt signs of demyelination or inflammation, are seen in the affected white matter. Axons are lost in severely affected areas, while they are relatively spared compared with myelin loss in less involved areas. Ultrastructural studies have revealed the occasional presence of vacuolated or "foamy" oligodendrocytes,

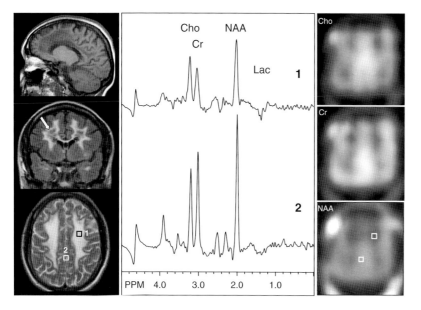

Figure 11.15. 38-year-old woman who presented with depression and ovaric insufficiency. She was diagnosed with genetically confirmed **adult-onset leukoencephalopathy with vanishing white matter**.
MRSI (PRESS: TR/TE = 1200/135 ms; 24 × 24 matrix; FOV = 200 × 200 × 15 mm^3) was acquired at the level of the centrum semiovale. Cho, Cr, and NAA maps with two selected spectra are illustrated.
The Cho, Cr, and NAA metabolite maps show diffuse reduction of signal in the centrum semiovale. The selected white matter spectrum (1) confirms moderate reduction of the main metabolites and mild elevation of Lac. The normal spectrum from the parasagittal parietal gray matter (2) is displayed for comparison. Diffuse signal hyperintensity is seen on the axial T_2-weighted MR image. The coronal FLAIR image shows initial formation of cavitations (isointense with the cerebrospinal fluid) in the deeper white matter (arrow). Diffuse hypointensity is seen in the sagittal T_1-weighted MR image.

while myelin vacuoles and intramyelinic edema are not features of this disease.[98]

MR imaging shows extensive white matter signal changes with replacement of involved white matter with fluid. The FLAIR sequence best demonstrates white matter rarefaction extending from the periventricular zone to the periphery over time. Other structures typically involved are the central tegmental tracts in the pons, the posterior limb of the internal capsules, and the external and extrema capsules. The corpus callosum is partially involved, whereas the basal ganglia and thalami are commonly spared. Often cavum septi pellucidum and vergae are patent: these normal variants are more frequently found than in the average population. The brain does not collapse and rarely shows signs of cortical atrophy, despite the highly rarefacted deep white matter. Only when the brain is removed at autopsy does the rarefied white matter collapse.

MRSI at the level of the centrum semiovale is valuable for the demonstratation of the loss of all main metabolites in the white matter with a relatively normal spectral profile in adjacent gray matter. In the early stages, NAA and Cr signal losses may be more prominent than Cho and detection of lactate may be inconsistent. In more advanced stages, the signal of all main metabolites is decreased, whereas lactate and glucose may accumulate in areas of cystic degeneration. No significant signal changes are found in the gray matter even in the most advanced stage.

In the adult onset subtype, MR imaging and spectroscopy features are similar to those in children. The metabolite maps are particularly useful to recognize the selective signal loss in the white matter of the centrum semiovale (Figure 11.15).

Megalencephalopathy with subcortical cyst

MLC is a rare autosomal recessive leukoencephalopathy with characteristic MR imaging features.[69] A mutation in the *MLC1* gene located on chromosome 22qtel is confirmed in 60–70% of patients with typical MR imaging findings.[99] In families without mutations in *MLC1* despite extended mutation analysis,[100] a mutation in another gene may be responsible for the disease. The *MLC1* gene encodes a membrane protein of unknown function that is highly conserved throughout evolution and is related to myelin. The protein is highly expressed in the astrocytic endfeet in subpial, perivascular and subependymal zones. This suggests that MLC1 may be involved in a transport process across the blood–brain barrier. The only autopsy study performed so far has shown a status spongiosis with many vacuoles in the white matter.[101] Myelin splitting and intramyelinic vacuoles only in the outer lamellae of the myelin sheaths have been found with electron microscopy. The cortex has been reported normal; no evidence of axonal degeneration was found.

Chapter 11: MRS in cerebral metabolic disorders

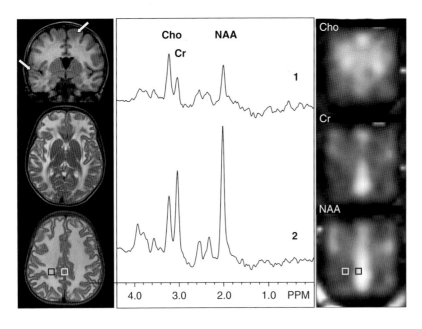

Figure 11.16. Six-year-old boy who presented with slow deterioration of motor function with ataxia, only mild cognitive impairment, and megalencephalopathy. He was diagnosed with **megalencephalic leukoencephalopathy with subcortical cysts (MLC)**.

MRSI (PRESS: TR/TE = 1200/135 ms; 24 × 24 matrix; FOV = 200 × 200 × 15 mm^3) was acquired at the level of the centrum semiovale. Cho, Cr, and NAA maps with two selected spectra are illustrated.

The Cr and NAA metabolite maps show diffuse symmetric reduction of signal in the centrum semiovale. The selected white matter spectrum (1) confirms reduction of the main metabolites. The gray matter spectrum from the parasagittal parietal cortex (2) is normal. On the axial T_2-weighted MR images diffuse signal hyperintensity in the white matter are seen. The cortex has a ribbon-like appearance due to the white matter swelling. The coronal FLAIR image (right top) confirms diffuse white matter hyperintensity; note focal isointensity (with the cerebrospinal fluid) in the subcortical white matter, where subcortical cysts may develop (arrows).

MLC was first described in Turkish and Indian people. A striking feature is the discrepancy between the diffuse white matter T_2 hyperintensity abnormalities and the mild clinical course.[89] Swelling of the white matter causes macrocephaly that may be already present at birth or develop during the first year of life. The ventricular size is normal with no evidence of hydrocephalus. Macrocephaly, a slowly progressive delay in motor development, and mild mental deterioration form the clinical triad of the disease.[102] Cognitive function is preserved for years after onset of the motor deficits. Walking may be delayed slightly and the child may fall frequently. Only several years later ataxia and other signs of motor deterioration may develop. Slowly, the child loses the ability to walk independently, and most children will become wheelchair-bound at the end of the first decade of life or during adolescence. Epilepsy may be an early feature of the disease; however, it is often controlled with medications. Speech is acquired, but later it becomes increasingly dysarthric. Mental deterioration is usually late and mild. Minor head trauma may be associated with temporary deterioration or coma in some patients.[103]

MRI shows diffusely abnormal and swollen white matter with occasional sparing of the most compact white matter tracts (cingulum, optic radiations, corpus callosum, and internal capsule). The cortex has normal signal, but appears thinner due to adjacent white matter swelling. The hallmark of MLC is the formation of large subcortical cysts, frequently found bilaterally in the temporal and parietal lobes. With time, the cysts may enlarge and increase in number. The swelling and the signal hyperintensity on T_2-weighted images is more pronounced in the first two years of life and may diminish with age. Diffusion imaging shows elevation of the ADC throughout the white matter. The differential diagnosis with MR imaging alone includes diseases with macrocephaly and extensive white matter involvement: AD, CD, L-2-OH-aciduria, and merosin-deficient congenital muscular dystrophy. MRSI may be useful narrowing these differential diagnoses. MLC shows mild signal loss of the main metabolites in the white matter of the centrum semiovale, whereas signals are normal in the adjacent cortical gray matter and deep nuclei (Figure 11.16). Mild elevation of lactate has been reported occasionally.[104] These findings are compatible with rarefaction of white matter and replacement with fluid (increased water content). The differential diagnosis for spectroscopy includes VWM and mitochondrial encephalopathies.

MRS to monitor disease outcome and response to therapy

Unfortunately, most metabolic disorders involving the CNS are difficult to treat, and few effective therapies exist. However, MRS has been proposed in trials of patients with GCL and x-ALD,[105] and NAA changes over time have been used as an endpoint after aspartoacylase gene transfer in CD.[106] These data add to those acquired in creatine deficiency,[5] mitochondrial disorders,[107] phenylketonuria,[108] and other rare conditions, where MRS has been used to monitor response to therapy.

Summary

^{1}H-MRS of the brain provides chemical–pathological information that has the potential to improve both diagnostic classification and management of patients with metabolic disorders affecting the CNS. In some instances, metabolic indices provided by MRS are sensitive indicators of early neurological involvement, and are relevant to patients' clinical status. A more extensive use of MRS (possibly with short echo time sequences) in combination with other nonconventional MR techniques might yield a more complete description of the dynamics responsible for pathological changes in this heterogeneous group of disorders, and may allow a more accurate evaluation of disease progression and response to therapeutical intervention.

References

[1] Bizzi A, Castelli G, Bugiani M, Barker PB, Herskovits EH, Danesi U, et al. Classification of childhood white matter disorders using proton MR spectroscopic imaging. *AJNR Am J Neuroradiol* 2008; **29**: 1270–5.

[2] Posse S, Otazo R, Caprihan A, Bustillo J, Chen H, Henry PG, et al. Proton echo-planar spectroscopic imaging of J-coupled resonances in human brain at 3 and 4 Tesla. *Magn Reson Med* 2007; **58**: 236–44.

[3] Schulze A. Creatine deficiency syndromes. *Mol Cell Biochem* 2003; **244**: 143–50.

[4] Stockler S, Hanefeld F, Frahm J. Creatine replacement therapy in guanidinoacetate methyltransferase deficiency, a novel inborn error of metabolism. *Lancet* 1996; **348**: 789–90.

[5] Stockler S, Holzbach U, Hanefeld F, Marquardt I, Helms G, Requart M, et al. Creatine deficiency in the brain: A new, treatable inborn error of metabolism. *Pediatr Res* 1994; **36**: 409–13.

[6] Bianchi MC, Tosetti M, Fornai F, Alessandri MG, Cipriani P, De Vito G, et al. Reversible brain creatine deficiency in two sisters with normal blood creatine level. *Ann Neurol* 2000; **47**: 511–3.

[7] Bizzi A, Bugiani M, Salomons GS, Hunneman DH, Moroni IME, et al. X-linked creatine deficiency syndrome: A novel mutation in creatine transporter gene SLC6A8. *Ann Neurol* 2002; **52**: 227–31.

[8] Martin E, Capone A, Schneider J, Hennig J, Thiel T. Absence of *N*-acetylaspartate in the human brain: impact on neurospectroscopy? *Ann Neurol* 2001; **49**: 518–21.

[9] Kreis R, Ernst T, Ross B. Absolute quantitation of water and metabolites in the human brain. II Metabolite concentrations. *J Magn Reson B* 1993; **102**: 9–19.

[10] Kreis R, Pietz J, Penzien J, Herschkowitz N, Boesch C. Identification and quantitation of phenylalanine in the brain of patients with phenylketonuria by means of localized in vivo 1H magnetic-resonance spectroscopy. *J Magn Reson B* 1995; **107**: 242–51.

[11] Leuzzi V, Tosetti M, Montanaro D, Carducci C, Artiola C, Carducci C, et al. The pathogenesis of the white matter abnormalities in phenylketonuria. A multimodal 3.0 tesla MRI and magnetic resonance spectroscopy (1H MRS) study. *J Inherit Metab Dis* 2007; **30**: 209–16.

[12] van der Knaap MS, Wevers RA, Struys EA, Verhoeven NM, Pouwels PJ, Engelke UF, et al. Leukoencephalopathy associated with a disturbance in the metabolism of polyols. *Ann Neurol* 1999; **46**: 925–8.

[13] Bizzi A, Danesi U, Moroni I, Castelli G, Bugiani M, Erbetta A, et al. Encefalomiopatie mitocondriali in eta' pediatrica: incidenza dell'accumulo di acido lattico documentato con immagini di spettroscopia RM del protone. [in Italian.] *Riv Neuroradiol* 2001; **14**: 149–52.

[14] Bizzi A, Danesi U, Bugiani M, Moroni I, Erbetta A, Savoiardo M, et al. Incidence of cerebral lactic acidosis in children with mitochondrial encephalomyopathy. Int Soc Magn Res in Medicine (ISMRM); 2002; Honolulu, Hawai'I, USA; 2002: 984.

[15] Brockmann K, Bjornstad A, Dechent P, Korenke CG, Smeitink J, Trijbels JM, et al. Succinate in dystrophic white matter: A proton magnetic resonance spectroscopy finding characteristic for complex II deficiency. *Ann Neurol* 2002; **52**: 38–46.

[16] Frahm J, Hanefeld F. Localized proton magnetic resonance spectroscopy of brain disorders in childhood. In: Bachelard H, ed. *Magnetic Resonance Spectroscopy and Imaging in Neurochemistry*. New York: Plenum Press; 1997: 329–401.

[17] Ghezzi D, Goffrini P, Uziel G, Horvath R, Klopstock T, Lochmüller H, et al. SDHAF1, encoding a LYR complex-II specific assembly factor, is mutated in SDH-defective infantile leukoencephalopathy. *Nat Genet* 2009; **41**: 654–6.

[18] Bugiani M, Lamantea E, Invernizzi F, Moroni I, Bizzi A, Zeviani M, et al. Effects of riboflavin in children with complex II deficiency. *Brain Dev* 2006; **28**: 576–81.

[19] Righini A, Ramenghi LA, Parini R, Triulzi F, Mosca F. Water apparent diffusion coefficient and T2 changes in the acute stage of maple syrup urine disease: Evidence of intramyelinic and vasogenic–interstitial edema. *J Neuroimaging* 2003; **13**: 162–5.

[20] Jan W, Zimmerman RA, Wang ZJ, Berry GT, Kaplan PB, Kaye EM. MR diffusion imaging and MR spectroscopy of maple syrup urine disease during acute metabolic decompensation. *Neuroradiology* 2003; **45**: 393–9.

[21] Battisti C, Tarugi P, Dotti MT, De Stefano N, Vattimo A, Chierichetti F. Adult onset Niemann–Pick type C disease: A clinical, neuroimaging and molecular genetic study. *Mov Disord* 2003; **18**: 1405–09.

[22] Sylvain M, Arnold DL, Scriver CR, Schreiber R, Shevell MI. Magnetic resonance spectroscopy in Niemann–Pick disease type C: Correlation with diagnosis and clinical response to cholestyramine and lovastatin. *Pediatr Neurol* 1994; **10**: 228–32.

[23] Canavan MM. Schilder's encephalitis periaxialis diffusa. *Arch Neurol Psychiatry* 1931; **25**: 299.

[24] Matalon R, Michals K, Sebesta D, Deanching M, Gashkoff P, Casanova J. Aspartoacylase deficiency and N-acetylaspartic aciduria in patients with Canavan disease. *Am J Med Genet* 1988; **29**: 463–71.

[25] Gambetti P, Mellman WJ, Gonatas NK. Familial spongy degeneration of the central nervous system (Van Bogaert–Bertrand disease). An ultrastructural study. *Acta Neuropathol (Berl)* 1969; **12**: 103–15.

[26] Grodd W, Krageloh-Mann I, Petersen D, Trefz FK, Harzer K. In vivo assessment of N-acetylaspartate in brain in spongy degeneration (Canavan's disease) by proton spectroscopy [Letter]. *Lancet* 1990; **336**: 437–8.

[27] Austin SJ, Connelly A, Gadian DG, Benton JS, Brett EM. Localized 1H NMR spectroscopy in Canavan's disease: A report of two cases. *Magn Reson Med* 1991; **19**: 439–45.

[28] Barker PB, Bryan RN, Kumar AJ, Naidu S. Proton NMR spectroscopy of Canavan's disease. *Neuropediatrics* 1992; **23**: 263–7.

[29] Righini A, Ramenghi LA, Parini R, Triulzi F, Mosca F. Water apparent diffusion coefficient and T2 changes in the acute stage of Maple Syrup Urine Disease: Evidence of intramyelinic and vasogenic-interstitial edema. *J Neuroimaging* 2003; **13**: 162–5.

[30] Hanefeld FA, Brockmann K, Pouwels PJ, Wilken B, Frahm J, Dechent P. Quantitative proton MRS of Pelizaeus–Merzbacher disease: Evidence of dys- and hypomyelination. *Neurology* 2005; **65**: 701–06.

[31] Bonavita SSR, Moore DF, Frei K, Choi B, Patronas MDN, Virta A, et al. Evidence for neuroaxonal injury in patients with proteolipid protein gene mutations. *Neurology* 2001; **56**: 785–8.

[32] Pizzini F, Fatemi AS, Barker PB, Nagae-Poetscher LM, Horska A, Zimmerman AW, et al. Proton MR spectroscopic imaging in Pelizaeus–Merzbacher disease. *Am J Neuroradiol* 2003; **24**: 1683–9.

[33] Toft PB, Geiss-Holtorff R, Rolland MO, Pryds O, Muller-Forell W, Christensen E, et al. Magnetic resonance imaging in juvenile Canavan disease. *Eur J Pediatr* 1993; **152**: 750–3.

[34] Varho T, Komu M, Sonninen P, Holopainen I, Nyman S, Manner T, et al. A new metabolite contributing to N-acetyl signal in 1H MRS of the brain in Salla disease *Neurology*. 1999; **52**(8):1668–72. [Published erratum appears in *Neurology* 1999; **53**(5): 1162.]

[35] Ruivo R, Sharifi A, Boubekeur S, Morin P, Anne C, Debacker C, et al. Molecular pathogenesis of sialic acid storage diseases: Insight gained from four missense mutations and a putative polymorphism of human sialin. *Biol Cell* 2008; **100**: 551–9.

[36] Morse RP, Kleta R, Alroy J, Gahl WA. Novel form of intermediate salla disease: Clinical and neuroimaging features. *J Child Neurol* 2005; **20**: 814–6.

[37] Shah DK, Tingay DG, Fink AM, Hunt RW, Dargaville PA. Magnetic resonance imaging in neonatal nonketotic hyperglycinemia. *Ped Neurol* 2005; **33**: 50–2.

[38] Heindel W, Kugel H, Roth B. Noninvasive detection of increased glycine content by proton MR spectroscopy in the brains of two infants with nonketotic hyperglycinemia. *Am J Neuroradiol* 1993; **14**: 629–35.

[39] Gabis L, Parton P, Roche P, Lenn N, Tudorica A, Huang W. In vivo 1H magnetic resonance spectroscopic measurement of brain glycine levels in nonketotic hyperglycinemia. *J Neuroimaging* 2001; **11**: 209–11.

[40] Di Mauro S, Schon EA. Mitochondrial respiratory-chain diseases. Review. *N Engl J Med* 2003; **348**: 2656–68.

[41] Bianchi MC, Sgandurra G, Tosetti M, Battini R, Cioni G. Brain magnetic resonance in the diagnostic evaluation of mitochondrial encephalopathies. Review. *Biosci Rep* 2007; **27**: 69–85.

[42] Farina L, Chiapparini L, Uziel G, Bugiani M, Zeviani M, Savoiardo M. MR findings in Leigh syndrome with COX deficiency and SURF-1 mutations. *Am J Neuroradiol* 2002; **23**: 1095–100.

[43] Moroni I, Bugiani M, Bizzi A, Castelli G, Lamantea E, Uziel G. Cerebral white matter involvement in children with mitochondrial encephalopathies. *Neuropediatrics* 2002; **33**: 79–85.

[44] Krägeloh-Mann I, Grodd W, Schöning M, Marquard K, Nägele T, Ruitenbeek W. Proton spectroscopy in five patients with Leigh's disease and mitochondrial enzyme deficiency. *Dev Med Child Neurol* 1993; **35**: 769–76.

[45] Bianchi MC, Tosetti M, Battini R, Manca ML, Mancuso M, Cioni G, et al. Proton MR spectroscopy of mitochondrial diseases: Analysis of brain metabolic abnormalities and their possible diagnostic relevance. *Am J Neuroradiol* 2003; **24**: 1958–66.

[46] Cross JH, Connelly A, Gadian DG, Kendall BE, Brown GK, Brown RM, et al. Clinical diversity of pyruvate dehydrogenase deficiency. *Ped Neurol* 1994; **10**: 276–83.

[47] Lin DD, Crawford TO, Barker PB. Proton MR spectroscopy in the diagnostic evaluation of suspected mitochondrial disease. *Am J Neuroradiol* 2003; **24**: 33–41.

[48] Matthews PM, Andermann F, Silver K, Karpati G, Arnold DL. Proton MR spectroscopic characterization of differences in regional brain metabolic abnormalities in mitochondrial encephalomyopathies. *Neurology* 1993; **43**: 2484–90.

[49] De Stefano N, Matthews PM, Ford B, Genge A, Karpati G, Arnold DL. Short-term dichloroacetate treatment improves indices of cerebral metabolism in patients with mitochondrial disorders. *Neurology* 1995; **45**: 1193–8.

[50] Krägeloh-Mann I, Grodd W, Niemann G, Haas G, Ruitenbeek W. Assessment and therapy monitoring of Leigh disease by MRI and proton spectroscopy. *Ped Neurol* 1992; **8**: 60–4.

[51] Grosso S, Balestri P, Mostardini R, Federico A, De Stefano N. Brain mitochondrial impairment in ethylmalonic encephalopathy. *J Neurol* 2004; **251**: 755–6.

[52] van der Knaap MS, van der Voorn P, Barkhof F, Van Coster R, Krageloh-Mann I, Feigenbaum A, et al. A new leukoencephalopathy with brainstem and spinal cord involvement and high lactate. *Ann Neurol* 2003; **53**: 252–8.

[53] Coburn B. A rare disorder, ethylmalonic encephalopathy, is caused by mutations in a mitochondrial protein. *Clin Genet* 2004; **65**: 460–2.

[54] Scheper GC, van der Klok T, van Andel RJ, van Berkel CG, Sissler M, Smet J, et al. Mitochondrial aspartyl-tRNA synthetase deficiency causes leukoencephalopathy with brain stem and spinal cord involvement and lactate elevation. *Nat Genet* 2007; **39**: 534–9.

[55] De Stefano N, Dotti MT, Mortilla M, Federico A. Magnetic resonance imaging and spectroscopic changes in brains of patients with cerebrotendinous xanthomatosis. *Brain* 2001; **124**: 121–31.

[56] Federico A, Dotti MT, Volpi N. Muscle mitochondrial changes in cerebrotendinous xanthomatosis [Letter]. *Ann Neurol* 1991; **30**: 734–5.

[57] Dotti MT, Manneschi L, Federico A. Mitochondrial enzyme deficiency in cerebrotendinous xanthomatosis. *J Neurol Sci* 1995; **129**: 106–08.

[58] Choi CG, Yoo HW. Localized proton MR spectroscopy in infants with urea cycle defect. *Am J Neuroradiol* 2001; **22**: 834–7.

[59] Franco LP, Anderson J, Okoh J, Pomper MG, Braverman N, Barker PB. Proton MR spectroscopy in hyperhomocysteinemia with elevated blood methionine levels. *J Magn Reson Imaging* 2006; **23**: 404–07.

[60] Bisschops RH, van der Graaf Y, Mali WP, van der Grond J. Elevated levels of plasma homocysteine are associated with neurotoxicity. *Atherosclerosis* 2004; **174**: 87–92.

[61] Bergman AJ, Van der Knaap MS, Smeitink JA, Duran M, Dorland L, Valk J, et al. Magnetic resonance imaging and spectroscopy of the brain in propionic acidemia: Clinical and biochemical considerations. Review. *Pediatr Res* 1996; **40**: 404–9.

[62] Trinh BC, Melhem ER, Barker PB. Multi-slice proton MR spectroscopy and diffusion-weighted imaging in methylmalonic acidemia: Report of two cases and review of the literature. *Am J Neuroradiol* 2001; **22**: 831–3.

[63] Commodari F, Arnold DL, Sanctuary BC, Shoubridge EA. 1H NMR characterization of normal human cerebrospinal fluid and the detection of methylmalonic acid in a vitamin B12 deficient patient. *NMR Biomed* 1991; **4**: 192–200.

[64] Yalçinkaya C, Dinçer A, Gündüz E, Fiçicioğlu C, Koçer N, Aydin A. MRI and MRS in HMG-CoA lyase deficiency. *Ped Neurol* 1999; **20**: 375–80.

[65] van der Knaap MS, Bakker HD, Valk J. MR imaging and proton spectroscopy in 3-hydroxy-3-methylglutaryl coenzyme A lyase deficiency. *Am J Neuroradiol* 1998; **19**: 378–82.

[66] Iles RA, Jago JR, Williams SR, Chalmers RA. 3-Hydroxy-3-methylglutaryl-CoA lyase deficiency studied using 2-dimensional proton nuclear magnetic resonance spectroscopy. *FEBS Lett* 1986; **203**: 49–53.

[67] D'Incerti L, Farina L, Moroni I, Uziel G, Savoiardo M. L-2-Hydroxyglutaric aciduria: MRI in seven cases. *Neuroradiology* 1998; **40**: 727–33.

[68] Moroni I, Bugiani M, D'Incerti L, Maccagnano C, Rimoldi M, Bissola L, et al. L-2-Hydroxyglutaric aciduria and brain malignant tumors: A predisposing condition? Review. *Neurology* 2004; **62**: 1882–4.

[69] van der Knaap MS, Breiter SN, Naidu S, Hart AA, Valk J. Defining and categorizing leukoencephalopathies of unknown origin: MR imaging approach. *Radiology* 1999; **213**: 121–33.

[70] van der Voorn JP, Pouwels PJ, Hart AA, Serrarens J, Willemsen MA, Kremer HP, et al. Childhood white matter disorders: Quantitative MR imaging and spectroscopy. *Radiology* 2006; **241**: 510–7.

[71] Bugiani M, Al Shahwan S, Lamantea E, Bizzi A, Bakhsh E, Moroni I, et al. GJA12 mutations in children with recessive hypomyelinating leukoencephalopathy. *Neurology* 2006; **67**: 273–9.

[72] Uhlenberg B, Schuelke M, Ruschendorf F, Ruf N, Kaindl AM, Henneke M, et al. Mutations in the gene encoding gap junction protein alpha 12 (connexin 46.6) cause Pelizaeus–Merzbacher-like disease. *Am J Hum Genet* 2004; **75**: 251–60.

[73] Orthmann-Murphy JL, Salsano E, Abrams CK, Bizzi A, Uziel G, Freidin MM, et al. Hereditary spastic paraplegia is a novel phenotype for GJA12/GJC2 mutations. *Brain* 2008 [Epub ahead of print].

[74] van der Knaap MS, Naidu S, Pouwels PJ, Bonavita S, van Coster R, Lagae L, et al. New syndrome characterized by hypomyelination with atrophy of the basal ganglia and cerebellum. *Am J Neuroradiol* 2002; **23**: 1466–74.

[75] Farina L, Bizzi A, Finocchiaro G, Pareyson D, Sghirlanzoni A, Bertagnolio B, et al. MR imaging and proton MR spectroscopy in adult Krabbe disease. *Am J Neuroradiol* 2000; **21**: 1478–82.

[76] De Stefano N, Dotti MT, Mortilla M, Pappagallo E, Luzi P, Rafi MA, et al. Evidence of diffuse brain pathology and unspecific genetic characterization in a patient with an atypical form of adult-onset Krabbe disease. *J Neurol* 2000; **247**: 226–8.

[77] Kruse B, Barker PB, van Zijl PC, Duyn JH, Moonen CT, Moser HW. Multislice proton magnetic resonance spectroscopic imaging in X-linked adrenoleukodystrophy. *Ann Neurol* 1994; **36**: 595–608.

[78] Oz G, Tkác I, Charnas LR, Choi IY, Bjoraker KJ, Shapiro EG, et al. Assessment of adrenoleukodystrophy lesions by high field MRS in non-sedated pediatric patients. *Neurology* 2005; **64**: 434–41.

[79] Marino S, De Luca M, Dotti MT, Stromillo ML, Formichi P, Galluzzi P, et al. Prominent brain axonal damage and functional reorganization in "pure" adrenomyeloneuropathy. *Neurology* 2007; **69**: 1261–9.

[80] Eichler FS, Barker PB, Cox C, Edwin D, Ulug AM, Moser HW, et al. Proton MR spectroscopic imaging predicts lesion progression on MRI in X-linked adrenoleukodystrophy. *Neurology* 2002; **58**: 901–07.

[81] Brenner M, Johnson AB, Boespflug-Tanguy O, Rodriguez D, Goldman JE, Messing A. Mutations in GFAP, encoding glial fibrillary acidic protein, are associated with Alexander disease. *Nat Genet* 2001; **27**: 117–20.

[82] Rodriguez D, Gauthier F, Bertini E, Bugiani M, Brenner M, N'guyen S, et al. Infantile Alexander disease: Spectrum of GFAP mutations and genotype–phenotype correlation. *Am J Hum Genet* 2001; **69**: 1134–40.

[83] Alexander WS. Progressive fibrinoid degeneration of fibrillary astrocytes associated with mental retardation in a hydrocephalic infant. *Brain* 1949; **72**: 373–81, 3 pl.

[84] Farina L, Pareyson D, Minati L, Ceccherini I, Chiapparini L, Romano S, et al. Can MR imaging diagnose adult-onset Alexander disease? *Am J Neuroradiol* 2008; **29**: 1190–6.

[85] van der Knaap MS, Naidu S, Breiter SN, Blaser S, Stroink H, Springer S, et al. Alexander disease: Diagnosis with MR imaging. *Am J Neuroradiol* 2001; **22**: 541–52.

[86] Grodd W, Krageloh-Mann I, Klose U. Metabolic and destructive brain disorders in children: Findings with localized proton MR spectroscopy. *Radiology* 1991; **181**: 173–81.

[87] Meins M, Brockmann K, Yadav S, Haupt M, Sperner J, Stephani U, et al. Infantile Alexander disease: A GFAP mutation in monozygotic twins and novel mutations in two other patients. *Neuropediatrics* 2002; **33**: 194–8.

[88] Brockmann K, Dechent P, Wilken B, Rusch O, Frahm J, Hanefeld F. Proton MRS profile of cerebral metabolic abnormalities in Krabbe disease. *Neurology* 2003; **60**: 819–25.

[89] van der Knaap MS, Barth PG, Stroink H, van Nieuwenhuizen O, Arts WF, Hoogenraad F, et al. Leukoencephalopathy with swelling and a discrepantly mild clinical course in eight children. *Ann Neurol* 1995; **37**: 324–34.

[90] Hanefeld F, Holzbach U, Kruse B, Wilichowski E, Christen HJ, Frahm J. Diffuse white matter disease in three children: An encephalopathy with unique features on magnetic resonance imaging and proton magnetic resonance spectroscopy. *Neuropediatrics* 1993; **24**: 244–8.

[91] Schiffmann R, Moller JR, Trapp BD, Shih HH, Farrer RG, Katz DA, et al. Childhood ataxia with diffuse

central nervous system hypomyelination. *Ann Neurol* 1994; **35**: 331–40.

[92] Tedeschi G, Schiffmann R, Barton NW, Shih HH, Gospe SM, Jr., Brady RO, *et al*. Proton magnetic resonance spectroscopic imaging in childhood ataxia with diffuse central nervous system hypomyelination. *Neurology* 1995; **45**: 1526–32.

[93] van der Knaap MS, Barth PG, Gabreels FJ, Franzoni E, Begeer JH, Stroink H, *et al*. A new leukoencephalopathy with vanishing white matter. *Neurology* 1997; **48**: 845–55.

[94] Leegwater PA, Konst AA, Kuyt B, Sandkuijl LA, Naidu S, Oudejans CB, *et al*. The gene for leukoencephalopathy with vanishing white matter is located on chromosome 3q27. *Am J Hum Genet* 1999; **65**: 728–34.

[95] van der Knaap MS, Leegwater PA, Könst AA, Visser A, Naidu S, Oudejans CB, *et al*. Mutations in each of the five subunits of translation initiation factor eIF2B can cause leukoencephalopathy with vanishing white matter. *Ann Neurol* 2002; **51**: 264–70.

[96] Hanefeld F, Kruse B, Bruhn H, Frahm J. In vivo proton magnetic resonance spectroscopy of the brain in a patient with L-2-hydroxyglutaric acidemia. *Pediatr Res* 1994; **35**: 614–6.

[97] Rodriguez D, Gelot A, della Gaspera B, Robain O, Ponsot G, Sarliève LL, *et al*. Increased density of oligodendrocytes in childhood ataxia with diffuse central hypomyelination (CACH) syndrome: Neuropathological and biochemical study of two cases. *Acta Neuropathol* 1999; **97**: 469–80.

[98] Wong K, Armstrong RC, Gyure KA, Morrison AL, Rodriguez D, Matalon R, *et al*. Foamy cells with oligodendroglial phenotype in childhood ataxia with diffuse central nervous system hypomyelination syndrome. *Acta Neuropathol* 2000; **100**: 635–46.

[99] Leegwater PA, Yuan BQ, van der Steen J, Mulders J, Könst AA, Boor PK, *et al*. Mutations of MLC1 (KIAA0027), encoding a putative membrane protein, cause megalencephalic leukoencephalopathy with subcortical cysts. *Am J Hum Genet* 2001; **68**: 831–8.

[100] Ilja Boor PK, de Groot K, Mejaski-Bosnjak V, Brenner C, van der Knaap MS, Scheper GC, *et al*. Megalencephalic leukoencephalopathy with subcortical cysts: An update and extended mutation analysis of MLC1. *Hum Mutat* 2006; **27**: 505–12.

[101] van der Knaap MS, Barth PG, Vrensen GF, Valk J. Histopathology of an infantile-onset spongiform leukoencephalopathy with a discrepantly mild clinical course. *Acta Neuropathol* 1996; **92**: 206–12.

[102] Topcu M, Saatci I, Topcuoglu MA, Kose G, Kunak B. Megalencephaly and leukodystrophy with mild clinical course: A report on 12 new cases. *Brain Dev* 1998; **20**: 142–53.

[103] Bugiani M, Moroni I, Bizzi A, Nardocci N, Bettecken T, Gärtner J, *et al*. Consciousness disturbances in megalencephalic leukoencephalopathy with subcortical cysts. *Neuropediatrics* 2003; **34**(4): 211–4.

[104] De Stefano N, Balestri P, Dotti MT, Grosso S, Mortilla M, Morgese G, *et al*. Severe metabolic abnormalities in the white matter of patients with vacuolating megalencephalic leukoencephalopathy with subcortical cysts. A proton MR spectroscopic imaging study. *J Neurol* 2001; **248**: 403–09.

[105] Moser HW, Barker PB. Magnetic resonance spectroscopy: A new guide for the therapy of adrenoleukodystrophy. *Neurology* 2005; **64**: 406–07.

[106] Leone P, Janson CG, Bilaniuk L, Wang Z, Sorgi F, Huang L. Aspartoacylase gene transfer to the mammalian central nervous system with therapeutic implications for Canavan disease. *Ann Neurol* 2000; **48**: 27–38.

[107] De Stefano N, Matthews PM, Arnold DL. Reversible decreases in N-acetylaspartate after acute brain injury. *Magn Reson Med* 1995; **34**: 721–7.

[108] Moats RA, Moseley KD, Koch R, Nelson Jr M. Brain phenylalanine concentrations in phenylketonuria: Research and treatment of adults. *Pediatrics* 2003; **112**: 1575–9.

Chapter 12

MRS in prostate cancer

Key points

- Prostate cancer has a high incidence, and is one of the leading causes of death in men.
- The sensitivity and specificity of diagnosing prostate cancer with conventional imaging methods (ultra sound, MRI) is relatively low.
- The normal prostate contains high levels of citrate (Cit) which can be detected in the proton spectrum at 2.6 ppm. Other compounds detectable in vivo include creatine, choline, spermine, and lipids.
- Citrate is a strongly coupled mutiple at 1.5 and 3.0 T. For optimum detection, careful attention to pulse sequence parameters (TR, TE) is required. TE 120 ms is commonly used at 1.5 T, and TE 75–100 ms at 3 T.
- Multiple studies have reported that prostate cancer is associated with decreased levels of citrate and increased levels of Cho, compared to both normal prostate and also benign prostatic hyperplasia (BPH).
- MRS and MRSI of the prostate is technically challenging: water- and lipid-suppressed 3D-MRSI is the method of choice for most prostate spectroscopy studies.
- Some studies report that adding MRSI to conventional MRI increases sensitivity and specificity of prostate cancer diagnosis.
- MRSI is traditionally performed with an endorectal surface coil, but acceptable quality data may be obtained at 3 T with external phased-array coils which are more comfortable for patients.

Introduction

Prostate cancer is the third most common cancer in the world among men, with 543,000 new cases every year according to the World Health Organization. In the majority of both developed and developing countries, prostate cancer is the most commonly diagnosed neoplasm affecting men beyond middle age.[1] In the United States, prostate cancer is the most frequently diagnosed cancer among men. According to the American Cancer Society, an estimated 192,280 new cases of prostate cancer (PCa) will occur, and there will be an estimated 27,360 deaths due to prostate cancer in the US during 2009.[2,3] The report also suggests that African American men are twice as vulnerable to prostate cancer compared to white men. Although the death rate has dropped over the last few years, it still remains the second leading cause of cancer deaths among men after respiratory system cancers including lung cancer in the US. The ACS recommends that the prostate-specific antigen (PSA) test and the digital rectal examination should be offered annually, beginning at age 50, to men who have a life expectancy of at least 10 years and those men that are at higher risk (African American men and those men with a strong family history of one or more first-degree relatives diagnosed with prostate cancer at an early age). The survival and successful treatment of PCa patients is dependent upon the early diagnosis of PCa. Further, the ability to monitor the progression and regression of malignancy is critical to the management of the disease. Currently the combination of digital rectal examination and PSA testing is the primary diagnostic procedure. Typically, an elevated PSA or a nodule detected on physical examination prompts an evaluation and an eventual transrectal ultrasound-guided (TRUS) biopsy may reveal cancer. However, in most cases, positive identification of PCa only becomes evident when malignancy has been established and the cancer has metastasized beyond the capsular region of the prostate. Staging tests are inaccurate most of the time and are unable to accurately estimate the cancer volume both within and outside the capsular region of the prostate. This makes the tracking of the progression and regression of malignancy following treatment very difficult.

Magnetic resonance imaging (MRI) in conjunction with endorectal coils provides superior visualization of zonal prostate anatomy compared with TRUS.[4] MRI by itself, however, has limited specificity, as various pathologies can mimic cancer with similar appearance on conventional images. In recent years, magnetic resonance spectroscopy (MRS) of the prostate has been shown to provide useful metabolic information, and overcome some of the limitations posed by conventional MRI, and has the potential for increasing the sensitivity and specificity for detection of prostate cancer.[5]

Prostate anatomy

The normal prostate gland is a conically shaped organ about the size of a walnut and is situated deep in the male pelvis. It is nestled between the symphysis pubis anteriorly, the rectum posteriorly, inferiorly by the levator ani muscles and superiorly by the inferior margin of the urinary bladder, and the lateral margins are covered by fat. The seminal vesicles are paired bilaterally between the prostate and the bladder. The prostate can be divided into five zones consisting of the non-glandular anterior fibromuscular stroma and the four glandular components of peripheral zone, central zone, transition zone and periurethral glandular tissue. With age, the periurethral glandular tissue and the transition zone, may hypertrophy considerably, and from a radiological point of view these five zones are essentially reduced to transition zone, central zone, and the peripheral zone, as shown in Figure 12.1. Each of the prostate zones has its own unique architecture, histology, and predisposition to characteristic local disease processes. All zones contain ducts and acini lined by mucin-containing secretory epithelium. It has been reported that 70% of the cancers arise in the peripheral zone, 10% in the central zone, and 20% in the transitional zone.[6]

Magnetic resonance imaging

MRI accurately depicts internal prostatic zonal anatomy and displays the physiologic complexity of the gland. Recent studies show that the combination of endorectal and phased array coils in conjunction with a high field strength MRI system (e.g. 3 Tesla) provides the highest image resolution possible for accurate assessment of prostate disease.[4] Over the past several years, the accuracy of MRI in the staging accuracy of cancer involving the peripheral zone has been consistently reported between 75% and 90%.[5] Most prostate cancer involves the peripheral zone of the gland, where cancer is identified as low signal abnormality on T_2-weighted imaging. Although MRI has allowed intra-prostatic evaluation of tumor location, results are often non-specific as chronic prostatitis, scar tissue, and hemorrhage may exhibit similar findings and potentially reduce the specificity in detection of prostate cancer.[7,8,9] Citrate produced by healthy prostate tissue is an anticoagulant and lengthens the time that the blood product persists. It has been shown that post biopsy blood products can persist for as long as 4–5 months.[10] One study also showed that MR imaging within 21 days after biopsy led to an overestimation of tumor presence and extracapsular extension, and that staging accuracy increased after 21 days. Hormonal ablation, for example, decreases the signal intensity of the normal peripheral zone, thus masking

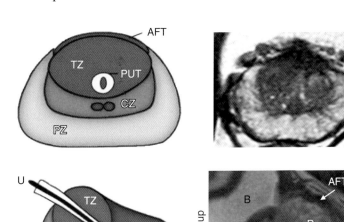

Figure 12.1. Schematic of prostate gland in the axial and sagittal planes along with the corresponding MR images. PZ, peripheral zone; CZ, central zone; TZ, transitional zone; U, urethra; AFT, anterior fibromuscular tissue; UT, periurethral tissue; ED, ejaculatory duct; NVB, neurovascular bundle; SV, seminal vesicles; B, bladder; P, prostate. Adapted with permission from [102].

the low signal intensity exhibited by prostate cancer.[11] Similarly, tumors arising in the central gland may be indistinct especially when low signal intensity stromal hyperplasia dominates. Extra-capsular extension may be seen with MRI as an irregularity or thickening of the capsule. Asymmetric enlargement of the neurovascular bundle as seen on T_1-weighted MRI can be an indication of tumor invasion. Seminal vesicle involvement is best detected as a hypointense signal on T_2-weighted images. A meta-analysis by Engelbrecht et al. that spanned MR imaging studies from 1984 to 2000 reported a joint maximum sensitivity and specificity of 71% for overall tumor staging, 64% for extracapsular extension, and 82% for seminal vesicle invasion. They also concluded that turbo spin echo, endorectal coil, and multiple imaging planes improved prostate cancer staging performance.[12]

Citrate metabolism

The metabolism of normal mammalian cells involves the complete oxidation of glucose and fat through the intermediary steps involving the synthesis and oxidation of citrate via the Krebs cycle.[13] Coupled with phosphorylation, this intermediary synthesis and oxidation of citrate is essential for the cells to generate their major supply of cellular energy through the production of ATP. The citrate synthesized during this process in the Krebs cycle forms the source for acetyl-CoA required for lipogenesis. The Krebs cycle, and the recycling of its intermediates, are essential for the various reactions of amino acid metabolism. These established pathways are essential to normal mammalian aerobic cell metabolism, cellular function, survival, growth, and reproduction.[14] The normal human prostate, on the other hand, does not go through the process of citrate oxidation, thus accumulating large amounts of citrate, which essentially is the end product of the intermediary metabolism. Cooper and Imfeld were the first to report that citrate levels were significantly decreased in prostate cancer tissue compared to the normal prostate or benign prostatic hyperplasia (BPH).[15] Shortly thereafter, the same group suggested that the biochemical alterations seen through altered citrate metabolism may well occur before any malignant changes are histologically obvious.[16] While these observations were made in the early 1960s, it is only in the last decade that scientists have been paying attention to the measurement of citrate levels within the prostate. Costello and Franklin along with their colleagues have further studied the altered citrate metabolism in prostate cancer and have shed some light on the role of zinc in the production of citrate.[17,18]

Table 12.1. Representative citrate and zinc levels in the human prostate per gram of tissue.

Tissue type	Citrate (nmoles g^{-1})	Zn ($\mu g\ g^{-1}$)
Normal central zone	5,000	209
Normal peripheral zone	13,000	
BPH (mixed tissue)	10,000–15,000	
BPH (glandular tissue)	20,000–50,000	589
PCa (mixed tissue)	1,000–3,000	55
Malignant tissue	<500 (estimated)	
Stromal tissue	150–300	
Other soft tissues	150–450	30
Blood plasma	90–110	1
Prostatic fluid	40,000–150,000	590

In addition to citrate accumulation, the normal and BPH prostate also accumulates high levels of zinc. The level of zinc in the normal prostate is about 150 $\mu g\ g^{-1}$ of tissue wet weight. However, the levels of zinc and citrate are not uniformly distributed throughout the prostate gland. For example, in the normal peripheral zone there is high level of zinc concomitant with high levels of citrate. In the normal central gland, the levels of zinc and citrate are at a lower concentration.[19] Table 12.1 lists the concentrations of citrate and zinc in different regions of the prostate in various pathological conditions. It is thought that in the presence of zinc, the mitochondrial aconitase activity that is responsible for citrate oxidation is severely limited in the normal prostate epithelial cells, which ultimately leads to the accumulation of citrate. As shown in Figure 12.2, the accumulation of citrate comes at the cost of ATP production, which is reduced by about 65% in the normal prostate epithelial cells (14 moles of ATP) compared to other normal mammalian cells (38 moles of ATP) that completely oxidize glucose. In prostate cancer, however, the ability of intramitochondrial accumulation of zinc diminishes and citrate is oxidized to produce ATP. Recent studies suggest that the zinc uptake transporter, ZIP1 is upregulated in normal prostate cells increasing zinc accumulation which inhibits cell growth and increases net citrate production.[20] This process is reversed where ZIP1 is down-regulated in PCa, thereby decreasing zinc accumulation in the cells. It is thought that such a decrease

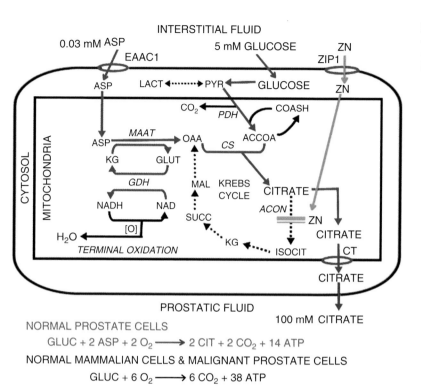

Figure 12.2. Metabolic pathways for citrate production in the normal and malignant prostate. Reproduced with permission from [18].

in the zinc level restores the m-aconitase activity that leads to increased citrate oxidation. This is coupled with ATP production essential for progression of the cells towards malignancy.[18,20,21,22] Although the "cause and effect" relationship between citrate and zinc has been established by Costello and Franklin, many other aspects of the zinc–citrate relationship including the exact mechanisms of zinc depletion during malignancy are still under active research investigation.

Magnetic resonance spectroscopy

Due to increased prostate cancer screening using serum PSA and TRUS-guided biopsy, prostate cancer patients are being identified at an earlier and potentially more treatable stage than previously.[3] However, the decision on how to manage prostate cancer once detected still poses a great dilemma for both patients and their clinicians. The dilemma stems from the fact that prostate cancers demonstrate a tremendous range in biologic malignancy and are treated with a broad spectrum of approaches from "watchful waiting" to aggressive surgical, radiation, and cryosurgical therapies. A number of clinical (digital rectal exam), pathologic (histologic grade from biopsy, and number and percentage of positive biopsies), and biochemical parameters in serum (PSA, PAP) can aid in assessing the extent and aggressiveness of the disease. However, these are often inaccurate or inadequate, particularly when used alone in individual patients. Recent studies suggest that the addition of MRI and MR spectroscopic imaging (MRSI) data to PSA and biopsy data can improve the characterization of prostate cancer in individual patients, and thereby aid in therapeutic selection. Specifically, studies in pre-prostatectomy patients have indicated that the metabolic information provided by MRSI combined with the anatomical information provided by MRI can significantly improve the assessment of cancer location and extent within the prostate, cancer spread beyond the prostate, and cancer aggressiveness.

MRSI of the prostate can provide useful biological information associated with several different metabolites.[23] Molecules that can be studied with MRS include water, lipids, choline, citrate, lactate, creatine, and amino acids.[23,24] The prostate gland is unique in the body in that it contains high levels of citrate.[25] As the normal glandular epithelial cells are replaced by cancer, the concentration of citrate and choline change in the transformation to a malignant state. Choline levels increase and citrate levels decrease in the presence of active cancer.[13] As mentioned above, the reason for the decline in the levels of citrate is the altered intermediate metabolism in the Krebs cycle.[14] Although the mechanism for the elevation of the choline peaks is less understood, just as in the case of brain spectroscopy, its elevation is thought to be associated with changes in cell membrane synthesis and degradation that is normally associated with cancer. The choline resonance observed in vivo at 3.22 ppm, sometimes referred to as total choline, arises from the methyl hydrogens of trimethylamines and is comprised of choline, phosphocholine (PC), glycerophosphocholine (GPC), phosphoethanolamine (PE), glycero-phosphoethanolamine (GPE), and ethanolamine.[8,29,31] These compounds are essential in the synthesis and hydrolysis of phosphatidylcholine and phosphatidylethanolamines that are an integral part of the characteristic bilayer structure of cells and regulate membrane integrity and function. Polyamines such as spermine can be visualized in prostate MRSI.[26] Polyamines are involved in many cellular processes, such as maintenance of DNA structure, RNA processing, translation, and protein activation.[27,28] Disruption to the synthesis of polyamines is known to modulate the genetic effects of these genes. Polyamines can be visualized in proton MRSI as a broad peak between choline and creatine. Normal prostate epithelial cells will demonstrate large amounts of citrate and polyamines. The malignant cells, on the other hand, exhibit low levels of citrate and polyamines to the extent that the choline and creatine resonances are resolved to the baseline. One limitation of prostate MRSI in its current form is the inability to monitor metabolites such as lactate and lipids in vivo due to the necessity for suppressing lipids to minimize contamination from the lipids surrounding the prostate gland. It has been shown in vitro that the citrate to lactate ratio can be used to discriminate prostate cancer from BPH and that the ratio can be used as an indicator of cancer aggressiveness.[16] It is hoped that future MRSI improvements will allow for the interrogation of lactate in the prostate gland.

A series of experiments by researchers in the late 1980s and early 1990s using animal models, cell lines, and tissue extracts had established that prostatic tissue citrate can be a marker for the differentiation of BPH from Pca.[29,30,31] Working with DU 145 xenografts (a poorly differentiated human prostatic adenocarcinoma cell line growth in nude mice), Kurhanewicz et al. found that the citrate concentrations in primary human adenocarcinoma were significantly lower than those observed for normal and BPH tissues.[29]

Following these investigations, Narayan and coworkers were among the first to show an improved SNR in the prostate image and the possibility of obtaining high-quality spectra from the prostate using an endorectal surface coil.[32,33] Initial studies from a normal prostate using a single-voxel ($2 \times 2 \times 2 \, cm^3$) technique showed high concentrations of citrate, whereas malignant prostates showed much lower levels of citrate and higher levels of lipids.[34] Following their findings from the DU145 xenografts study, Kurhanewicz et al. performed in vivo ^{1}H spectroscopy on 28 patients and 5 volunteers. They used the STEAM sequence to obtain water-suppressed spectra from regions of normal prostate peripheral zone, BPH, and Pca.[35] They then correlated their spectra with T_2-weighted MR images and histological studies of the step-sectioned gland after surgery. Consistently they found lower levels of citrate in PCa compared to both BPH and normal prostate peripheral zone using citrate/[creatine+choline] peak area ratios and these results correlated with the histological findings. Furthermore, they did not find any significant difference in the citrate levels in the regions of normal peripheral zone among the patients and the age-matched volunteers. This study further confirmed the role of citrate as an in vivo marker for discriminating PCa from surrounding normal peripheral zone and BPH.[35] Liney et al. reported a strong correlation between the quantitative T_2 maps and citrate concentration in the normal prostate.[36] Their subsequent study on patients with prostate carcinoma showed that both the T_2 maps and the citrate concentration as determined by ^{1}H-MRS could be useful in differentiating PCa from benign disease and normal tissue.[37] Although these studies established the feasibility of MRS in the prostate, they were far from practical, as the spatial resolution was low and spectra from several voxels had to be obtained, which increased the scan time significantly.

MRSI techniques

Although significant developments have been made with MRSI of the brain, the translation of this technology to other body parts including the prostate gland has proven to be far from trivial. In the case of the prostate gland, its small size, deep location, and the possible movement during the MRSI acquisition, as well as the dominating triglyceride signals from the surrounding adipose tissues, all pose a challenge in obtaining reliable spectra. Initial studies employing

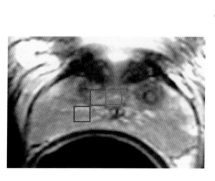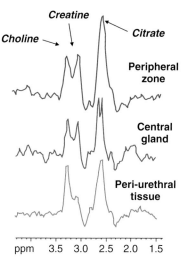

Figure 12.3. Variation in the metabolite concentration in various regions of the prostate. Note the nearly 60% decrease in citrate concentration in the central gland in comparison to the peripheral zone and a decrease in the citrate and elevation of choline in the periurethral tissue. Courtesy of Dr John Kurhanewicz, UCSF.

prostate spectroscopy used single-voxel techniques such as STEAM (Stimulated Echo Acquisition Method) and PRESS (Point Resolved Spectroscopy) using the body coil.[35,38,39,40] Usually the voxel size was large and encompassed both the peripheral zone and the central gland. Although these techniques showed the feasibility for performing proton spectroscopy of the prostate, their use in the clinical setting was limited due to long scan times, poor resolution, and the poor signal-to-noise ratio (SNR) of the spectra. Improvements in MRI hardware, and the arrival of 2D and 3D MRSI techniques have clearly made an impact on the application of MRSI in the diagnosis of prostate cancer.[41,42,43,44] However, the direct use of techniques developed for brain MRSI on the prostate gland proved to be less effective and non-diagnostic. Several improvements have been suggested and implemented for the accurate localization and visualization of relevant metabolites following the suppression of large signals from both water and lipids.[45,46,47,48]

Kuhranewicz et al. were among the first to use a 3D-chemical shift imaging (or 3D MRSI) technique for the prostate.[49] A combination of the body-phased array coil and the endorectal coil enabled spectra from the entire prostate at a volumetric resolution of $0.24\,cm^3$ to be obtained with reasonable SNR. The 3D MRSI technique was able to cover the entire prostate in a single scan and provide quantitative information on the tumor burden within the prostate based on spatial maps of individual metabolites and their relevant ratios. Further, they were able to demonstrate the zonal variation of the metabolites within the normal prostate, and more significantly they found higher levels of choline in regions of cancer compared to BPH and normal peripheral zone. Figures 12.3 and 12.4 demonstrate the variation in metabolites in the normal prostate and the elevation of choline and reduction of citrate in the diseased prostate. Although other techniques of obtaining MRSI data from the entire prostate have been demonstrated, such as multisection proton MRSI using a single spin-echo, the 3D MRSI technique remains the most popular technique.[50] Since the introduction of the 3D MRSI technique, several studies have been performed to assess its role in prostate cancer. Wefer et al. demonstrated that while the accuracy of MRI/MRSI was better than digital rectal exam and comparable to that of sextant TRUS-guided biopsy, it was more accurate for cancers in the apex.[51] The anterior peripheral zone and the transition zone are not palpable during digital rectal exam. In such cases, it has been shown that the addition of MRSI can be useful to target biopsy, especially on patients exhibiting elevated PSA but with negative biopsy results. [52,53] The combined use of MRI/MRSI may help increase the specificity of cancer detection, thereby minimizing unnecessary biopsies.[54]

Investigators have used MRS for treatment planning to specifically target regions of the prostate with higher tumor burden by optimizing the radiation dose delivered to those regions. Zelefski et al. ported the spectroscopy information into an intraoperative computer-based optimization planning system for prostate cancer patients who underwent permanent interstitial implantation. They successfully demonstrated escalation of up to 139–192% of the 144-Gy prescription dose to the MRS-positive voxels while keeping the urethral and rectal doses within the prescription range.[55,56] Similar techniques have been described for intensity-modulated radiation therapy.[57,58] DiBiase et al. implanted ^{125}I seeds, boosting the dose by 130% to focal tumors that showed abnormal (i.e. low) citrate levels, while maintaining the normal radiation tolerances to adjacent tissues with morbidity comparable to normally treated patients during short-term follow-up.[59] Although such approaches may have an impact on the long-term outcomes especially among patients with organ confined prostate cancer, they should be viewed in the context of the limitation of MRI and MRSI in assessing accurate tumor volumes. [60,61] Further, such image-guided treatments will have to take into consideration the amount of distortion caused by the endorectal coil prior to optimization of the treatment plan and performing brachytherapy using ultrasound/CT imaging.[62,63]

The ability of MRI to detect residual or recurrent prostate cancer following radiation therapy is limited due to post-treatment changes, volume changes, and low signal intensity on T_2-weighted images. Preliminary studies have shown that MRSI alone can be very sensitive (89%) and specific (82%) for the diagnosis of local recurrence following external beam radiation therapy when only the number of suspicious voxels (in this case three consecutive voxels) was used to define diagnostic threshold.[64] In another study, MRSI had a sensitivity and specificity of 77% and 78%, respectively, following external beam therapy, which was confirmed with step-section pathological findings. [65] MRSI may be a useful adjunct as an early tool for evaluating the treatment response for patients treated with external beam therapy, permanent prostate implantation, or hormonal therapy by characterizing the time to metabolic atrophy as this may be an early indicator compared to the PSA. It has been shown that metabolic atrophy occurs months earlier compared to PSA nadir.[66,67] It should be noted that total metabolic atrophy results in spectra with no peaks that looks like noise and may be falsely attributed to a failed exam as shown in Figure 12.4. When evaluating patients for recurrence of prostate cancer to assess metabolic atrophy, expanding the spectral window to view the residual water peak is advisable to confirm that the lack of metabolic peaks is indeed due to metabolic atrophy, rather than technical failure.

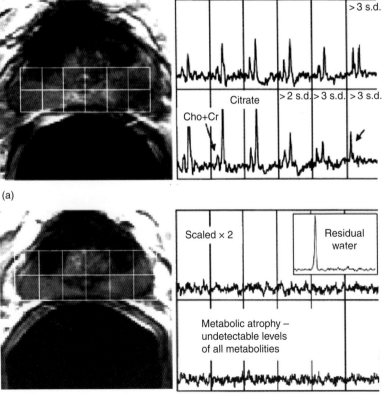

Figure 12.4. Metabolite concentrations (a) before and (b) ~3 years after external beam radiation therapy. Note the increase in (Cho+Cr)/Cit ratio in the left peripheral zone. Complete metabolic atrophy is seen after radiation therapy which may be mistaken for failed exam. The quality of the exam can be confirmed by verifying the linewidth of the residual water peak which confirms metabolic atrophy. From [67], with permission.

MRI/MRSI protocol

A conventional prostate MR exam is best conducted using a combination of the body-phased array receive coil and a balloon-covered endorectal receiver coil. It is necessary to verify the position of the coil once it is placed in the rectum to ensure that the coil is centered to the prostate and that the face of the coil is well aligned with the posterior margins of the prostate. This minimizes any artifacts that may result from the exacerbation of the B_0-inhomogeneity introduced by the endorectal coil. The balloon of the coil is inflated with 80–100 cc of air to ensure good coupling with the prostate and to minimize motion of the prostate; however, the presence of the air causes unfavorable magnetic susceptibility effects which decrease field homogeneity. More recent studies have shown improved spectral linewidths using perfluorocarbon balloon inflation.[102] To minimise prostate motion due to bowel peristalsis, many investigations have administered about 1 mg of glucagon intra-muscularly. High resolution axial T_1- and T_2-weighted, and coronal T_2-weighted, images with a slice thickness of about 3 mm through the prostate and the seminal vesicles are typically acquired. A spectroscopic imaging volume is then prescribed using the axial T_2-weighted images in a manner that would maximize the coverage of the prostate while minimizing the inclusion of periprostatic fat and rectal air. To facilitate accurate interpretation of the MRSI data, care should be taken such that the spectroscopic volume (usually defined using a PRESS sequence) is aligned with the axial slices obtained for anatomical information. In this way, depending on the slice thickness used for the anatomical scans, two or three of the anatomical slices will exactly match the thickness of each of the 3D-MRSI slices. To further suppress unwanted signals from outside the prostate, spatial saturation pulses with very sharp cut-off may be positioned around the prostate, as shown in Figure 12.5. Saturating the much larger water and fat signals is necessary in order to visualize the metabolites and to minimize lipid contamination. This can be done in several ways; a common method is to use band-selective excitation with gradient dephasing (known as "BASING"), which provides excellent lipid and water suppression while sparing some water signal for quantification purposes.[45] Initial 3D MRSI experiments were as long as 17 min for a $16 \times 8 \times 8$ phase-encoded spectra array that provided a nominal spatial resolution as low as 0.24 cm³.[49,51,68,69,70] In recent years, the scan time for this technique has been reduced through the use of elliptical encoding to about 9 min.[71,72] As with any spectroscopic technique, shimming over the volume of interest is necessary prior to acquiring spectroscopic data. Linewidths of around 12–20 Hz should provide good quality spectra at 1.5 T. One can expect larger linewidths in the range of 20–30 Hz at 3.0 T. Ideally, the total MRI/MRSI exam time including the prep time should be no more than 60–75 min.

Spectral interpretation

Interpretation resulting from a combined evaluation of the MR images and metabolic changes observed through MRSI has been shown to be up to 98% specific for diagnosing prostate cancer.[51] Decreased signal intensity on T_2-weighted images in conjunction with decreasing levels of citrate and polyamines and a concomitant increase in the levels of choline increases the specificity in the diagnosis of prostate cancer. More recently, an additional peak corresponding with various polyamine compounds, particularly spermine, has received closer attention. Similar to citrate, polyamines are accumulated and secreted by well-differentiated benign prostate tissue.[73] Polyamines appear as a broad peak between the choline and creatine resonances. Since the choline and creatine resonances are inseparable for quantification purposes, most investigators use [Choline+Creatine]/ Citrate (CC/C) for spectral analysis. A standardized scoring method was developed by Jung et al. which is based on the deviation of the CC/C ratio from its normal value of 0.22±0.013.[74] A voxel CC/C value within one standard deviation of this normal value was given a score of 1, a value between 1 and 2 standard deviations was given a score of 2, a value between 2 and 3 standard deviations was given a score of 3, a value between 3 and 4

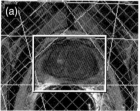 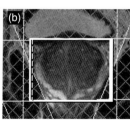 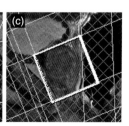

Figure 12.5. Multiple spatial saturation pulses are used in the (a) axial, (b) coronal, and (c) sagittal plane to conform to the shape of the prostate to minimize off-resonance artifacts in the resultant spectra. The white rectangular box shows the bounding box for the prostate and the hashed lines represent the saturation slabs that are used to shape the prostate.

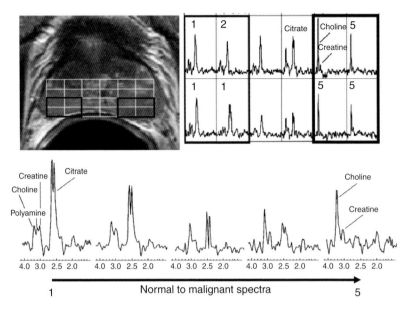

Figure 12.6. Scoring methodology employed by the University of San Francisco researchers for the interpretation of the prostate spectra that takes into consideration the role of polyamines. A score of 1 is considered normal prostate tissue, whereas a score of 5 indicates malignant prostate tissue whose [Cho+Cr]/Cit ratio is greater than four standard deviations from the normal ratio of 0.22 ± 0.13. From [74], with permission.

standard deviations was given a score of 4, and a value greater than 4 standard deviations was assigned a score of 5. Additional adjustments were made to the score to account for the elevation of choline over creatine, reduced polyamines, and poor SNR due to metabolite atrophy (Figure 12.6). In this way each voxel obtained a score between 1 and 5, which was designated to an interpretative scale of likely benign, probably benign, equivocal, probably malignant and likely malignant, corresponding to a voxel score from 1 to 5, respectively. Using this standardized five-point scale, they were able to show good accuracy and excellent interobserver agreement. It should be noted that 3D-MRSI produces large amounts of spectroscopic data, and a standardized scale such as the one developed by Jung et al. is likely to make the task of spectral interpretation less difficult.[74] Zakian et al. performed a study on 123 patients to determine whether MRSI can be used to predict aggressiveness of prostate cancer in comparison with step-section histopathology.[75] In this study, the investigators used a four-point scale similar to the above scale except that only CC/C ratio was used with no additional adjustments. The study demonstrated that detectability of the tumors depended on the Gleason score (a histological measure of tumor differentiation which correlates with prognosis). The sensitivity of MRSI increased to 56% for tumor detection compared to 44% for MRI alone in Gleason score 3 + 3 and increased to 89% on lesions with Gleason score ≥4 + 4. Development of such standardized scales may therefore allow characterization of tumor aggressiveness.

J-coupling

Quantification of the citrate resonance needs to account for the J-coupling interactions of the methylene protons and the chemical shift difference between them (see Chapter 1). Because the J-coupling (J = 16.1 Hz) and the chemical shift difference (9.4 Hz at 1.5 T; 18.8 Hz at 3.0 T) are of the same order of magnitude at the clinical field strengths of 1.5 and 3.0 T, the corresponding spectral shape of citrate is similar to an AB-type multiplet centered around 2.6 ppm as shown in Figure 12.7. This multiplet is dependent not only on J-coupling and the chemical shift, but is also modulated by the timing of the RF pulses used in the pulse sequence. At 1.5 T, van der Graaf et al. reported that when using the PRESS sequence the optimal echo time was 120, with the $\tau 1$ being 11 ms (time between the 90° and the 180° pulse) and $\tau 2$ being 60 ms (time between the two 180° pulses). At this echo time, the two outer multiples are minimized and the inner lines have the maximum absorptive intensity and are in phase with the other metabolites.[76] As demonstrated in Figure 12.7, the optimum TE at 3.0 T could be in the range of 75–100 ms if inverted (negative) magnetization is acceptable. A TE of 140–145 ms could also provide an absorptive spectrum with the inner and outer lines visible and in phase with the rest of the metabolites from the prostate. Scheenan et al. argued that the optimal spectral shape is not the only factor of interest, as the TR has to be taken into account.[72] Given the short T_2 values of citrate (~0.17 ± 0.05 s), they recommend

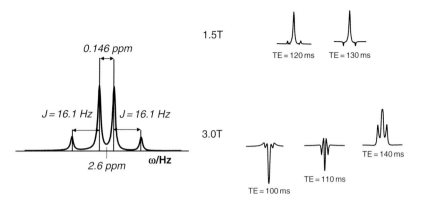

Figure 12.7. J-coupling interactions of the methylene protons in citrate can make spectral quantification problematic. The optimum echo time for citrate spectrum to be in absorptive mode are at 120 ms at 1.5 T and 140 ms at 3.0 T. Adapted from [103].

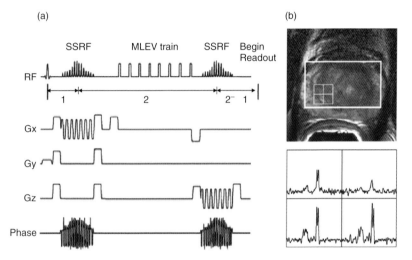

Figure 12.8. A conventional PRESS sequence using spectral spatial RF (SSRF) refocusing pulses. The Michael–Levittt (MLEV) train of pulses between the two refocusing pulses provide the J-refocusing thus minimizing the effects of J-modulation. From [77], with permission.

using a shorter TE of 75 ms and a TR of 750 ms with weighted averaging. However, most spectra reported in the literature to date on 3 T have been obtained at a TE of 140–145 ms.

New pulse sequences have been proposed that control the J-modulation of citrate to obtain pure absorption spectra from citrate at reasonable echo times. Cunningham et al. designed a new pulse sequence (Figure 12.8) that used short spectral–spatial refocusing pulses to minimize the J-modulation along with a phase-cycled train of nonselective Michael–Levitt (MLEV) pulses to manipulate the J-modulation of the citrate spectrum. An optimum echo time of 85 ms was arrived at from their simulations followed by phantom studies and in vivo validation as shown in Figure 12.8.

Two-dimensional J-resolved PRESS sequences have been proposed at higher field strengths. In this technique, acquisitions can be obtained at incremental echo times to obtain metabolite coupling information. This can be particularly advantageous in the case of prostate cancer, as it allows for the separation of the polyamine metabolite from the choline and creatine resonance compared to normal acquisition techniques which are unable to separate them. Further, because J-resolved spectroscopy steps through several echo times, one can obtain T_2 values of the metabolites. The first in vivo studies of the prostate using 2D J-resolved spectroscopy was by Yue et al. where they demonstrated the ability to resolve choline-containing compounds and spermine unequivocally, and further they were able to distinguish between the benign and malignant tissue. [78] However, the time to acquire the spectra from a single voxel was as long as 17 min and not practical for routine clinical use. Later, this technique was further refined by Kim et al., who employed spiral readout MRSI instead of the traditional phase-encoded method of acquiring spectroscopy data.[79,80] They used 16 spatial interleaves of the spirals to obtain a 32×32 spatial matrix for a voxel resolution of 0.59 cc. Sixteen different echo times starting from 35 to 285 ms were used to collect the data at spectral resolution of 4 Hz for a total acquisition time of 17 min at a TR of

2 s. Although this technique makes it more practical for clinical use, the coverage provided is still limited compared to the 3D-MRSI techniques currently available. Future hardware and software improvements may enable this technique to be more practical in the clinic.

1.5 T vs. 3.0 T

To date, most of the clinical spectroscopy on the prostate has been performed at 1.5 T. With the increasing availability of clinical 3.0 T scanners, MRSI studies of the prostate are beginning to appear at this higher field strength. Advantages of the higher field include a theoretical twofold increase in SNR, and increased spectral resolution, while increased magnetic susceptibility effects, potential relaxation time changes, and RF deposition may be disadvantageous. There is about a 20% increase in the T_1 value (1597±42 ms at 3 T) and a 16% decrease in the T_2 value (74±9 ms) of the prostate tissue at 3 T compared to 1.5 T, suggesting some optimization of MRI acquisition parameters is necessary to obtain optimum contrast and sensitivity.[81] Initial studies comparing MRI of prostate cancer at 1.5 T with the endorectal coil and 3.0 T using the body-phased array coil alone showed that the 1.5 T images with the endorectal coil had superior image quality and provided better delineation of prostate cancer. [82] Other investigators have found no significant difference between the 1.5 T with endorectal coil and 3.0 T with the body-phased array coil alone, and note that the added advantage of the undistorted images allows for accurate volume assessment and visualization of posterior border and seminal vesicles.[83,84] This suggests that prostate imaging may benefit from the use of an endorectal coil even at the higher field strength of 3.0 T. Another study compared endorectal coil images from normal volunteers at both 1.5 and 3.0 T and found that the image quality and the resolution achievable by 3.0 T far exceeded that of 1.5 T.[85]

Kim et al. developed a transmit/receive pelvic phase array coil and were among the first to show diagnostic quality spectra from the prostate using a surface coil at 3.0 T.[88] Scheenan et al. used 3D-MRSI without endorectal coil on 45 patients at 3.0 T and compared the results with histopathology (Figure 12.9).[86] Using the CC/C ratio, they obtained an accuracy (as determined by the area under the ROC curve, A_z) of 0.84 in the peripheral zone, which was significantly higher compared to the accuracy in the central gland of 0.69. They concluded that it was feasible to perform

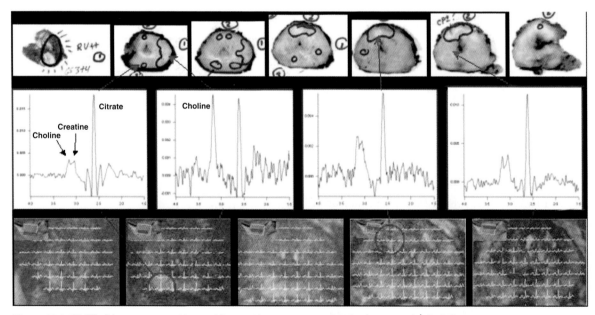

Figure 12.9. 3D-CSI of the prostate on a 64-year-old man with prostate cancer (PSA level, 6.86 ng ml^{-1}; final Gleason score, 3+4; stage, pT2c) using phased-array body coil without the endorectal coil (TR/TE = 750/145 ms) at 3 T. Top: photographs of histopathologic findings from apex to midgland in transverse slices. Arrows = areas of specimen that correspond to location of spectra and images below. Middle: spectra from in vivo prostate. Left to right: healthy peripheral zone, cancer in peripheral zone, cancer in central gland, and healthy central gland. Scale is adjusted for each spectrum. Bottom: transverse T_2-weighted MR images of three of 16 sections of 3D 1H-MRS data from apex to midgland overlaid with corresponding spectral maps (range 2.0–3.5 ppm). True size and location of voxels of which spectra are shown in middle row are indicated with circles. CP±? = possible capsular penetration, RV++ = positive resection margins. From [86], with permission.

3D MRSI with an external surface coil, to discriminate cancer from healthy tissue. Futterer *et al.* scanned 10 patients on both 1.5 and 3.0 T and reported excellent spectral resolution along with spatial and temporal resolution at 3.0 T, and suggested that this may result in improved accuracy in delineating and staging prostate cancer.[87] Kaji and colleagues used an external transceive coil for anatomic and metabolic assessment of the prostate on healthy volunteers. They were able to visualize duct-like structures at high resolution and obtain consistently good quality spectra showing both the outer and inner citrate peaks.[89]

Taken together, the above studies indicate that the higher field strength of 3.0 T can be beneficial for the detection of prostate cancer. The use of external surface coils to obtain spectroscopic information benefits those patients needing the exam shortly after surgery, such as after abdominal perineal resection for rectal cancers or after radiation therapy of the pelvis. Similarly, 3.0 T without the endorectal coil may be recommended for those patients undergoing prostate brachytherapy or external beam therapy, where metabolic and structural information may be obtained without any distortion to the prostate. Further, the surface coil method may be recommended for those patients that are "watchful waiting" to monitor the progression of the disease. Future research, however, will focus on pushing the boundaries of spatial and spectral resolution with the use of an endorectal coil that may allow us to have a more accurate assessment of the disease.

The combination of MRI and MRSI in conjunction with the endorectal and phased-array body coil is emerging as a useful tool for anatomic and metabolic evaluation of the prostate gland.[5,73,74] Improvements in pulse sequences and MR technology have enabled the acquisition of the metabolic information from the entire prostate at high resolution within a reasonable time of 10 min or less. Proton MRI/MRSI may be of great value for patients who are at increased risk for prostate cancer, for patients who have chosen watchful waiting, for longitudinal follow up from therapy, and in guiding various localized therapeutic treatments.[59,62,90]

Although the majority of the literature provides a positive outlook on MRSI of the prostate, it should be viewed in the context of at least two studies that have not seen any benefit from MRSI. Wetter *et al.* examined 50 patients to evaluate a routine protocol for combined MR and spectroscopic imaging of the prostate for staging accuracy.[91] The combination of the two techniques only provided a marginal improvement in staging performance that was not statistically significant. A recently concluded multisite ACRIN study that enrolled 134 patients scheduled for radical prostatectomy did not detect any incremental benefit from MRI and MRSI, compared to MRI alone, in sextant tumor localization in the peripheral zone.[92] While the reason for this discrepancy is not clear, patient selection criteria, technical factors, and interpretation of the data may have contributed to lack of extra value of MRSI compared to MRI alone. A recent meta-analysis of all published studies (meeting certain inclusion/exclusion criteria, 31 in total) related to the use of MRSI in the diagnosis of prostate cancer was cautious in its implications for clinical practice, but did suggest that combined MRI and MRSI may have a role for ruling-in cancer among patients who otherwise would be considered low-risk.[103]

Overall, MRSI is a sensitive tool for the evaluation of morphological and metabolic changes within the prostate. Combined with MRI, it can be used for detecting cancer, assessing tumor aggressiveness and staging. MRI/MRSI of the prostate gland is likely to benefit from the recent trend towards ultra-high field magnet systems and the use of multichannel parallel imaging. Further, newer techniques such as diffusion and perfusion are likely to increase the sensitivity and specificity of prostate cancer detection and characterization.[93,94,95,96,97,98,99,100,101]

References

[1] Stewart BW, Kleihus P, Eds. *World Cancer Report*. Lyon: IARC Press, 2003.

[2] American Cancer Society website: www.cancer.org; Cancer Facts and Figures 2006.

[3] American Cancer Society. *Cancer Facts & Figures 2009*. Atlanta: American Cancer Society; 2009.

[4] Hricak H, White S, Vigneron D, Kurhanewicz J, Kosco A, Levin D, *et al.* Carcinoma of the prostate gland: MR imaging with pelvic phased-array coils versus integrated endorectal-pelvic phased-array coils. *Radiology* 1991; **193**: 703–09.

[5] Kurhanewicz J, Vigneron DB, Males RG, Swanson MG, Yu K, Hricak H. The prostate: MR imaging and spectroscopy. Present and future. *Radiol Clin North Am* 2000; **38**: 115–38, viii–ix.

[6] McNeal JE. Normal anatomy of the prostate and changes in benign hypertrophy and carcinoma. *Semin US CT MR* 1988; **9**: 329–34.

[7] Perrotti M, Han KR, Epstein RE, Kennedy EC, Rabbani F, Badani K, et al. Prospective evaluation of endorectal magnetic resonance imaging to detect tumor foci in men with prior negative prostatic biopsy: A pilot study. *J Urol* 1999; **162**: 1314–7.

[8] Schiebler M, Miyamoto KK, White M, Maygarden SJ, Mohler JL. In vitro high resolution 1H-spectroscopy of the human prostate: Benign prostatic hyperplasia, normal peripheral zone and adenocarcinoma. *Magn Reson Med* 1993; **29**: 285–91.

[9] Torricelli P, Iadanza M, De Santis M, Pollastri CA, Cesinaro AM, Trentini G, et al. Magnetic resonance with endorectal coil in the local staging of prostatic carcinoma. Comparison with histologic macrosections in 40 cases. *Radiol Med (Torino)* 1999; **97**: 491–8.

[10] White S, Hricak H, Forstner R, Kurhanewicz J, Vigneron DB, Zaloudek CJ, et al. Prostate cancer: Effect of postbiopsy hemorrhage on interpretation of MR images. *Radiology* 1995; **195**: 385–90.

[11] Chen M, Hricak H, Kalbhen CL, Kurhanewicz J, Vigneron DB, Weiss JM, et al. Hormonal ablation of prostatic cancer: Effects on prostate morphology, tumor detection, and staging by endorectal coil MR imaging. *Am J Roentgenol* 1995; **166**: 1157–63.

[12] Engelbrecht MR, Jager GJ, Laheij RJ, Verbeek ALM, van Lier HJ, Barentsz JO. Local staging of prostate cancer using magnetic resonance imaging: A meta-analysis. *Eur Radiol* 2002; **12**: 2294–302.

[13] Costello LC, Franklin RB, Narayan P. Citrate in the diagnosis of prostate cancer. *The Prostate* 1999; **38**: 237–45.

[14] Costello LC, Franklin RB. The intermediary metabolism of the prostate: A key to understanding the pathogenesis and progression of prostate malignancy. *Oncology* 2000; **59**: 269–82.

[15] Cooper JF, Imfeld H. The role of citric acid in the physiology of the prostate: A preliminary report. *J Urol* 1959; **81**: 157–63.

[16] Cooper JF, Farid I. The role of citric acid in the physiology of the prostate: Lactic/citrate rations in benign and malignant prostatic homogenates as an index of prostatic malignancy. *J Urol* 1984; **92**: 533–6.

[17] Costello LC, Liu Y, Franklin RB, Kennedy MC. Zinc inhibition of mitochondrial aconitase and its importance in citrate metabolism of prostate epithelial cells. *J Biol Chem* 1997; **46**: 28875–81.

[18] Costello LC, Franklin RB. Novel role of zinc in the regulation of prostate citrate metabolism and its implications in prostate cancer. *The Prostate* 1998; **35**: 285–96.

[19] Zaichick VY, Sviridova TV, Zaichick SV. Zinc concentration in human prostatic fluid: Normal, chronic prostatitis, adenoma and cancer. *Int Urol Nephrol* 1996; **28**: 687–94.

[20] Franklin RB, Ma J, Zou J, Kukoyi BI, Feng P, Costello LC. hZIP1 is a major zinc uptake transporter for the accumulation of zinc in prostate cells. *J Inorg Biochem* 2003; **96**: 435–42.

[21] Zaichick VY, Sviridova TV, Zaichick SV. Zinc in the human prostate gland: Normal hyperplasia, cancerous. *Int Urol Nephrol* 1997; **29**: 565–74.

[22] Costello LC, Franklin RB, Liu Y, Kennedy MC. Zinc causes a shift toward citrate at equilibrium of the m-aconitase reaction of prostate mitochondria. *Inorg Biochem* 2000; **78**: 161–5.

[23] Negendank W. Studies of human tumors by MRS: A review. *NMR Biomed* 1992; **5**: 303–24.

[24] Gillies RJ, Morse DL. In vivo magnetic resonance spectroscopy in cancer. *Annu Rev Biomed Eng* 2005; 7: 287–326.

[25] Costello LC, Franklin RB. Bioenergetic theory of prostate malignancy. *Prostate* 1994; **25**: 162–6.

[26] van der Graaf M, Schipper RG, Oosterhof GO, Schalken JA, Verhofstad AA, Heerschap A. Proton MR spectroscopy of prostatic tissue focused on the detection of spermine, a possible biomarker of malignant behavior in prostate cancer. *Magma* 2000; **10**: 153–9.

[27] Childs AC, Mehta DJ, Gerner EW. Polyamine-dependent gene expression. *Cell Mol Life Sci* 2003; **60**: 1394–406.

[28] Babban N, Gerner EW. Polyamines as modifiers of genetic risk factors in human intestinal cancers. *Biochem. Soc Trans* 2003; **31**: 388–92.

[29] Kurhanewicz J, Dahiya R, Macdonald JM, Chang LH, James TL, Narayan P. Citrate alterations in primary and metastasis human prostatic Aden carcinomas: 1H magnetic resonance spectroscopy and biochemical study. *Magn Reson Med* 1993; **29**: 149–57.

[30] Fowler AH, Pappas AA, Holder, JC, Finkbeiner AE, Dalrymple GV, Mullins MS, et al. Differentiation of human prostate cancer from benign hypertrophy by in vitro 1H NMR. *Magn Reson Med* 1992; **25**: 140–7.

[31] Cornel, EB, Smits GAHJ, Oosterhof GON, Karthaus HFM, Debruyne FMJ, Schalken JA, et al. Characterization of human prostate cancer, benign prostatic hyperplasia and normal prostate by in vitro ^{1}H and ^{31}P magnetic resonance spectroscopy. *J Urol* 1993; **150**: 2019–24.

[32] Narayan D, Vigneron DB, Jajodia CM, Anderson CM, Hedgcock MW, Tanagho EA, et al. Transrectal probe for 1H MRI and 31P MR spectroscopy of the prostate gland. *J Magn Reson* 1989; **11**: 209–20.

[33] Narayan P, Kurhanewicz J. Magnetic resonance spectroscopy in prostate disease: Diagnostic possibilities and future developments. *Prostate* 1992; **4**: 43–50.

[34] Thomas MA, Narayan P, Kurhanewicz J, Jajodia P, Weiner MW. 1H MR spectroscopy of normal and malignant human prostate in vivo. *J Magn Reson* 1990; **87**; 610–9.

[35] Kurhanewicz J, Vigneron DB, Nelson SJ, Hricak H, MacDonald JM, Konety B, et al. Citrate as an in vivo marker to discriminate prostate cancer from benign prostatic hyperplasia and normal prostate peripheral zone: Detection via localized proton spectroscopy. *Urology* 1995; **45**; 459–66.

[36] Liney GP, Lowry M, Turnbull LW, Manton DJ, Knowles AJ, Blackband SJ, et al. Proton MR T2 maps correlate with the citrate concentration in the prostate. *NMR Biomed* 1996; **9**: 59–64.

[37] Liney GP, Turnbull LW, Lowry M, Turnbull LS, Knowles AJ, Horsman A. In vivo quantification of citrate concentration and water T2 relaxation time of the pathologic prostate gland using 1H MRS and MRI. *Magn Reson Imag* 1997; **15**: 1177–86.

[38] Frahm J, Bruhn H, Gyngell ML, Merbolt KD, Hanicke W, Sauter R. Localized high-resolution proton NMR spectroscopy using echoes: Initial applications to human brain in vivo. *Magn Reson Med* 1989; **9**: 79–93.

[39] Bottomley PA. Spatial localization in NMR spectroscopy in vivo. *Ann NY Acad Sci* 1987; **508**: 333–48.

[40] Heerschap A, Jager G, de Koster A, Barentsz J, de la Rosette J, Debruyne F, et al. 1H MRS of prostate pathology. In *Proceedings of Soc of Magn Reson Med*, 12th annual meeting, New York, 1993, p. 213.

[41] Brown TR. Practical applications of chemical shift imaging. *NMR Biomed* 1992; **5**: 238–43.

[42] Brown TR, Kincaid BM, Ugurbil K. NMR chemical shift in three dimensions. *Proc Natl Acad Sci USA* 1982; **79**: 3523–6.

[43] Maudsley AA, Hilal SK, Simon HE, Wittekoek S. In vivo MR spectroscopic imaging with P-31. Work in progress. *Radiology* 1984; **153**: 745–50.

[44] Luyten PR, Marien AJ, den HJ. Acquisition and quantitation in proton spectroscopy. *NMR Biomed* 1991; **4**: 64–9.

[45] Star-Lack J, Nelson SJ, Kurhanewicz J, Huang LR, Vigneron DB. Improved water and lipid suppression for 3D PRESS CSI using RF Band selective inversion with gradient dephasing (BASING). *Magn Reson Med* 1997; **38**: 311–21.

[46] Tran T-KC, Vigneron DB, Sailasuta N, Tropp J, Le Roux P, Kurhanewicz J, et al. Very selective suppression pulses for clinical MRSI studies of brain and prostate cancer. *Magn Reson Med* 2000; **43**: 23–33.

[47] Star-Lack J, Vigneron DB, Pauly J, Kurhanewicz J, Nelson SJ. Improved solvent suppression and increased spatial excitation bandwidths for three-dimensional PRESS CSI using phase-compensating spectral/spatial spin-echo pulses. *J Magn Reson Imaging* 1997; **7**: 745–57.

[48] Males RG, Vigneron DB, Star-Lack J, Falbo SC, Nelson SJ, Hricak H, et al. Clinical application of BASING and spectral/spatial water and lipid suppression pulses for prostate cancer staging and localization by in vivo 3D 1H magnetic resonance spectroscopic imaging. *Magn Reson Med* 2000; **43**: 17–22.

[49] Kurhanewicz J, Vigneron DB, Hricak H, Narayan P, Carroll P, Nelson SJ. Three-dimensional H-1 MR spectroscopic imaging of the in situ human prostate with high (0.24–0.7 cm^3) spatial resolution. *Radiology* 1996; **198**: 795–805.

[50] van der Graaf M, van den Boogert HJ, Jager GJ, Barentsz JO, Heerschap A. Human prostate: Multisection proton MR spectroscopic imaging with a single spin-echo sequence – preliminary experience. *Radiology* 1999; **213**: 919–25.

[51] Wefer AE, Hricak J, Vigneron DB. Sextant localization of prostate cancer; comparison of sextant biopsy, magnetic resonance imaging and magnetic resonance spectroscopic imaging with step section histology. *J Urol* 2000; **164**: 400–04.

[52] Zakian KL, Eberhardt S, Hricak H, Shukla-Dave A, Kleinman S, Muruganandham M, et al. Transition zone prostate cancer: Metabolic characteristics at 1H spectroscopic imaging – initial results. *Radiology* 2003; **229**: 241–7.

[53] Beyersdorff D, Taupitz M, Winkelmann B, Fischer T, Lenk S, Loening SA, et al. Patients with a history of elevated prostate-specific antigen levels and negative transrectal US-guided quadrant or sextant biopsy results: Value of MR imaging. *Radiology* 2002; **224**: 701–06.

[54] Terris MK. Prostate biopsy strategies: Past, present, future. *Urol Clin North Am* 2002; **29**: 205–12.

[55] Zelefsky MJ, Cohen G, Zakian KL, Dyke J, Koutcher JA, Hricak H, et al. Intraoperative conformal optimization for transperineal prostate implantation using magnetic resonance spectroscopic imaging. *Cancer J* 2000; **6**: 249–55.

[56] Zaider M, Zelefsky MJ, Lee EK, Zakian KL, Amols HI, Dyke J, et al. Treatment planning for prostate implants using magnetic resonance spectroscopy imaging. *Int J Radiation Oncology Biol Phys* 2000; **47**: 1085–96.

[57] Pickett B, Vigneault E, Kurhanewicz J, Verhey L, Roach M. Static field intensity modulation to treat a dominant intra-prostatic lesion to 90 Gy compared to seven field

[57] ...3-dimensional radiotherapy. *Int J Radiat Oncol Biol Phys* 1999; **44**: 921–9.

[58] Xia P, Pickett B, Vigneault E, Verhey LJ, Roach M 3rd. Forward or inversely planned segmental multileaf collimator IMRT and sequential tomotherapy to treat multiple dominant intraprostatic lesions of prostate cancer to 90 Gy. *Int J Radiat Oncol Biol Phys* 2001; **51**: 244–54.

[59] DiBiase SJ, Hosseinzadeh K, Gullapalli RP, Jacobs SC, Naslund MJ, Sklar GN, et al. Magnetic resonance spectroscopic imaging-guided brachytherapy for localized prostate cancer. *Int J Radiat Oncol Biol Phys* 2002; **52**: 429–38.

[60] Coakley FV, Kurhanewicz J, Lu Y, Jones KD, Swanson MG, Chang SD, et al. Prostate cancer tumor volume: Measurement with endorectal MR and MR spectroscopic imaging. *Radiology* 2002; **223**: 91–7.

[61] Hom JJ, Coakley FV, Simko JP, Qayyum A, Lu Y, Schmitt L, et al. Prostate cancer: Endorectal MR imaging and MR spectroscopic imaging – distinction of true-positive results from chance-detected lesions. *Radiology* 2006; **238**: 192–9.

[62] Mizowaki T, Cohen GN, Fung AY, Zaider M. Towards integrating functional imaging in the treatment of prostate cancer with radiation: The registration of the MR spectroscopy to ultrasound/CT images and its implementation in treatment planning. *Int J Radiat Oncol Biol Phys* 2002; **54**: 1558–64.

[63] Wu X, Dibiase SJ, Gullapalli RP, Yu CX. Deformable image registration for the use of magnetic resonance spectroscopy in prostate treatment planning. *Int J Radiat Oncol Biol Phys* 2004; **58**: 1577–83.

[64] Coakley FV, The HS, Qayyum A, Swanson MG, Lu Y, Roach M 3rd, et al. Endorectal MR imaging and MR spectroscopic imaging for locally recurrent prostate cancer after external beam radiation therapy: preliminary experience. *Radiology* 2004; **233**: 441–8.

[65] Pucar D, Shukla-Dave A, Hricak H, Moskowitz CS, Kuroiwa K, Olgac S, et al. Prostate cancer: Correlation of MR imaging and MR spectroscopy with pathologic findings after radiation therapy-initial experience. *Radiology* 2005; **236**: 545–53.

[66] Pickett B, Ten Haken RK, Kurhanewicz J, Qayyum A, Shinohara K, Fein B, et al. Time to metabolic atrophy after permanent prostate seed implantation based on magnetic resonance spectroscopic imaging. *Int J Radiat Oncol Biol Phys* 2004; **59**: 665–73.

[67] Pickett B, Kurhanewicz J, Coakley F, Shinohara K, Fein B, Roach M 3rd. Use of MRI and spectroscopy in evaluation of external beam radiotherapy for prostate cancer. *Int J Radiat Oncol Biol Phys* 2004; **60**: 1047–55.

[68] Yu KK, Scheidler J, Hricak H, Vigneron DB, Zaloudek CJ, Males RG, et al. Prostate cancer: Prediction of extracapsular extension with endorectal MR imaging and three-dimensional proton MR spectroscopic imaging. *Radiology* 1999; **213**: 481–8.

[69] Coakley FV, Kurhanewicz J, Liu Y, Jones KD, Swanson MG, Chang SD, et al. Prostate cancer tumor volume: Measurement by endorectal MR imaging and MR spectroscopic imaging. *Radiology* 2002; **223**: 91–7.

[70] Kurhanewicz J, Vigneron DB, Nelson SJ. Three-dimensional magnetic resonance spectroscopic imaging of brain and prostate cancer. *Neoplasia* 2000; **2**: 166–89.

[71] Scheme TWJ, Klomp DWJ, Roll SA, Futterer JJ, Barentsz JO, Heerschap A. Fast acquisition-weighted three-dimensional proton MR spectroscopic imaging of the human prostate. *Magn Reson Med* 2004; **52**: 80–8.

[72] Scheme TWJ, Gambarota G, Weiland E, Klomp DWJ, Futterer JJ, Barentsz, Heerschap A. Optimal timing for in vivo 1H-MR spectroscopic imaging of the human prostate at 3 T. *Magn Reson Med* 2005; **53**: 1268–74.

[73] Swanson MG, Vigneron DB, Tran T-KC, Kurhanewicz J. Magnetic resonance imaging and spectroscopic imaging of prostate cancer. *Cancer Invest* 2001; **19**: 510–23.

[74] Jung JA, Coakley FV, Vigneron DB, Swanson MG, Qayyum A, Weinberg V, et al. Prostate depiction at endorectal MR spectroscopic imaging: Investigation of a standardized evaluation system. *Radiology* 2004; **233**: 701–08.

[75] Zakian KL, Sircar K, Hricak H, Chen H-N, Shukla-Dave A, Eberhardt S, et al. Correlation of proton MR spectroscopic imaging with Gleason score based on step-section pathological analysis after radical prostatectomy. *Radiology* 2005; **234**: 804–14.

[76] van der Graaf M, Jager GJ, Heerschap A. Removal of the outer lines of the citrate multiplet in proton magnetic resonance spectra of the prostatic gland by accurate timing of a point-resolved spectroscopy pulse sequence. *MAGMA* 1994; **5**: 65–9.

[77] Cunningham CH, Vigneron DB, Marjanska M, Chen AP, Xu D, Hurd RE, et al. Sequence design for magnetic resonance spectroscopic imaging of prostate cancer at 3 T. *Magn Reson Med* 2005; **53**: 1033–9.

[78] Yue K, Marmot A, Bines N, Thomas MA. 2D JPRESS of human prostates using an endorectal receiver coil. *Magn Reson Med* 2002; **47**: 1059–64.

[79] Kim D-H, Henry R, Spielman DM. Fast multi-voxel two-dimensional spectroscopic imaging at 3 T. *Magn Reson Imaging* 2007; **25**: 1144–61.

[80] Kim D-H, Margolis D, Xing L, Daniel B, Spielman D. In vivo prostate magnetic resonance spectroscopic imaging using two-dimensional J-resolved PRESS at 3 T. *Magn Reson Med* 2005; **49**: 1177–82.

[81] de Baseline CMJ, Duane GD, Risky NM, Alsip DC. MR imaging relaxation times of abdominal and pelvic tissues measured in vivo at 3.0 T: Preliminary results. *Radiology* 2004; **230**: 652–9.

[82] Beyersdorff D, Taymoorian K, Knosel T, Schnorr D, Felix R, Hamm B, et al. MRI of prostate cancer at 1.5 and 3.0 T: Comparison of image quality in tumor detection and staging. *Am J Roentgenol* 2005; **185**: 1214–20.

[83] Sosna J, Rofsky NM, Gaston SM, DeWolf WC, Lenkinski RE. Determinations of prostate volume at 3-Tesla using an external phased array coil: Comparison to pathologic specimens. *Acad Radiol* 2003; **10**: 846–53.

[84] Sosna J, Pedrosa I, Dewolf WC, Mahallati H, Lenkinski RE, Rofsky NM. MR imaging of the prostate at 3 Tesla: Comparison of an external phased-array coil to imaging with an endorectal coil at 1.5 Tesla. *Acad Radiol* 2004; **11**: 857–62.

[85] Bloch BN, Rofsky NM, Baroni RH, Marquis RP, Lenkiski RE. 3-Tesla magnetic resonance imaging of the prostate with combined pelvic phased-array and endorectal coils initial experience. *Acad Radiol* 2004; **11**: 863–7.

[86] Scheenan TWJ, Heijmink SWTPJ, Roell SA, Hulsbergen-Van de Kaa, Knipscheer BC, Witjes JA, et al. Three-dimensional proton MR spectroscopy of human prostate at 3 T without endorectal coil: Feasibility. *Radiology* 2007; **245**: 507–16.

[87] Futterer JJ, Scheenen TWJ, Huisman HJ, Klomp DWJ, van Dorsten FA, Hulsbergen-van de Kaa CA, et al. Initial experience of 3 Tesla endorectal coil magnetic resonance imaging and 1H-spectroscopic imaging of the prostate. *Invest Radiol* 2004; **39**: 671–80.

[88] Kim H-W, Buckley DL, Peterson DM, Duensing GR, Caserta J, Fitzsimmons J, et al. In vivo prostate magnetic resonance imaging and magnetic resonance spectroscopy at 3 Tesla using a transceiver pelvic phased array coil. *Invest Radiol* 2003; **38**: 443–51.

[89] Kaji Y, Kuroda K, Maeda T, Kitamura Y, Fujiwara T, Matsuoka Y, et al. Anatomical and metabolic assessment of prostate using a 3-Tesla MR scanner with a custom-made external transceiver coil: Healthy volunteer study. *J Magn Reson Imaging* 2007; **25**: 517–26.

[90] D'Amico AV, Whittington R, Malkowicz B, Schnall M, Schultz D, Cote K, et al. Endorectal magnetic resonance imaging as a predictor of biochemical outcome after radical prostatectomy in men with clinically localized prostate cancer. *J Urol* 2000; **164**: 759–63.

[91] Wetter A, Engl TA, Nadjmabadi D, Fliessback K, Lehnert T, Gurung J, et al. Combined MRI and MR spectroscopy of the prostate before radical prostatectomy. *Am J Roentgenol* 2006; **187**: 724–30.

[92] Weinreb JC, Coakley F, Blume J, Wheeler T, Cormack J, Kurhanewicz J. ACRIN 6659: *MRI and MRSI of prostate cancer prior to radical prostatectomy: A prospective multi-institutional clinicopathological study.* Chicago: RSNA, 2006

[93] Gibbs P, Pickles MD, Turnbull LW. Diffusion imaging of the prostate at 3.0 Tesla. *Invest Radiol* 2006; **41**: 185–8.

[94] Pickles MD, Gibbs P, Sreenivas M, Turnbull LW. Diffusion-weighted imaging of normal and malignant prostate tissue at 3.0 T. *J Magn Reson Imaging* 2006; **23**: 130–4.

[95] Shimofusa R, Fujimoto H, Akamata H, Motoori K, Yamamoto S, Ueda T, et al. Diffusion-weighted imaging of prostate cancer. *J Comput Assist Tomogr* 2005; **29**: 149–53.

[96] Padhani AR, Gapinski CJ, Macvicar DA, Parker GJ, Suckling J, Revell PB, et al. Dynamic contrast enhanced MRI of prostate cancer: Correlation with morphology and tumor stage, histological grade and PSA. *Clin Radiol* 2000; **55**: 99–109.

[97] Hara N, Okuizumi M, Koiki H, Kawaguchi M, Bali V. Dynamic contrast-enhanced magnetic resonance imaging (DCE-MRI) is a useful modality for the precise detection and staging of early prostate cancer. *The Prostate* 2004; **62**: 140–7.

[98] Huisman HJ, Engelbrecht MR, Barentsz JO. Accurate estimation of pharmacokinetic contrast enhanced dynamic MRI parameters of the prostate. *J Magn Reson Imaging* 2001; **13**: 607–14.

[99] Engebrecht MR, Huisman HJ, Laheij RJF, Jager GJ, van Leenders GJLH, Hulsbergenvan de Kaa CA, et al. Discrimination of prostate cancer from normal peripheral zone and central gland tissue by using contrast-enhanced MR imaging. *Radiology* 2003; **229**: 248–54.

[100] Buckley D, Roberts C, Parker GJM, Hutchinson CE. Prostate cancer: Evaluation of vascular characteristics with dynamic contrast-enhanced T1-weighted MR imaging – initial experience. *Radiology* 2004; **233**: 709–15.

[101] van Dorsten FA, van der Graaf M, Engelbrecht MR, van Leenders GJLH, Verhofstad A, Rjpkema M, et al. Combined quantitative dynamic contrast enhanced MR imaging and 1H MR spectroscopic imaging of human prostate cancer. *J Magn Reson Imaging* 2004; **20**: 279–87.

[102] Prando A, Kurhanewicz J, Borges AP, Oliveira EM Jr, Figueiredo E. Prostatic biopsy directed with endorectal MR spectroscopic imaging findings in patients with elevated prostate specific antigen levels and prior negative biopsy findings: Early experience. *Radiology* 2005; **236**: 903–10.

[103] Umbehr M, Bachmann LM, Held U, Kessler TM, Sulser T, Weishaupt D, *et al*. Combined magnetic resonance imaging and magnetic resonance spectroscopy imaging in the diagnosis of prostate cancer: A systematic review and meta-analysis. *Eur Urol* 2009; **55**: 575–91.

Chapter 13: MRS in breast cancer

Key points

- MRS of the breast is more technically demanding than that in the brain.
- Cho levels have been reported to be higher in malignant breast cancer than in benign lesions and normal breast tissue.
- Early decreases in Cho signal intensity may be seen in lesions that respond to treatment.
- MRS is limited by sensitivity to lesions at least 1 cm^3.
- Inadequate sensitivity may lead to false negatives, and both false positives and negatives may arise due to insufficient water and lipid suppression, or other artifacts.

Introduction: MRS of breast tissues

Although the vast majority of magnetic resonance spectroscopy (MRS) studies in humans have been performed to date in the central nervous system, there is growing interest in the application of MRS to other organ systems in the body. This is particularly true for areas such as breast cancer, where conventional diagnostic techniques have relatively limited sensitivity and/or specificity. MRS of the breast presents a number of technical challenges (described in detail later in this chapter) which are gradually being overcome, allowing clinical research studies to be performed. Early MRS studies of human breast cancer focused on the phosphorus (^{31}P) nucleus,[1] since localized, water-suppressed proton spectroscopy was not available at that time. However, with the development of improved gradient hardware, spatial localization, and water suppression techniques, ^{31}P spectroscopy has largely been replaced by proton (^{1}H) MRS. The much higher sensitivity of the proton nucleus allows spectra with higher signal-to-noise ratios (SNR) to be recorded from smaller volumes of tissue compared to ^{31}P.

The normal breast consists of glandular and adipose tissue, which in the proton spectrum exhibit large signals from water and fat (lipid) resonances, respectively (Figure 13.1A). Typically, without water or lipid suppression techniques, these are the only visible resonances in the spectrum. Water resonates at approximately 4.7 ppm in the spectrum, while multiple lipid resonances occur, the largest ones at 0.9 (CH$_3$) and 1.3 (–CH$_2$)$_n$–) ppm, but other somewhat smaller lipid resonances may also be detected at 2.2 and 2.5 ppm (due to –CH$_2$–C=O, and diallylic CH$_2$ protons, respectively), and unsaturated (olefinic) protons from the glycerol backbone at 5.2 ppm. When water and lipid suppression pulses are turned on, it becomes possible to observe much smaller metabolite signals, either in normal breast tissue or in breast lesions (Figure 13.1B). The assignments of these resonances has been investigated by the use of "magic-angle spinning" (MAS) spectroscopy of tissue samples at high magnetic fields (Figure 13.1C), as well as two-dimensional spectroscopy of perchloric acid extracts.[2] Signals have been assigned to various compounds including choline (Cho, 3.2 ppm), creatine (Cr, 3.0 and 3.9 ppm), taurine (Tau, 3.3 ppm), glycine (Gly, 3.5 ppm), and alanine (Ala, 3.8 ppm), amongst others.[3] It should be noted that these signals are generally only clearly observed under high signal-to-noise (SNR) conditions, i.e. using high-strength magnetic fields and state-of-the-art methodology; more commonly, in normal breast tissue using lower field strength (e.g. 1.5 T) systems, the only observable signals may be residual water and lipids. For this reason, many studies to date have used the visible presence of any Cho signal in the spectrum as an indication of cancer, since under these conditions at lower magnetic fields no Cho is usually seen. However, this may not be the case under all circumstances; for instance, it has been reported that Cho and lactose signals (~3.8 ppm) can be observed in women who are lactating (Figure 13.1D). The sensitivity and specificity

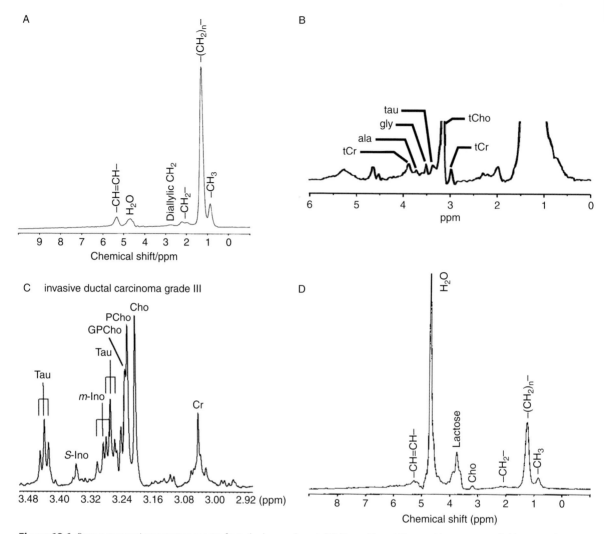

Figure 13.1. Proton magnetic resonance spectra from the human breast: (A) Normal breast tissue, without water or lipid suppression, demonstrates spectra from water and lipids (reproduced with permission from [58]). (B) High field (4 T) spectrum from a breast cancer showing peaks from Cho and other compounds (reproduced with permission from [3]). (C) High-resolution magic-angle spinning (MAS) spectrum of an invasive ductal carcinoma showing signals from Cho, PC, and GPC, as well as other signals from creatine (Cr), taurine (Tau) and scyllo- and myo-inositol (reproduced with permission from [2]). (D) Normal breast tissue in a lactating patient, showing the presence of lactose at 3.8 ppm, as well as a detectable Cho signal (reproduced with permission from [29]).

for Cho for breast cancer detection will be discussed in detail below.

Clinical background: breast cancer screening and diagnosis

Breast cancer is the most frequently diagnosed type of cancer in women and is the third leading cause of death.[4] Incidence rates increased by approximately 21% between 1973 and 1990, but have begun to decrease in recent years, with relatively constant mortality rates until the last few years, where annual decreases have been seen. At least one reason for the improvement in mortality rates has been improved screening using X-ray mammography, although exact recommendations for initiation and frequency of mammographic screening remain controversial.[4,5]

Although early detection of breast cancer is critical for successful treatment, the current most widely used screening technique, X-ray mammography, has appreciable shortcomings both in terms of specificity and sensitivity. It has been estimated that between 5% and 15% of tumors are missed on mammography, and that the overall yield of breast cancers per biopsy (recommended on the basis of screening mammography) has been reported to be between only 10% and 50%.[6] While this yield may be moderately increased by the use of ultrasound for the differentiation of cystic vs. solid masses, there continues to be a search for alternative or adjunctive screening techniques with both higher sensitivity and specificity.[6,7] Various techniques show promise, ranging from radionucleotide imaging methods (PET, SPECT), elastography, optical imaging techniques, electrical impedance imaging, to contrast-enhanced MRI.[7] Clearly, in order to be a viable screening modality, these techniques should be non-invasive, low-risk, have good sensitivity and specificity, and cost-effective. While some of the lower-cost non-invasive imaging techniques listed above may have some role in the future as *screening* modalities, more realistically most of the more complicated and expensive procedures are likely to be used in the general population for *diagnostic* purposes only.

Breast imaging methods based on in vivo magnetic resonance (imaging or spectroscopy) are therefore most likely to be applied for diagnosis of lesions identified by other screening techniques (except for MRI screening in certain high-risk populations[8,9,10]). While breast MRI has been shown to have very high sensitivity, its specificity for distinguishing malignant from benign lesions has been reported to be highly variable (20–100%).[11] The vast majority of breast lesions, and virtually all breast cancers (with the exception of some (≈30%) ductal carcinomas in situ (DCIS))[12] enhance on administration of GdDTPA contrast agent, and are best visualized on high-resolution, subtracted (post- minus pre-contrast) fat-suppressed T_1-weighted images. Once detected, a number of approaches can help establish whether a lesion is benign or malignant: these include analysis of lesion morphology, location, signal intensity on other MRI sequences (T_1, T_2, etc.), and the rate of contrast uptake/washout in dynamic MR sequences.[11] However, even when all of these factors are considered, there often remains uncertainty as to whether a lesion is malignant or benign, and a biopsy cannot be avoided.[12] Therefore, it would be valuable to add additional sequences or other modalities which could increase the diagnostic confidence (in particular, specificity), either as adjunctive methods,[13] as part of a multi-parametric model,[14,15,16] or as an independent technique. Magnetic resonance spectroscopy has the potential to fulfill this role, although at the current time it largely remains mainly an experimental procedure. The rest of this chapter discusses the potential clinical applications of MRS in breast cancer, as well as some of the special technical challenges (and their solutions) that occur in MRS of the breast.

Magnetic resonance spectroscopy of breast cancer

Over the last few years, there has been appreciable interest in the use of proton magnetic resonance spectroscopy for the non-invasive diagnosis of breast lesions. One of the spectroscopic hallmarks of the neoplastic process appears to be presence of a choline (Cho) signal (actually, a composite resonance at 3.2 ppm in the in vivo spectrum (Figure 13.1C), predominantly consisting of phosphocholine (PC), glycerophosphocholine (GPC), and some free choline itself),[2,17] usually in higher concentrations than in normal tissue or in non-malignant lesions. This appears to be generally true, whether the lesion occurs in the brain, prostate, breast, liver, extremity, or the head and neck.[18] Studies performed in breast cancer specimens and cell culture preparations indicate high levels of PC in malignant breast cancer cells,[19,20,21] and up-regulation of phospholipase C, the enzyme which converts phosphatidylcholine into PC. The same pattern also seems to hold for lesions in vivo. For example, Figure 13.1C shows an (in vitro) MAS spectrum from an invasive ductal carcinoma specimen, showing prominent signals from both Cho and PC, larger than the GPC signal and other resonances in the spectrum.

Breast cancer has been studied somewhat less than some of these other organ systems in vivo (notably brain and prostate, mainly because of the additional technical challenges of performing breast spectroscopy), but the results available to date are nevertheless promising for the diagnosis of breast cancer by MRS.[22,23] In one of the earliest MRS studies in vivo, Roebuck *et al.*[24] found that 7 out of 10 malignant lesions (in situ and infiltrating ductal and lobular carcinomas) had detectable levels of Cho (estimated at a threshold detection level of 0.2 mM, based on measurements of an external Cho phantom) using 1–2 cm^3 voxel sizes. In contrast, six benign lesions (fibroadenomas and fibrocystic changes, adenosis) showed no detectable Cho signals, and only one lesion considered benign (a rare tubular adenoma) had a detectable Cho signal. In another study which also used water-suppressed single-voxel (SV) proton MRS

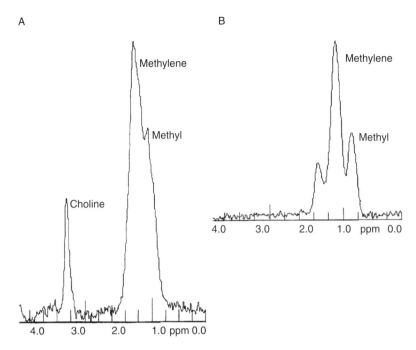

Figure 13.2. (A) and (B) Localized proton spectra from the human breast recorded using the PRESS pulse sequence (1.5 T, TR 2 sec, TE 350 msec, 256 averages, 3.8 cm^3 voxel size). (A) Infiltrating ductal carcinoma, (B) fibroadenoma. Lipid methyl and methylene signals are observed in both spectra, while an elevated choline peak at 3.2 ppm is only detected in the carcinoma (A). A peak at 1.5 ppm is observed in the fibroadenoma which was tentatively assigned to the β-protons of CH$_2$-CCO in fatty acid chains. Reproduced with permission from [25]).

(typical voxel sizes ranging from approximately 2 to 8 cm^3), Kvistad et al.[25] found that a Cho signal was detected in 9 of 11 carcinomas, whereas only 2 of 11 benign lesions (fibroadenomas, fibrocystic disease) had a detectable Cho signal (Figures 13.2A and B). In addition, in one patient undergoing neoadjuvant chemotherapy, the Cho signal was observed to decrease dramatically following treatment[25] (see below for more detail regarding treatment response studies).

A meta-analysis of the available MRS studies published prior to 2002 (5 studies, 153 lesions in total, 100 malignant and 53 benign)[23] indicated a sensitivity and specificity of 83% and 85%, respectively, for MRS. The sensitivity and specificity were even higher if the age range was restricted to young patients only. Since the meta-analysis performed in 2002, additional papers on the use of proton MRS in breast cancer have been published.[22] Generally, these studies have continued to suggest that MRS is a promising technique for diagnosis of breast lesions,[22] and have indicated how MRS may be added to breast MRI in order to improve diagnostic accuracy.[13] For instance, Haddadin et al. showed that MRS-based measurements of Cho concentrations, when provided to four radiologists, improved diagnostic accuracy and sensitivity over MRI alone, using a cutoff value of approximately 1 mmol kg^{-1} to distinguish benign from malignant lesions.[13] Techniques for quantifying lesion choline concentrations (and their potential variability) are discussed in more detail below; however, it should be noted that quantitative (concentration) analysis methods are not currently commercially available, so only selective research groups with sufficient technical expertise can perform this type of analysis at present.

Various factors should be kept in mind when evaluating spectra from breast lesions, with perhaps the most important issue being the lesion size. Since the signal-to-noise ratio (SNR) of a spectrum depends linearly on the voxel size (amongst other factors), smaller lesions will be less likely to show detectable Cho signals than larger ones. Haddadin et al.[3] found a less than 50% likelihood of detecting Cho in lesions less than 1 cm^3 using a 4 T magnet system, which would suggest that at lower field strengths (e.g. 1.5 T) the likelihood of detection is even lower, probably limiting MRS evaluations to lesions 2 cm^3 or larger. It should also be kept in mind that, under high SNR conditions, it is possible to detect Cho signals from normal fibroglandular tissue (Figure 13.3A), so that the presence of a Cho signal does not necessarily imply the presence of a lesion or malignancy. Inspection of Figure 13.3B, however, suggests that normal tissue, and many fibroadenomas, should not exceed a Cho concentration of 1 mmol kg^{-1}, so a concentration higher than this value is suspicious for malignancy.

Finally, in addition to evaluating primary breast lesions, MRS has also shown promise as a means of

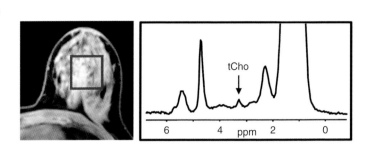

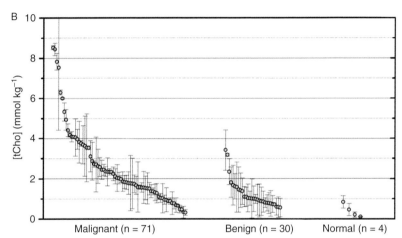

Figure 13.3. (A) 3 T breast MRS showing spectrum containing Cho signal from normal glandular tissue, and (B) ordered distribution of Cho concentrations from 71 malignant lesions, 30 benign lesions, and 4 normal subjects (reproduced with permission from [3]).

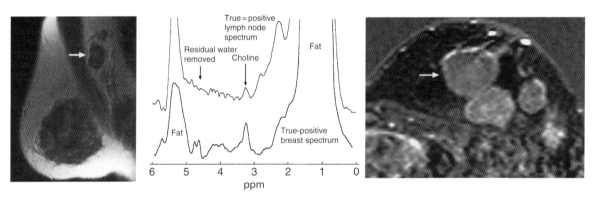

Figure 13.4. 1.5 T spectra (TE 135 msec) showing Cho signal at 3.2 ppm in both the breast lesion (12 cm^3 voxel size) and axillary lymph node (7.5 cm^3 voxel size). Post-contrast, fat-suppressed MRI shows multiple malignant lesions in the breast, and an enlarged (2 cm) lymph node (reproduced with permission from [27]).

investigating metastatic spread of breast cancer into the axillary lymph nodes, either using MRS of biopsy samples[26] or in vivo (if the nodes are large enough) MRS.[27] Figure 13.4 shows one example of a Cho signal detected in an enlarged axillary lymph node. In the study of Yeung et al.,[27] a sensitivity of 82% and specificity of 100% was reported for MRS detection of axillary lymph metastases (using ultrasound-guided fine-needle aspiration biopsy as the gold standard).

Monitoring treatment response

As mentioned above,[25] another area of interest for breast MRS is the detection of response to therapy in patients undergoing neoadjuvant chemotherapy.[28]

With conventional imaging techniques and/or clinical examination, it may take several weeks (based on tumor size) to determine if a tumor is responding to a particular chemotherapy regime. An earlier indicator of tumor response (or lack thereof) could in principle be used to alter chemotherapeutic agent or dosage, rather than performing several weeks of ineffective therapy.

It has been hypothesized that biochemical measurements of tumor Cho will decrease as cells are killed, prior to reduction in tumor size. Jagannathan et al. reported that Cho decreased or disappeared in 89% of patients undergoing chemotherapy.[29] The Minnesota group[28] found that reduced Cho could be detected within 24 h of the first treatment of locally advanced breast cancer with doxorubicin, and correlated well with decrease in tumor size measured after 4 cycles of chemotherapy. The change in the Cho signal was statistically significant between responders and non-responders, and responders actually showed higher Cho levels at baseline, indicating that pre-treatment Cho levels may have a role in predicting response. Figure 13.5 shows an example of a treatment responder, before and after 1 and 4 cycles of chemotherapy, respectively, and the summary of the Cho levels in responders and non-responders from the study by Haddadin et al.[3].

One issue with using MRS to estimate treatment response is that the Cho signal is harder to detect when the lesion becomes smaller. It could be argued that a decrease in lesion size on MRI is a primary indicator of response, and therefore MRS is not needed if the size changes; however, this issue does raise questions as to the appropriate MRS analysis method to be used (i.e. should the Cho signal be normalized to the size of the lesion, or lesion water content, or not?). It would appear that the main utility of MRS in monitoring neoadjuvant chemotherapy is in predicting early response prior to gross anatomical changes.

Technical aspects of breast spectroscopy

The first four chapters of this book have covered the fundamental aspects of the physics of MR spectroscopy, and have also described the techniques most commonly used for proton MR spectroscopy of the human brain. While many of these same techniques and principles are used for MRS of the breast, there are a number of additional challenges in breast spectroscopy that are beyond those encountered in the brain. Some examples of these challenges include spatial resolution and SNR (e.g. in characterizing small breast ($\sim 1\,cm^3$) lesions), the presence of huge lipid signals, optimization of field homogeneity, the effects of respiration, and how to quantify Cho signals in the absence of any convenient reference signals. These issues and some of the solutions to them are discussed below.

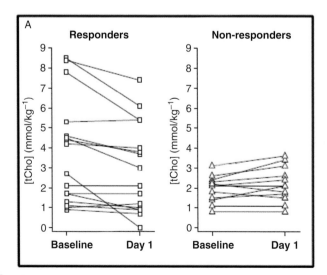
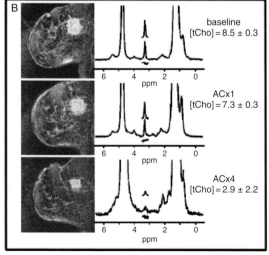

Figure 13.5. (A) Total choline levels in treatment responders and non-responders at baseline and after one cycle of chemotherapy. Responders have higher initial Cho levels and show a significantly greater decrease in Cho after treatment compared to non-responders. (B) An example of a decreasing Cho in a responder after 1 and 4 cycles of neoadjuvant chemotherapy (reproduced with permission from [3]).

Field homogeneity

Spectral resolution and SNR are critically dependent on magnetic field homogeneity, which can be challenging to optimize in the breast, even at field strengths of 1.5 T, and particularly at higher field strengths (3 T and above). Magnetic field homogeneity is a function of both the homogeneity of the main magnetic field itself (which may be somewhat lower than expected, since breast imaging is not performed at magnetic isocenter) and susceptibility effects of the breast tissue and torso. At 3 T, Maril *et al.* demonstrated that magnetic susceptibility effects cause quite a strong anterior–posterior gradient, with strong local inhomogeneity in the vicinity of the nipple, as well as second-order gradients in the left–right dimension.[30] Field map,[31,32] (or projection [33]) based, high-order shimming algorithms (with sufficient strength shim coils and amplifiers) are required to correct these inhomogeneities. The optimization method must be designed such that the water–fat composition of the breast does not influence the result (e.g. if using the field map-based algorithms, the water and fat signals must be in phase so that they do not give erroneous field strength values). Figure 13.6 shows an example of a field map at 3 T, before and after corrections using shimming up to the second order. It should also be remembered that field homogeneity and correct setting of center frequency (on the water resonance) are also important for the frequency-based fat suppression techniques that are required for breast MRI.

Field homogeneity and center frequency can also be affected by respiration – although motion of the breast itself secondary to respiration is small when the subject is prone (and properly positioned, including the use of compression), susceptibility effects from the lungs can cause significant scan to scan variations, particularly in the proximity of the chest wall.[34] Since this is a magnetic susceptibility effect, it will increase linearly with magnetic field strength. One approach to correcting this problem in SV-MRS is to store each free-induction decay (FID) separately in computer memory, and perform phase and frequency corrections prior to time-averaging.[34] This approach has been shown at 4 T to produce spectra with significantly better SNR and resolution compared to those collected without correction (Figure 13.7).[34]

Water and lipid suppression

Other factors may also affect the ability of MRS to detect a small Cho signal in the presence of much larger water and lipid signals. Unlike the brain, MRS of the breast is complicated by the presence of large lipid signals from adipose tissue, which have the potential to obscure the much smaller Cho signal. It has been shown that small magnetic field instabilities (such as gradient-induced vibrations, causing eddy currents) can modulate the large lipid signals typically found in the breast to produce discrete sidebands, which can overlap with or mimic small Cho signals.[35] One solution to this problem is to change the experimental conditions from scan to scan to alter the phase of the sidebands. One way to do this is to alter the echo time (TE); TE-averaged spectra have been shown to

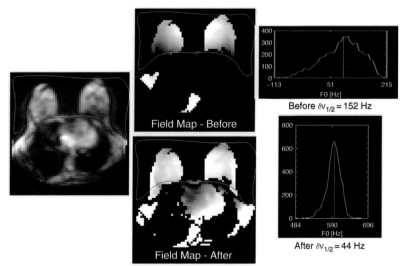

Figure 13.6. An example of field–map-based shimming for optimization of field inhomogeneity in the breast bilaterally at 3 T. Field map before shim correction shows regional inhomogeneity and a width at half height of 152 Hz. Second-order shim correction provides more uniform magnetic field map and a global linewidth of 44 Hz. Data provided courtesy of Dr Michael Schär.

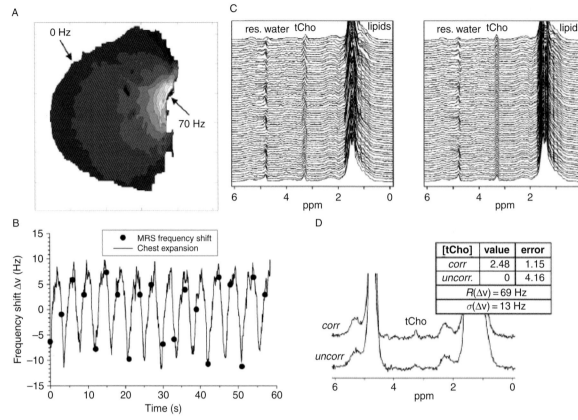

Figure 13.7. (A) Difference field map of the breast showing frequency difference between full inspiration and expiration at 4 T. Paramagnetic air in the lungs causes an appreciable frequency shift in the breast, particularly for the posterior regions adjacent to the chest wall. (B) MRS frequency shift (TR 3 sec) and chest expansion plotted versus time showing a good correlation, (C) Individual spectra recorded as a function of time showing frequency variations due to respiration (left), and (right) after application of frequency correction. (D) Averaged spectra from the left and right columns of figure (C). The frequency corrected spectra show a much more prominent Cho peak compared to the uncorrected. Reproduced with permission from [34]).

have much reduced sideband intensity[35] and produce cleaner spectra at 4 T. This methodology is more difficult to apply in magnetic resonance spectroscopic imaging (MRSI) experiments (see below), since proton MRSI does not usually involve any time-averaging, and therefore multiple TE acquisitions would involve unacceptably long scan times. While TE-averaged MRSI may be possible by combining phase-encoded steps with different echo times, perhaps a better approach is to develop improved water and lipid suppression schemes (and also therefore suppress unwanted modulation sidebands from water and lipid).

Various approaches to lipid suppression have been proposed, based either on differences in T_1 or frequency between Cho and lipids.[36,37] Typically, frequency-selective water and lipid suppression pulses are applied for breast spectroscopy, similar to the brain (for water). By using multiple pulses with carefully chosen flip angles,[38] quite high suppression factors can be obtained (e.g. ~100–1000). In the future, it is likely that simultaneous, dual-frequency suppression techniques will be used in order to maximize water and lipid suppression.[39,40] While, conventionally, pre-saturation is used for water or lipid suppression, frequency-selective pulses can also be applied during the echo time of the pulse sequence in order to increase suppression factors by selective dephasing of the water and lipid signals (so-called MEGA/BASING type experiments [40,41]). Of course, it should be noted that all frequency-selective suppression schemes require sufficient B_0 field homogeneity across the breast in order to function correctly.

Magnetic resonance spectroscopic imaging of the breast

Most of the MRS studies of human breast cancer published to date used single-voxel spectroscopy; however,

Chapter 13: MRS in breast cancer

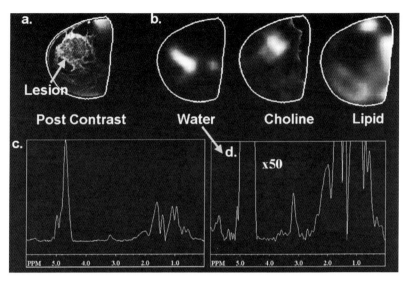

Figure 13.8. Proton MRI and MRSI of a 41-year-old patient with infiltrating ductal carcinoma of the breast. (a) Post-contrast T_1-weighted images of the breast lesion. (b) MRSI images of water, Cho, and lipids. Representative spectrum of elevated Cho within the lesion (c) and magnified (× 50) region (d) demonstrates a detectable Cho signal. Reproduced with permission from [36].

a few studies have used proton MRSI techniques. [14,16,36,42] There are several potential advantages of MRSI over SV: increased spatial coverage, usually also increased spatial resolution, information on distribution of the Cho signal, the retrospective ability to select voxel locations (e.g. those only identified post-hoc by the radiologist or computer-aided detection (CAD) software), and the ability to investigate multiple or serendipitous lesions. An example of a breast MRSI study studied at 1.5 T in a patient with an invasive ductal carcinoma is given in Figure 13.8.

However, MRSI of the breast also presents many technical challenges, and for this reason is only used for research purposes at the current time. Whereas SV spectroscopy allows the optimization of field homogeneity, water and lipid suppression on a small, targeted volume, MRSI requires that these parameters be optimized on a much larger volume of tissue. Generally, the result is that the spectrum from a typical voxel selected from an MRSI scan has lower quality than a corresponding, targeted SV acquisition, because of inhomogeneities in the magnetic and radiofrequency fields. Other problems also occur, including the effects of motion and aliasing of large, unwanted water and lipid signals into the region of interest in the breast. Perhaps surprisingly, a significant amount of signal can be excited from the chest wall and torso, particularly using the body transmit coil. Although this should not be detected with PRESS localization (confined to the breast), in practice even a small amount of residual out-of-volume magnetization (particularly since it originates from water and fat in regions of poor field homogeneity, where frequency-selective suppression methods will fail) can fold-over into the MRSI data, causing artifacts. Fold-over from the contralateral breast may also be a problem, depending on the acquisition conditions. These signals can be suppressed by optimal use of crusher field gradients in the PRESS sequence,[43] and also by the use of outer-volume suppression (OVS) saturation pulses.[44]

Scan time is also an issue for MRSI. Whereas a SV acquisition (including the necessary prescan steps, e.g. shimming, adjusting suppression pulses, etc.) can typically be completed within about 10 min, for MRSI (particularly with extended coverage and/or high spatial resolution) scan times are usually longer. This is because a large number of phase-encoding steps are involved, particularly if 3D phase-encoding is performed (which is required for whole breast coverage).[45] Using conventional phase-encoding techniques, scan time becomes prohibitively long for routine clinical application, even using spherical phase-encoding schemes and reduced field-of-views.[46,47] Therefore, to date, all published MRSI studies of the human breast have used slice-single, 2D-MRSI methodology.[36,42] In the future, approaches to fast MRSI that have been developed in the brain will need to be applied to the breast if high spatial resolution and coverage is required. Some examples of fast MRSI methods that may be used include sensitivity-encoded (SENSE)-MRSI (probably with SENSE-acceleration in two or even three dimensions [39,48]), or echo-planar MRSI (EPSI).[49]

Spectroscopic quantitation techniques

Most of the early MRS studies in breast cancer simply used visual inspection to assess the spectrum for presence or absence of Cho. However, the detectability of a signal depends on many technical factors, not least the sensitivity of the particular scanner and protocol being used, and visual interpretation is subjective and can vary from one observer to another. A better approach is to estimate lesion Cho concentrations using spectral quantitation techniques.

Some of the quantitation techniques developed for SV-MRS of the breast include the use of an external standard, or the use of tissue water as an internal intensity reference.[50,51] While both of these approaches have been demonstrated to work, they may be susceptible to systematic errors. For instance, external standards are sensitive to variations in both the B_0 and B_1 fields, and in fact the presence of the external standard placed next to the breast may cause artifacts because of its magnetic susceptibility properties. Internal references (such as water) may vary as a function of the water content of the lesion, as well as possible partial volume with adipose and ductal tissue within the voxel, so care has to be taken to correct for these factors. Typically, the water signal will also have to be corrected for relaxation time effects, which may vary from one lesion to another.[3] Another approach is the so-called "phantom replacement" method, which involves comparing the in vivo Cho signal with that recorded from a suitable phantom scanned either before or after the patient scan.[52] Various correction factors have to be applied (perhaps the most important being for radiofrequency coil loading), but the method (in the brain, at least) appears to be quite reproducible. In the future this technique may be applied to breast spectroscopy, either for SV-MRS or MRSI. It is particularly appealing for MRSI, since unlike the internal or external standard methods, it does not involve any additional *patient* scan time. However, to date, only the water referencing method for SV-MRS has been routinely applied for the evaluation of breast cancer.[3]

Breast spectroscopy at high magnetic fields

One of the limitations of MRS is its low SNR, due to the low concentration of Cho in the breast, and this limits the minimum level of Cho that can be detected, and also the minimum usable voxel size. One way to increase SNR is to use higher magnetic field strengths (since, in vivo, SNR is predicted to increase linearly with field strength [53]). Use of higher magnetic field strengths also increases the chemical shift dispersion (measured in Hz) of the water, Cho, and lipid signals. Development of high-field single-voxel (SV) MRS of the breast and other organs (e.g. brain) has been performed by the Minnesota group.[34,35,51] For instance, it has been shown that the use of a 4 T magnet with good field homogeneity can produce better resolved spectra, with higher SNR, than is possible at lower field strengths.[34,54]

Figure 13.9 shows an example of a 4 T breast MRI and SV-MRS study of an invasive ductal carcinoma – the good spectral resolution (i.e. lack of overlap between water, Cho, and lipids) and high SNR is readily apparent. It is hoped that continued improvements in SNR and resolution will be possible at even higher magnetic field strengths such as 7 T, which, at the current time, are becoming available for human studies. However, high field systems also present technical challenges in terms of magnet and radiofrequency field homogeneity, RF power deposition, and other factors, which will need to be overcome in order to realize the expected improvements in SNR and resolution.

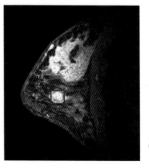
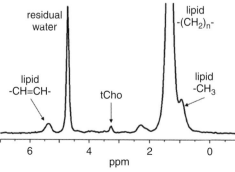

Figure 13.9. High-field MRS: 4 Tesla breast MRI and SV-MRS from an invasive ductal carcinoma, showing signals from Cho, lipids, and residual water. The excellent spectral resolution and SNR attainable at high field is demonstrated. Reproduced with permission from [22].

Effect of contrast agents (gadolinium-chelates) on MRS of the breast

While many breast lesions are often visible on non-contrast-enhanced T_1- or T_2-weighted images, in clinical practice the contrast enhanced T_2-weighted images (usually, subtracted post- minus pre-gadolinium (Gd)) are by far the most important for lesion detection and diagnosis. Therefore, in order to correctly identify lesion location, most SV-MRS scans must by necessity be performed post-Gd administration. It is therefore an important question as to what effect Gd may have on the Cho signal. As a paramagnetic compound, Gd has the potential to shorten Cho relaxation times, and also perturb the magnetic field homogeneity. Relatively few studies have examined the effect of Gd on breast Cho signals, although there have been a few studies in brain tumors performed. Generally, in the brain, Gd has been reported to either have a small or no effect on Cho signals.[55,56] Where some effect was measured, it was typically a ~20% increase in linewidth and a ~10–15% decrease in peak area. However, usually no adjustment of shim currents was performed pre- and post-Gd, and it's possible that reshimming may restore the linewidth to the pre-Gd value. In the breast, similar effects have been reported.[57] Generally, effects of Gd on Cho are likely to be small, since Cho is believed to be predominantly from the intracellular compartment, while Gd should be in the extracellular space. However, if using the water signal for quantitation referencing, it should be remembered that Gd has a substantial impact on the water T_1, and so, if possible, the water signal should be recorded pre-Gd.

Recommended protocol for MRS of the breast

At the current time, at most institutions, MRS of the breast is only routinely possible using SV-MRS, and MRSI still requires appreciable technique development. Since the Cho signal is typically very small, sensitive phased-array receiver coils should be used, and high field magnet systems are preferable (e.g. 3.0 T). Voxel size should be adjusted to the size of the lesion, but a minimum voxel size of $1-3 \text{ cm}^3$ is recommended (linear dimension ~1–1.5 cm), since SNR will be too low if attempting to go below this resolution. Typically, scan time should be at least ~5 min to generate enough SNR (e.g. ~200–256 averages with a TR of 1.5 s). If possible, each average should be stored separately and TE-averaging performed in order to reduce sidebands and phase and frequency instabilities.[34,35] The TE value chosen depends on various factors; while SNR should be better at short TE (for instance, 20–35 ms, because of fewer T_2 losses), in practice cleaner spectra may be attainable at longer TEs (e.g. 130–280 ms) because water and lipid suppression are better at long TE (water and lipids have shorter T_2's than Cho). If TE averaging is performed, typical ranges of TE values might be from 45 to 200 ms in increments of 1–2 ms. Cho T_2 relaxation times become shorter at higher field, so shorter TEs should be used as the field strength increases. Both water and lipid suppression are recommended.

High-order shimming is also valuable, particularly at higher magnetic field strengths, and it is important to record a quick spectrum (with perhaps 4 or 8 averages only) without any water or lipid suppression, which can be used to estimate field homogeneity (from the width of the water peak), center frequency, and used as a quantitation reference. The SNR and linewidth of this reference scan should provide some likelihood of the minimum detectable Cho signal; for instance, if the linewidth (width at half maximum) is greater than 0.15 ppm (~20 Hz at 3 T, or 10 Hz at 1.5 T), the ability to detect Cho is degraded compared to narrower linewidths.

Summary

In summary, proton MRS is a promising diagnostic technique for breast cancer, which in particular shows promise for improving the specificity of breast MRI examinations. It can be readily performed within the same session as a breast MRI exam, although voxel placement may be difficult unless there has been an opportunity to fully review the contrast-enhanced MRI scans (in this regard, the development of MRSI methods with increased breast coverage is important). MRS is low-risk in that it is non-invasive, and does not involve radiation. It promises to be cost-effective and can be performed in radiographically dense breasts. While there are currently only a relatively small number of MRS studies in the literature, nearly all of them are encouraging in terms of sensitivity and specificity. Nevertheless, it is debatable whether at present MRS is ready for routine clinical application, for several reasons, including the commercial availability of MRS methods which are robust enough to detect small Cho signals in the presence of much larger water and lipid resonances, and other artifacts. The limited sensitivity of MRS must be recognized, such that small lesions (<1 cm diameter) cannot be evaluated because of inadequate signal-to-noise ratios.

References

[1] Sijens PE, Wijrdeman HK, Moerland MA, Bakker CJ, Vermeulen JW, Luyten PR. Human breast cancer in vivo: H-1 and P-31 MR spectroscopy at 1.5 T. *Radiology* 1988; **169**: 615–20.

[2] Sitter B, Sonnewald U, Spraul M, Fjosne HE, Gribbestad IS. High-resolution magic angle spinning MRS of breast cancer tissue. *NMR Biomed* 2002; **15**: 327–37.

[3] Haddadin IS, Mcintosh A, Meisamy S, Corum C, Snyder AL, Powell NJ, *et al.* Metabolite quantification and high-field MRS in breast cancer. *NMR Biomed* 2009; **22**: 65–76.

[4] Smith RA, Mettlin CJ, Davis KJ, Eyre H. American Cancer Society guidelines for the early detection of cancer. *CA Cancer J Clin* 2000; **50**: 34–49.

[5] Abeloff MD, Lichter AS, Niederhuber JE, Pierce LJ, Love RR. Breast. In Abeloff MD, Armitage JO, Lichter AS, Eds. *Clinical Oncology*. 2nd edn. Oxford: Churchill Livingstone, 1999.

[6] Sabel M, Aichinger H. Recent developments in breast imaging. *Phys Med Biol* 1996; **41**: 315–68.

[7] Nass S, Henderson I, Lashof J. *Mammography and Beyond*. Washington, DC: National Academy Press, 2001.

[8] Gundry KR. The application of breast MRI in staging and screening for breast cancer. *Oncology (Williston Park)* 2005; **19**: 159–69; discussion 170, 173–4, 177.

[9] Kriege M, Brekelmans CT, Boetes C, Besnard PE, Zonderland HM, Obdeijn IM, *et al.* Efficacy of MRI and mammography for breast-cancer screening in women with a familial or genetic predisposition. *N Engl J Med* 2004; **351**: 427–37.

[10] Lehman CD, Blume JD, Weatherall P, Thickman D, Hylton N, Warner E, *et al.* Screening women at high risk for breast cancer with mammography and magnetic resonance imaging. *Cancer* 2005; **103**: 1898–905.

[11] Schnall MD, Blume J, Bluemke DA, Deangelis GA, Debruhl N, Harms S, *et al.* Diagnostic architectural and dynamic features at breast MR imaging: Multicenter study. *Radiology* 2006; **238**: 42–53.

[12] Bluemke DA, Gatsonis CA, Chen MH, Deangelis GA, Debruhl N, Harms S, *et al.* Magnetic resonance imaging of the breast prior to biopsy. *J Am Med Assoc* 2004; **292**: 2735–42.

[13] Meisamy S, Bolan PJ, Baker EH, Pollema MG, Le CT, Kelcz F, *et al.* Adding in vivo quantitative 1H MR spectroscopy to improve diagnostic accuracy of breast MR imaging: Preliminary results of observer performance study at 4.0 T. *Radiology* 2005; **236**: 465–75.

[14] Jacobs MA, Barker PB, Argani P, Ouwerkerk R, Bhujwalla ZM, Bluemke DA. Combined dynamic contrast enhanced breast MR and proton spectroscopic imaging: A feasibility study. *J Magn Reson Imaging* 2005; **21**: 23–8.

[15] Jacobs MA, Barker PB, Bluemke DA, Maranto C, Arnold C, Herskovits EH, *et al.* Benign and malignant breast lesions: Diagnosis with multiparametric MR imaging. *Radiology* 2003; **229**: 225–32.

[16] Jacobs MA, Ouwerkerk R, Wolff AC, Stearns V, Bottomley PA, Barker PB, *et al.* Multiparametric and multinuclear magnetic resonance imaging of human breast cancer: Current applications. *Technol Cancer Res Treat* 2004; **3**: 543–50.

[17] Barker PB, Breiter SN, Soher BJ, Chatham JC, Forder JR, Samphilipo MA, *et al.* Quantitative proton spectroscopy of canine brain: In vivo and in vitro correlations. *Magn Reson Med* 1994; **32**: 157–63.

[18] Lee J, Yamaguchi T, Abe A, Shizukuishi K, Uemura H, Miyagi E, *et al.* Clinical evaluation of choline measurement by proton MR spectroscopy in patients with malignant tumors. *Radiat Med* 2004; **22**: 148–54.

[19] Aboagye EO, Bhujwalla ZM. Malignant transformation alters membrane choline phospholipid metabolism of human mammary epithelial cells. *Cancer Res* 1999; **59**: 80–4.

[20] Cheng LL, Chang IW, Smith BL, Gonzalez RG. Evaluating human breast ductal carcinomas with high-resolution magic-angle spinning proton magnetic resonance spectroscopy. *J Magn Reson* 1998; **135**: 194–202.

[21] Mountford CE, Somorjai RL, Malycha P, Gluch L, Lean C, Russell P, *et al.* Diagnosis and prognosis of breast cancer by magnetic resonance spectroscopy of fine-needle aspirates analysed using a statistical classification strategy. *Br J Surg* 2001; **88**: 1234–40.

[22] Bolan PJ, Nelson MT, Yee D, Garwood M. Imaging in breast cancer: Magnetic resonance spectroscopy. *Breast Cancer Res* 2005; **7**: 149–52.

[23] Katz-Brull R, Lavin PT, Lenkinski RE. Clinical utility of proton magnetic resonance spectroscopy in characterizing breast lesions. *J Natl Cancer Inst* 2002; **94**: 1197–203.

[24] Roebuck JR, Cecil KM, Schnall MD, Lenkinski RE. Human breast lesions: Characterization with proton MR spectroscopy. *Radiology* 1998; **209**: 269–75.

[25] Kvistad KA, Bakken IJ, Gribbestad IS, Ehrnholm B, Lundgren S, Fjosne HE, *et al.* Characterization of

neoplastic and normal human breast tissues with in vivo (1)H MR spectroscopy. *J Magn Reson Imaging* 1999; **10**: 159–64.

[26] Seenu V, Pavan Kumar MN, Sharma U, Gupta SD, Mehta SN, Jagannathan NR. Potential of magnetic resonance spectroscopy to detect metastasis in axillary lymph nodes in breast cancer. *Magn Reson Imaging* 2005; **23**: 1005–10.

[27] Yeung DK, Yang WT, Tse GM. Breast cancer: In vivo proton MR spectroscopy in the characterization of histopathologic subtypes and preliminary observations in axillary node metastases. *Radiology* 2002; **225**: 190–7.

[28] Meisamy S, Bolan PJ, Baker EH, Bliss RL, Gulbahce E, Everson LI, *et al*. Neoadjuvant chemotherapy of locally advanced breast cancer: Predicting response with in vivo (1)H MR spectroscopy – a pilot study at 4 T. *Radiology* 2004; **233**: 424–31.

[29] Jagannathan NR, Kumar M, Seenu V, Coshic O, Dwivedi SN, Julka PK, *et al*. Evaluation of total choline from in-vivo volume localized proton MR spectroscopy and its response to neoadjuvant chemotherapy in locally advanced breast cancer. *Br J Cancer* 2001; **84**: 1016–22.

[30] Maril N, Collins CM, Greenman RL, Lenkinski RE. Strategies for shimming the breast. *Magn Reson Med* 2005; **54**: 1139–45.

[31] Blamire AM, Rothman DL, Nixon T. Dynamic shim updating: A new approach towards optimized whole brain shimming. *Magn Reson Med* 1996; **36**: 159–65.

[32] Sukumar S, Johnson MO, Hurd RE, Van Zijl PC. Automated shimming for deuterated solvents using field profiling. *J Magn Reson* 1997; **125**: 159–62.

[33] Gruetter R. Automatic, localized in vivo adjustment of all first- and second-order shim coils. *Magn Reson Med* 1993; **29**: 804–11.

[34] Bolan PJ, Henry PG, Baker EH, Meisamy S, Garwood M. Measurement and correction of respiration-induced B0 variations in breast 1H MRS at 4 Tesla. *Magn Reson Med* 2004; **52**: 1239–45.

[35] Bolan PJ, Delabarre L, Baker EH, Merkle H, Everson LI, Yee D, *et al*. Eliminating spurious lipid sidebands in 1H MRS of breast lesions. *Magn Reson Med* 2002; **48**: 215–22.

[36] Jacobs MA, Barker PB, Bottomley PA, Bhujwalla Z, Bluemke DA. Proton magnetic resonance spectroscopic imaging of human breast cancer: A preliminary study. *J Magn Reson Imaging* 2004; **19**: 68–75.

[37] Yongbi MN, Ding S, Dunn JF. Fat suppression at 7 T using a surface coil: Application of an adiabatic half-passage chemical shift selective radiofrequency pulse. *J Magn Reson Imaging* 1995; **5**: 768–72.

[38] Ogg RJ, Kingsley PB, Taylor JS. WET, a T1- and B1-insensitive water-suppression method for in vivo localized 1 H NMR spectroscopy. *J Magn Reson B* 1994; **104**: 1–10.

[39] Smith M, Gillen J, Barker PB, Golay X. Simultaneous water and lipid suppression for in vivo brain spectroscopy in humans. *Magn Reson Med* 2005; **54**: 691–6.

[40] Star-Lack J, Nelson SJ, Kurhanewicz J, Huang LR, Vigneron DB. Improved water and lipid suppression for 3D PRESS CSI using RF band selective inversion with gradient dephasing (BASING). *Magn Reson Med* 1997; **38**: 311–21.

[41] Mescher M, Merkle H, Kirsch J, Garwood M, Gruetter R. Simultaneous in vivo spectral editing and water suppression. *NMR Biomed* 1998; **11**: 266–72.

[42] Hu J, Vartanian SA, Xuan Y, Latif Z, Soulen RL. An improved 1 H magnetic resonance spectroscopic imaging technique for the human breast: Preliminary results. *Magn Reson Imaging* 2005; **23**: 571–6.

[43] Ernst T, Chang L. Elimination of artifacts in short echo time H MR spectroscopy of the frontal lobe. *Magn Reson Med* 1996; **36**: 462–8.

[44] Shungu DC, Glickson JD. Sensitivity and localization enhancement in multinuclear in vivo NMR spectroscopy by outer volume presaturation. *Magn Reson Med* 1993; **30**: 661–71.

[45] Barker PB, Smith M, Gillen JS, Jacobs MA. A protocol for quantitative 3D MR spectroscopic imaging of the human breast. *ENC*. Asilomar, CA, 2004.

[46] Maudsley AA, Matson GB, Hugg JW, Weiner MW. Reduced phase encoding in spectroscopic imaging. *Magn Reson Med* 1994; **31**: 645–51.

[47] Golay X, Gillen J, Van Zijl PCM, Barker PB. Scan time reduction in proton magnetic resonance spectroscopic imaging of the human brain. *Magn Reson Med* 2002; **47**: 384–7.

[48] Dydak U, Weiger M, Pruessmann KP, Meier D, Boesiger P. Sensitivity-encoded spectroscopic imaging. *Magn Reson Med* 2001; **46**: 713–22.

[49] Posse S, Tedeschi G, Risinger R, Ogg R, Le Bihan D. High speed 1 H spectroscopic imaging in human brain by echo planar spatial-spectral encoding. *Magn Reson Med* 1995; **33**: 34–40.

[50] Bakken IJ, Gribbestad IS, Singstad TE, Kvistad KA. External standard method for the in vivo quantification of choline-containing compounds in breast tumors by proton MR spectroscopy at 1.5 Tesla. *Magn Reson Med* 2001; **46**: 189–92.

[51] Bolan PJ, Meisamy S, Baker EH, Lin J, Emory T, Nelson M, *et al*. In vivo quantification of choline compounds in the breast with 1 H MR spectroscopy. *Magn Reson Med* 2003; **50**: 1134–43.

[52] Soher BJ, Van Zijl PC, Duyn JH, Barker PB. Quantitative proton MR spectroscopic imaging of the human brain. *Magn Reson Med* 1996; **35**: 356–63.

[53] Hoult DI, Richards RE. The signal-to-noise ratio of the nuclear magnetic resonance experiment. *J Magn Reson* 1976; **24**: 71–85.

[54] Tkac I, Andersen P, Adriany G, Merkle H, Ugurbil K, Gruetter R. In vivo 1H NMR spectroscopy of the human brain at 7 T. *Magn Reson Med* 2001; **46**: 451–6.

[55] Sijens PE, Van Den Bent MJ, Nowak PJ, Van Dijk P, Oudkerk M. 1H chemical shift imaging reveals loss of brain tumor choline signal after administration of Gd-contrast. *Magn Reson Med* 1997; **37**: 222–5.

[56] Smith JK, Kwock L, Castillo M. Effects of contrast material on single-volume proton MR spectroscopy. *Am J Neuroradiol* 2000; **21**: 1084–9.

[57] Joe BN, Chen VY, Salibi N, Fuangtharntip P, Hildebolt CF, Bae KT. Evaluation of 1H-magnetic resonance spectroscopy of breast cancer pre- and postgadolinium administration. *Invest Radiol* 2005; **40**: 405–11.

[58] Jagannathan NR, Singh M, Govindaraju V, Raghunathan P, Coshic O, Julka PK, *et al.* Volume localized in vivo proton MR spectroscopy of breast carcinoma: Variation of water–fat ratio in patients receiving chemotherapy. *NMR Biomed* 1998; **11**: 414–22.

Chapter 14: MRS in musculoskeletal disease

Key points

- ^{31}P-MRS allows the detection of phosphate-containing metabolites that are central to energy metabolism, and therefore is particularly suitable for studying muscle physiology and its disorders in vivo.
- Time-resolved signals from inorganic phosphates, phosphocreatine, phosphodiesters/monoesters, and intermediates of ATP reflect physiologic changes in muscles during rest, exercise, and recovery.
- Quantitative analysis of metabolites allows estimates of cytosolic ADP based on a number of assumptions, and the recovery of ADP has been used as a measure of in vivo mitochondrial function.
- In pathologic states including metabolic (mitochondrial or glycolytic pathway) dysfunction, hereditary and acquired myopathies, ^{31}P-MRS shows biochemical alterations (reduced PCr, increased Pi, slow ADP recovery) that tend to overlap between pathologies.
- Glycogenolytic disorders (such as McArdle's disease) may show paradoxical alkalosis during exercise.
- Muscle ^{31}P-MRS is valuable in monitoring therapeutic response in a number of neuromuscular disorders.
- ^{1}H-MRS currently has a limited role in the clinical evaluation of musculoskeletal disease, but has been used as a research tool to assess intramyocellular lipid, which has been implicated in skeletal muscle insulin resistance and type 2 diabetes mellitus.

Introduction

Magnetic resonance spectroscopy (MRS) of skeletal muscle has been studied over several decades. In particular, muscle MRS has been utilized to study carbohydrate metabolism (by 13-carbon (^{13}C) MRS), lipid metabolism (by proton (^{1}H) MRS) and, more widely, energy metabolism (by 31-phosphorus (^{31}P) MRS).

The observation of ^{13}C-MRS requires equipment that is typically not available on clinical MR scanners. These studies are very expensive (particularly if using isotopic enrichment of ^{13}C) and hampered by the low sensitivity of the ^{13}C nucleus which limits spatial resolution. As a result, studies have been limited to research teams, with practically no application on clinical grounds.[1]

The use of ^{1}H-MRS in skeletal muscle studies is, in principle, much easier since it can be performed on most conventional MRI scanners without any special hardware. The high sensitivity of the proton allows well-localized spectra to be recorded from small, well-defined voxels. Despite this, relatively little attention has been paid to clinical ^{1}H-MRS studies of muscle, with most studies to date having focused on the assessment of levels of intramyocellular lipids and their relevance to insulin resistance.[2]

^{31}P-MRS was first used to study human muscle disease more than two decades ago,[3,4] and, although this technique is not utilized routinely in neuromuscular or musculoskeletal clinics, its impact on the field of muscular disorders can now be evaluated with some perspective. While the sensitivity of the ^{31}P nucleus is relatively low, resulting in large voxel sizes or non-localized spectra being recorded, the concentrations of phosphorus containing metabolites in muscle are quite high (e.g. total phosphate content of the order of 40 mM), so that spectra can typically be recorded in time periods as short as a few seconds.

In this chapter, we will mainly review the ^{31}P-MRS observations in neuromuscular disorders, with emphasis on those disorders associated with impairment of energy metabolism. The use of ^{1}H-MRS will be also briefly discussed.

^{31}P-MRS of muscle

This is a particularly suitable technique for detecting signals from phosphate-containing metabolites central to energy metabolism in vivo. The methodology is not particularly challenging, but requires some hardware modifications (RF coils, transmitter and receiver) to MR scanners in order for them to operate at the resonance frequency of phosphorus (approximately 25.8 MHz in a 1.5 T scanner). Many studies of pathologies have found relatively little specific information in metabolite levels in resting muscle, but greater insights into mechanisms of metabolic dysfunction are often obtained from temporally-resolved spectra recorded during exercise and recovery. With appropriate analysis techniques and modeling, such studies provide quantitative information on rates of both oxidative phosphorylation and glycolytic metabolism.[5]

Basic principles

^{31}P-MRS studies of human muscle usually employ a surface coil (typically 10–12 cm diameter) for the recording of spectra. The sample volume is usually effectively on the order of a few (10–20) cubic centimeters. The signals recorded represent an average from the overall volume studied, weighted by factors related to the radiofrequency pulses and the surface coil used. Within this volume, the signal strength from tissue closer to the coil is stronger than the signal from tissue farther away. Thus, the spectrum obtained represents a weighted average of the energy state of the volume of muscle under the coil and may contain several inhomogeneous compartments (e.g. healthy and dying cells, active and resting cells, and even different muscles that may be recruited to different degrees in response to the exercise).

Not all compounds containing phosphorus produce a signal that is visible in the MR spectrum. In living human muscle, only metabolites that are unbound and present at concentrations of at least 1 mM give rise to peaks that are sufficiently narrow and have sufficient signal-to-noise ratio to be visible.

^{31}P-MRS in normal muscle

The normal phosphorus spectrum consists of five major peaks (Figure 14.1): three from adenosine triphosphate (α-, β-, and γ-ATP), the central intermediate of energy metabolism; one from phosphocreatine (PCr), a high energy buffer compound; and one from inorganic phosphate (Pi), a product of ATP breakdown. Two additional peaks from phosphomonoesters (PME) and phosphodiesters (PDE) may be often observed.

The single Pi peak observed in the spectrum is really a summed peak of HPO_4^- and $H_2PO_4^-$. The

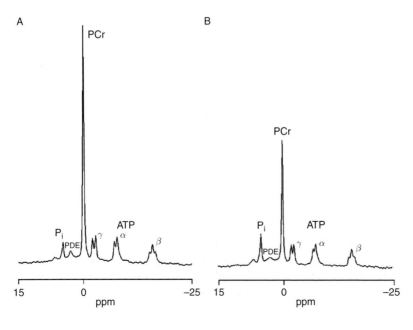

Figure 14.1. (A) Normal resting ^{31}P-MR spectrum of muscle. The major peaks of the ^{31}P-MR spectrum are: phosphocreatine (PCr), inorganic phosphates (Pi), phosphodiesters (PDE), and the three peaks (γ, α, β) of ATP. The chemical shift (in parts per million [ppm]) of the Pi is used for calculation of intracellular, cytosolic pH. (B) Resting spectrum from a patient with mitochondrial disease. Note the high Pi and low PCr resonances. Data recorded in the laboratory of Professor D. L. Arnold, Montreal Neurological Institute, McGill University.

acidic form of Pi resonates at one position in the spectrum and the basic form resonates at a different position, further away from the PCr peak. However, because the protonation/deprotonation reaction is extremely fast on the MRS timescale, these individual peaks are not separated. This results in a single peak, the position of which moves between these two extremes depending on the relative concentrations of the acidic and basic forms. The titration curve of the Pi is in the physiological pH range, thus its position reports on the pH environment of the Pi molecules. pH is calculated from the chemical shift of inorganic phosphate (δ_i) using

$$pH = 6.77 + \log_{10}[(\delta_i - 3.29)/(5.68 - \delta_i)] \quad (14.1)$$

As the intracellular water space is at least five times larger than the extracellular water space in normal skeletal muscle, and the intracellular phosphate concentration is approximately twice the extracellular phosphate concentration, the vast majority of the Pi (> 90%) in the phosphorus MR spectrum originates from cytoplasmic Pi and the pH determined from the Pi position can be considered as the cytoplasmic pH. Thus, the pH reported by the Pi peak will represent an average intracellular pH (pHi) in the volume of muscle under the coil. In normal muscle, pHi is approximately 7.0, much lower than extracellular pH.[6]

Adenosine diphosphate (ADP), the concentration of which regulates the rates of mitochondrial ATP synthesis, does not produce a peak that is visible in the spectrum, because the amount of ADP not bound to myofibrils is too low for direct measurement. However, the concentration of this free, metabolically active ADP can be calculated using the creatine kinase (CK) equilibrium equation, measurement of relative PCr and ATP concentrations, and pHi according to

$$[ADP] = \frac{[ATP] \times [Cr]}{K_{eq} \times [H] \times [PCr]} \quad (14.2)$$

where K_{eq} is the creatine kinase equilibrium constant (1.66×10^9 mol^{-1}) and $[H] = 10^{-pHi}$. Since Cr is usually not measured (not visible in the ^{31}P spectrum since it has no phosphorus atom), it is normally assumed that the total creatine (Cr + PCr) concentration remains constant.[7] ADP is an important regulator of mitochondrial (oxidative) metabolism, and its levels can be used to assess the energy state of the cell independent of Pi.

During exercise, transient imbalances in energy supply and demand are buffered by shifts in the CK reaction. Thus, ^{31}P-MRS of working normal muscle typically shows that the concentration of ATP remains stable while the PCr concentration declines as it transfers its energy phosphate to ADP to form ATP. At the same time, the concentration of cytoplasmic Pi stoichiometrically increases due to PCr hydrolysis. The calculated cytosolic ADP also rises initially, but it may be depressed subsequently if intracellular acidosis becomes severe. Pi and ADP must be transported into the mitochondria for the mitochondria to rephosphorylate ADP to make ATP. The transport of Pi into the mitochondria is mediated by a carrier and is pH-dependent.[8] During exercise, acute changes in muscle pHi presumably reflect a balance between early proton consumption by PCr hydrolysis and lactate production by glycolysis. Intense exercise may produce severe intracellular acidosis (pHi < 6). The type of muscle exercise that is performed in the magnet (isometric, concentric, or eccentric) and whether it is ischemic or non-ischemic are important factors in determining the magnitude of metabolic changes recorded.[5] It is also observed that phosphomonoesters (most likely glucose-6-phosphate) may increase in the ^{31}P-MRS spectrum during exercise in normal subjects.

During recovery from exercise (Figure 14.2), when no mechanical work is being done, glycolysis ceases but mitochondrial oxidative phosphorylation continues at an accelerated rate to replenish high-energy phosphate stores utilized during the exercise. Thus, when muscle contraction stops, PCr gradually increases, Pi and ADP decrease, and pH returns to its resting level. ^{31}P-MRS can assess recovery rates for each of these metabolites and produce valuable quantitative information about mitochondrial function. Due to the complexities involved in the analysis of data obtained during exercise in relation to work performed, it is probably much easier and more practical to evaluate data from recovery after exercise when work and associated ATP consumption have stopped.[5] In particular, the calculation of cytosolic free ADP recovery has been proposed as a sensitive measure of in vivo mitochondrial function.[9] Rephosphorylation of ADP is commonly expressed as the time required for ADP concentration to reach the halfway recovery point (ADP $t_{1/2}$) and reflects the initial rate of oxidative phosphorylation (Figure 14.3).

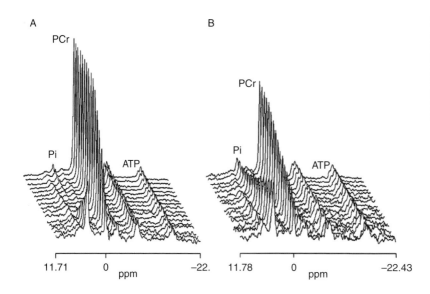

Figure 14.2. Two sets of ^{31}P-MR spectra of the gastrocnemius muscle during post-exercise recovery, time progresses from bottom to top. A. Normal control; B. Patient with mitochondrial myopathy. Note the slower metabolic recovery in B. Data recorded in the laboratory of Professor D. L. Arnold, Montreal Neurological Institute, McGill University.

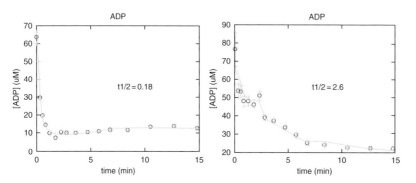

Figure 14.3. Pattern of ADP changes during recovery from intense ischemic exercise in a normal control (left panel) and a patient with mitochondrial myopathy (right panel). Note the much slower ADP kinetic of the mitochondrial patient ($t_{1/2} = 2.6$ min) in comparison to the normal control ($t_{1/2} = 0.18$ min).

^{31}P MRS exercise-recovery paradigms: practical considerations

Typically, dynamic ^{31}P-MRS paradigms involve exercise performed directly in the magnet (e.g. hand-grip or foot-flexion exercises) using custom-built, MR-compatible exercise devices. The subject's maximum voluntary contraction force should be measured prior to MRS, and then a fixed percentage effort should be applied (e.g. 60%) for a given period of time (e.g. 90 s). Monitoring of the subject's muscle force production using an MR compatible force transducer is important in order to achieve reproducible results. Other paradigms used involve patients' exercising until a fixed PCr depletion (e.g. 50%) has been achieved. A typical ^{31}P-MRS series might last ~6–7 min (e.g. 20 s rest, 90 s exercise, 5 min recovery) with continuous scanning at a time resolution of a few seconds per spectrum. With a 1 s TR, data can be presented as a 10–20 s "sliding average" to increase SNR.

Automated curve-fitting or peak integration is commonly used to measure PCr and Pi peak areas over the study time course. In addition to the rate of ADP recovery mentioned above (as a measure of oxidative phosphorylation), other metrics are quite commonly extracted from these time courses, including the initial rate of PCr (V_{PCr}) and Pi recovery which should be similar, and both reflect the rate of ATP synthesis at the beginning of the recovery period due to oxidative phosphorylation. The "maximum mitochondrial turnover" (Q_{max}) can also be calculated from V_{PCr} assuming Michaelis–Menten control of the mitochondria by ADP. The rate of PCr depletion during exercise can be used to estimate the forward creatine kinase flux, and, with various other assumptions, the flux of the anaerobic glycolysis pathway (for ATP synthesis) can also be estimated.

It should also be noted that ^{31}P MRS can be used to estimate tissue magnesium concentrations based on ATP chemical shift measurements. ATP binds to magnesium, and there is fast exchange between the species MgATP and Mg$_2$ATP, resulting in the dependency of the α- and β-ATP chemical shifts on magnesium levels. With appropriate calibrations and modeling, intracellular magnesium concentrations can be estimated.

Assumptions and interpretation of ^{31}P-MRS data analysis

It must be stressed that the analysis of ^{31}P-MRS data depends on a number of assumptions.[5] This should be recognized in order to have a better understanding of the strengths and weaknesses of the technique. Calculations of metabolite concentrations from their MRS signals require calibration with an internal standard of known concentration. Most studies have used the ATP peak as a convenient internal standard, assuming an unchanged average ATP concentration of approximately 8 mM in the intracellular water of muscle.[7] This concentration is a reasonable estimate in normal persons, but may be incorrect for diseased muscle cells. Muscle ATP has been measured directly in muscle biopsies using biochemical methods and found to be normal in a few patients with mitochondrial disorders or with other metabolic disorders, but it is not entirely clear that diseased cells always retain the normal concentration of ATP.[5] Recent work suggests that in healthy human muscle, the concentrations of PCr and Pi can be safely taken from ^{31}P-MRS measurements using the standard assumptions. In muscle affected by disease, however, it remains to be seen whether or not ^{31}P-MRS measurements give reliable results.[10]

Another assumption is that the cytosolic concentration of ATP does not change significantly during exercise or recovery. This can be confirmed by comparison of the acquired ATP signal at each time point to the resting spectra as a reference value and by calculation of metabolite concentrations during recovery using the ATP value from the same spectrum, thus correcting for any possible change in ATP signal due to technical factors.

Moreover, the interpretation of ^{31}P-MRS data implies that the observed metabolic system is a closed one with no loss of phosphate metabolites or total creatine (Cr + PCr) during exercise and recovery. In this case, this could be verified by calculating the total sum of phosphate metabolites in the spectra.

Finally, when interpreting ^{31}P-MRS results from healthy persons and patients, it is important to realize that multiple and different factors can influence the spectral pattern. Factors such as age, gender, nutrition or hydration, level of training or deconditioning, and even the presence of cardiovascular and respiratory disease can all significantly affect the results.[5]

^{31}P-MRS in muscle diseases

In most forms of metabolic myopathies, there is impairment of anaerobic or aerobic metabolism (or both) that can be assessed precisely by ^{31}P-MRS. As a result, this has been used extensively in most of these disorders.

Failure of muscle acidification during exercise by patients with defective glycolysis was the first diagnostic use of MR spectroscopy in a patient with McArdle's disease.[3] This was soon followed by the use of MRS to diagnose muscle disorders with impairment of the mitochondrial oxidative metabolism.[9]

Glycogenolytic and glycolytic disorders

Deficiency of glycogenolytic or glycolytic enzymes leads to impaired anaerobic synthesis of ATP in association with reduced production of lactate. This represents the biochemical basis for a distinctive muscle ^{31}P-MRS pattern in glycogenolytic or glycolytic enzyme defects, characterized by a rapid depletion of PCr, cytosolic alkalosis, or reduced acidification during muscle exercise, and slow rate of PCr resynthesis during recovery after exercise.[5]

On this basis, patients with McArdle's disease (myophosphorylase deficiency) show rapid PCr depletion and paradoxical increase in muscle pHi (due to proton consumption by PCr hydrolysis) during exercise (Figure 14.4).[3] Moreover, in the muscles of patients with McArdle's disease, there is no synthesis of glucose-6-phosphate or other sugar phosphates of the glycolytic pathway, so there is no increase in the PME peak. This might be used to differentiate this disorder from disorders of glycolysis such as enzymatic blocks at or distal to phosphofructokinase (PFK): the exercise ^{31}P-MR spectra show the accumulation of glycolytic intermediates (phosphorylated sugars observed in the PME region) exclusively in the latter disorders.[11,12,13,14]

In glycolytic enzyme defects more distal to the PFK reaction, a mild acidosis during extreme exercise can be observed. This may be the result of either residual activity of the deficient phosphoglycerate mutase or phosphoglycerate kinase enzyme enabling some continuous glycolytic activity, or the accumulation of H+ ions produced by the preceding step of glycolysis

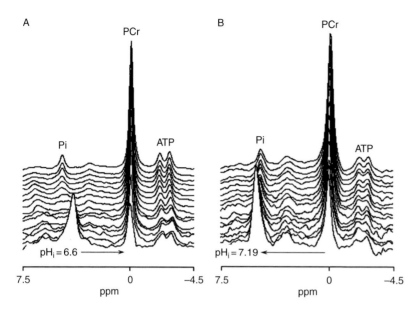

Figure 14.4. Two sets of ^{31}P-MR spectra of the gastrocnemius muscle during recovery from brief ischemic exercise, acquired using early high time-resolution protocol: (A) normal control, (B) patient with McArdle's disease. Note the acidosis in the normal control (pHi = 6.6) and the paradoxical alkalosis in the patient with McArdle's disease after exercise (quantified from the shift between Pi and PCr). This returns to normal values in both subjects by the end of the study. Data recorded in the laboratory of Professor D. L. Arnold, Montreal Neurological Institute, McGill University.

proximal to the block.[14] As mentioned before, the PME level rises continuously with increased aerobic work in muscles with impaired glycolysis and may be an indicator of the degree of the enzymatic block.[14] It should be noted that PME levels are not increased at rest in such muscles, suggesting that mechanisms for recovery or compensation are adequate at rest.

Several therapies have been attempted to improve muscle energy availability in patients with these disorders. Dietary creatine supplementation was monitored using ^{31}P-MRS of muscle and resulted in an increase in work capacity associated with increased PCr breakdown.[15] Very recently, changes in parameters of exercise physiology as assessed by muscle ^{31}P-MRS have been used as an endpoint in a double-blind, randomized, placebo-controlled trial assessing the efficacy of an angiotensin-converting enzyme (ACE) inhibitor in patients with McArdle's disease.[16] This tested the hypothesis that the severity of exercise capacity and disability in patients affected by this disease is associated with the ACE insertion/deletion haplotype, with patients of this genotype associated with higher ACE activity expressing the most severe phenotype. Both studies are practical examples that ^{31}P-MRS of human skeletal muscle is a suitable tool for the in vivo monitoring of response to treatment in different metabolic conditions.

Mitochondrial disorders

Mitochondrial disorders (MDs) are a heterogeneous group of diseases caused by molecular defects affecting one or more of the many biochemical steps in the oxidation of substrates: substrate transport into mitochondria, tricarboxylic acid cycle, electron transport chain, and oxidation and phosphorylation coupling. Electron transport chain defects due to mutations in mitochondrial DNA (mtDNA) (either rearrangements or point mutations) account for the majority of mitochondrial diseases.[17] The proportion of mutant versus wild-type mtDNA within affected cells or tissues determines the biochemical phenotype. However, the link between the biochemical phenotype and the clinical phenotype is not always clear.[18]

The diagnosis of a mitochondrial disorder is often very challenging.[17] Some patients have obvious myopathies, but others who come to neurological attention have primarily central nervous system (CNS) disorders. ^{31}P-MRS of muscle has been extensively used to investigate patients with and without clinical involvement of skeletal muscle. Despite the diversity of biochemical lesions and clinical presentations, mtDNA-related disorders share a characteristic pattern of ^{31}P-MRS detectable abnormality, which reflects the common deficit of mitochondrial ATP synthesis. However, this is presently performed only in specialized centers. In these cases, the evaluation of muscle by ^{31}P-MRS may precede muscle biopsy and genetic analysis in the diagnostic protocols of mitochondrial disorders and has demonstrated higher sensitivity in the detection of abnormal oxidative metabolism than functional evaluations such as blood lactate or bicycle exercise test.

There are many studies of patients with mitochondrial disorders examined by muscle ^{31}P-MRS.[9,19,20] Despite the differences in protocols, muscles interrogated, and patient selection, the major qualitative findings in these series are similar.[20] A ^{31}P-MR spectrum of resting muscle from a patient with a mitochondrial myopathy most often shows an increase in Pi. Less often, there is a decrease in PCr (and in the calculated cytosolic ADP). As a result, many patients with mitochondrial encephalomyopathies have abnormally low PCr/Pi at rest (Figure 14.1). Although resting abnormalities are common in patients with mtDNA defects, low PCr and high Pi cannot be considered specific indices of defective mitochondrial respiration per se, as similar changes can be described in patients with other types of muscle pathology.[21,22,23,24]

Under aerobic work conditions, patients with mitochondrial disease tend to exhibit a more rapid rate of PCr depletion per unit work than healthy individuals, consistent with limited oxidative metabolism.[9,25] The consequence is an increase in non-oxidative energy production (i.e. increased PCr consumption and glycolytic activity). Noteworthy is the fact that this rapid decline in energy state occurs without an accompanying severe intracellular acidosis. This is somewhat unexpected, given the increased lactic acid occurring in these conditions, and has been attributed to stimulated buffering systems or enhanced lactic acid extrusion from the cell in response to chronic overproduction of lactate.[9,25] Standardization of muscle exercise is an unsolved problem, making it somewhat difficult to define specific diagnostic criteria for ^{31}P-MRS of muscle.

Studies of the recovery rates of PCr and ADP, as well as other derived measures, provide valuable quantitative indices of mitochondrial dysfunction. The initial rate of recovery can provide a measure of maximal oxidative rate in the tissue and thus a sensitive and reliable index of mitochondrial dysfunction in vivo.[26] When recovery data were calculated from comparable end-exercise conditions or only for the initial phase of recovery, most of the patients with mitochondrial myopathies had abnormal recovery kinetics (Figures 14.2 and 14.3). This is less so, however, if patients have primarily brain rather than muscle involvement.[5] However, in many cases subclinical muscle involvement can be better detected by ^{31}P-MRS of muscle after stressing the mitochondria with brief anaerobic exercise.[19,27] Thus, the analysis of ^{31}P-MRS of the muscle during the initial phase of the recovery from exercise can help in the screening of patients with mitochondrial dysfunctions, and can be used to follow the clinical course of these patients.

^{31}P-MRS of skeletal muscle has been widely used to objectively evaluate the response to specific treatments in different forms of mitochondrial myopathies.[28,29,30] The positive effect of moderately intense aerobic training alone for 8 weeks on muscle oxidative metabolism was confirmed in 10 patients with various mtDNA abnormalities, in whom improvements in exercise performance and blood lactate concentrations (both 30%) were associated with an even greater (60%) improvement in ^{31}P-MRS-assessed ADP recovery rate.[31] In contrast, ADP recovery did not improve after a similar training regimen in patients with various forms of MDs or in sedentary control subjects.[32] A clinically effective therapy for mtDNA disorders has yet to be identified. ^{31}P-MRS of muscle is certainly an ideal tool to assess any potential new treatment approach.

Muscular dystrophies

Muscular dystrophies comprise a large and heterogeneous group of inherited disorders characterized by progressive muscle weakness and wasting. Structural alterations lead to fiber atrophy, necrosis, and fatty replacement of muscle tissue. These changes are easily detectable by computed tomography or MR imaging. ^{31}P-MRS of skeletal muscle can detect some of the biochemical abnormalities that underlie structural changes.

A large number of ^{31}P-MRS studies have investigated muscle metabolism in patients with dystrophinopathies such as Duchenne or Becker diseases.[21,33,35,36] ^{31}P-MRS results of the resting muscle show an increase in pH and Pi and reduced PCr. As a result, the calculated cytosolic ADP concentration is increased. These abnormalities are usually more pronounced in Duchenne than in Becker disease. These findings can be considered as evidence for secondary mitochondrial dysfunction in muscle cells undergoing necrosis or due in any case to an increased ATP turnover. A progressive increase in the PDE peak at rest has also been observed in some patients, possibly as a result of membrane breakdown.[34] Finally, a high muscle pHi at rest has been observed in numerous patients with muscular dystrophy, but the cause of this intracellular alkalosis is unclear.[23,34,35,37]

During exercise, a premature drop in PCr (or PCr/Pi) was observed in Becker dystrophy.[35,36,37] This leads to a higher than normal cytosolic ADP during exercise. There is a reduced acidosis during exercise which is, again, not well explained; neither a modified buffering capacity nor a change in proton efflux can totally account for the relative cytoplasmic alkalosis during exercise.[37]

During recovery from exercise, the maximum rate of mitochondrial ATP production was normal in patients with Becker[37] and limb girdle muscular dystrophy.[23] These observations are not necessarily in conflict with the abnormalities at rest, as it is likely that fibers undergoing necrosis do not contract during exercise.

The recent development of the new mouse model for dystrophinopathies, which exhibits many of the pathophysiologic features that are present in these disorders, may help to better understand the mechanisms leading to progressive muscle necrosis. The first in vivo ^{31}P-MRS study of these mice has shown metabolic abnormalities in skeletal muscle at rest, during exercise, and during recovery similar to those present in Duchenne and Becker diseases.[38]

Inflammatory myopathies

Inflammatory myopathies are a heterogeneous group of acquired muscle disorders characterized by muscle weakness and chronic inflammatory infiltrates. For the three major forms – polymyositis, dermatomyositis, and inclusion body myositis (IBM) – different pathogenic mechanisms have been proposed.[39]

Several studies have been performed in patients with dermatomyositis and polymyositis.[40,41,42,43,44,45] A high Pi/PCr at rest and larger depletion of PCr/Pi in the muscle of patients with dermatomyositis than in normal subjects for similar percentage of maximal exercise load was generally found. Furthermore, the abnormalities correlated with the degree of muscle involvement when the amyopathic (without weakness) form of dermatomyositis was compared with the myopathic form. Exercise and recovery muscle abnormalities reported in dermatomyositis and polymyositis patients are consistent with a deficit of oxidative phosphorylation secondary to impaired blood supply and suggest that the mitochondrial abnormalities reported in some dermatomyositis and polymyositis biopsy studies are probably secondary to the inflammatory process.[46,47] In some studies, abnormalities were partly reversed with clinical improvement after steroids and immunosuppressant therapy,[43,45] suggesting that ^{31}P-MRS findings in the inflammatory myopathies have potential for monitoring therapy.

The pathology of sporadic IBM includes inflammatory infiltrates, intracellular degenerative changes, and abnormal mitochondria (with inclusion bodies). The disease is characterized by weakness and atrophy affecting both proximal and distal muscle groups, early involvement of the quadriceps and deep finger flexors, and lack of response to conventional immunotherapy. [39] The pathogenesis of the intracellular changes is unclear. The finding of multiple deletions of mitochondrial DNA in this disorder raised the possibility that a defect of oxidative metabolism might play a role in the pathogenesis of IBM and might be involved in the progressive skeletal muscle degeneration.[48] This question has been addressed in two ^{31}P-MRS studies, [22,24] both of which showed that the calf muscle mitochondrial ATP production rate as measured from post-exercise recovery data is normal in IBM patients. The conclusions were that the accumulation of mtDNA deletions and the presence of other mitochondrial abnormalities are secondary processes and do not contribute to the pathogenesis of the muscle weakness and atrophy.

^{31}P-MRS of muscle in CNS disorders

Occasionally, ^{31}P-MRS of muscle has been used in central nervous system (CNS) disorders with the aim to better understand the pathogenesis of these diseases, to help in the diagnostic workup, or to monitor therapeutic treatment.

As also mentioned before, this is certainly the case in patients with primary mitochondrial disorders with prevalent CNS involvement where ^{31}P-MRS of muscle can be used to assess and monitor clinical or subclinical mitochondrial impairment.[19,28,49]

However, studies have been performed in a variety of neurological disorders where mitochondrial dysfunction has been suspected. These include migraine,[50,51] Parkinson's disease,[52,53] alternating hemiplegia,[54] and some hereditary ataxias. [55,56]. Particularly in Friedreich ataxia, ^{31}P-MRS of both cardiac and skeletal muscles has been used as a useful tool for monitoring response to therapy with antioxidants.[57,58] Results showed significant improvements in both cardiac and skeletal muscle bioenergetics that were maintained throughout the 4 years of antioxidant therapy.[58]

^{1}H-MRS of muscle

Although the proton is the most frequently used nucleus for brain MRS examinations, its application to skeletal muscle was virtually ignored for a long time, and is still rarely used for clinical purposes today.[1] Originally, this was due to the fact that the two large peaks of water and fat were masking the resonances of all the potentially clinically relevant metabolites. More recently, the improved performance of MR scanners in terms of gradients, homogeneity, volume selection, and water suppression, has made the detection of ^{1}H-MR spectra routinely possible.

In particular, it is routinely possible to observe intra- (IMCL, at 1.28 ppm) and extramyocellular lipids (EMCL, shifted approximately of 0.2 ppm) using conventional ^{1}H MRS techniques (Figure 14.5). These were first observed by Schick and coworkers as two separate groups of resonances for lipids in voxels of skeletal muscle, despite the fact that both compartments contain fatty acids or triglycerides of similar compositions.[59] The difference in resonance frequency observed in the spectrum is believed to be due to magnetic susceptibility effects due to the different geometries of the intra- and extramyocellular environments. The methylene resonances of EMCL are shifted and broadened due to bulk magnetic susceptibility effects while the signals from IMCL remain unaffected.[2,60] Improved resolution of the lipid signals of muscle tissue was obtained using long echo times, TE > 200 ms. The existence of two fatty acid compartments was supported by measurements of the relaxation times and line shape analysis.[2]

Other resonances are also observed in the proton spectrum of normal muscle. These include resonances from creatine and trimethylamines (TMA, also commonly referred to as "choline-containing compounds") as also seen in brain spectra, as well as smaller resonances from taurine, acetylcarnitine (generally not seen at rest, but elevated after heavy exercise), and, downfield from water, carnosine (whose resonances are pH-sensitive) and deoxyhemoglobin (detectable when muscle oxygenation decreases during exercise). While lactate is expected to be elevated during exercise, it is generally obscured by the much larger lipid resonances, and can only be detected using spectral-editing techniques.

Careful inspection of muscle ^{1}H MR spectra reveals "fine structure" not normally seen in other organ systems, or from spectra in solution. For instance, the Cr methyl group at 3 ppm often appears as a triplet (Figure 14.5) in many muscle groups, whereas in other organs it is a single resonance. The origin of this effect was investigated, and it was found that this splitting (fine structure) was due to orientation-dependent residual dipolar couplings. This occurs when rotational averaging of the creatine molecule is not complete, presumably due to its restricted movement within muscle fibers.[60] The splitting is dependent on the angle between the muscle fiber and the main magnetic field, and decreases to zero at the "magic angle" of 54.7°. The magnitude of the residual dipolar coupling can in fact be used to estimate fiber orientations in different muscle groups.

IMCLs are stored in droplets in the cytoplasm of muscle cells and can be in readily accessible energy storage form, in particular during long-term exercise. The depletion of IMCL after exercise can be replenished quickly and ^{1}H-MRS is able to follow this depletion and replenishment with a good inter- and intra-individual reproducibility.[2] In contrast, the quantitative interpretation of EMCL is not straightforward. While IMCL is evenly distributed in skeletal muscle and is largely independent of the choice of the voxel position, EMCL is concentrated in subcutaneous fat and fibrotic structures and can change considerably even for tiny shifts of the voxel position. Owing to these limitations, ^{1}H-MRS should not be used for the determination of EMCL levels.[1]

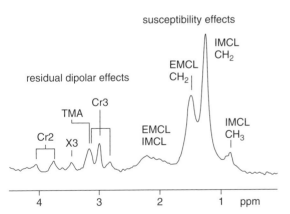

Figure 14.5. ^{1}H-MR spectrum of a PRESS acquisition of a single voxel in the anterior m. tibialis (TR 3000 ms, TE 20 ms). In the region of 1–2 ppm, bulk magnetic susceptibility effects lead to a shift of the extramyocellular lipid (EMCL) resonances relative to the chemically very similar intramyocellular lipids (IMCL). Residual dipolar coupling leads to a splitting of resonances, e.g. of creatine – CH$_2$ at 3.96 ppm (Cr2). Additional abbreviations: X3 = tentatively assigned to taurine; TMA = trimethylammonium-containing compounds; Cr3 = creatine–CH$_3$. (From [2], figure 1, with permission.)

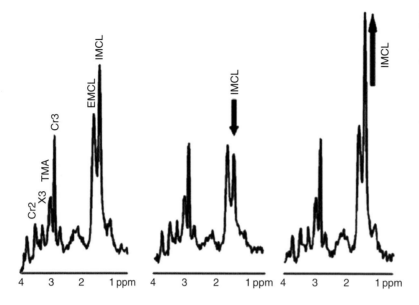

Figure 14.6. Three ^{1}H-MR spectra illustrate the influence of exercise and nutrition upon IMCL levels in skeletal muscle. The spectrum to the left was before strenuous exercise that resulted in a reduced IMCL level in the spectrum in the middle. Subsequent nutrition led to a replenishment of the IMCL stores typically after about 48 h, in the spectrum to the right. (From [2], figure 6, with permission.)

The potential for observing IMCL noninvasively has been utilized in exercise physiology studies.[60,61] IMCL is used during strenuous exercise and pre- and post-marathon spectra show a distinct reduction of IMCL during the activity.[61] Later, the observation of increased IMCL in insulin-resistant subjects has increased significantly the interest in the use of ^{1}H-MRS to assess IMCL.[62,63,64,65] These data were interpreted as a possible role of IMCL in the pathogenesis of skeletal muscle insulin resistance and type 2 diabetes mellitus. ^{1}H-MRS was also proposed as a simple measurement of insulin sensitivity for patients with diabetes. However, a number of reports showed that IMCL levels are dependent on many other factors, such as diet, exercise, and muscle type (Figure 14.6). [1, 2,65,66]

Thus, ^{1}H-MRS of muscle, although it may allow non-invasive and repeated studies of muscle metabolism, has limited use in clinic and is presently utilized as a research tool by research groups.[1] In the future, ^{1}H MRS may become more widely used for other applications, for instance evaluating tumors of the musculoskeletal system, where measures of choline may be useful in distinguishing malignant from benign lesions, or recurrent tumor from treatment effects.[67] However, more work is required before such applications can make the transition from research studies to clinical practice.

Summary

The anticipation that MRS of skeletal muscle, in particular ^{31}P-MRS, would become a major tool for diagnosing myopathies and elucidating muscular physiology has been only partially maintained. This application is not widely used in clinic and remains a tool for expert research groups interested in muscle physiology. However, ^{31}P-MRS is sensitive in detecting abnormalities in energy metabolism and might be taken into account in the clinical workup of patients with suspected muscle disorders, particularly if a metabolic dysfunction is suspected. More importantly, the ability of ^{31}P-MRS to provide a non-invasive, repetitive assessment makes it a powerful monitoring tool of therapeutic efficacy. This might become particularly useful in an era in which molecular medicine holds great promise in identifying modern therapies for muscle disorders.

References

[1] Boesch C. Musculoskeletal spectroscopy. *J Magn Reson Imaging* 2007; **25**: 321–38.

[2] Boesch C, Machann J, Vermathen P, Schick F. Role of proton MR for the study of muscle lipid metabolism. *NMR Biomed* 2006; **19**: 968–88.

[3] Ross BD, Radda GK, Gadian DG, Rocker G, Esiri M, Falconer-Smith J. Examination of a case of suspected

McArdle's syndrome by ^{31}P nuclear magnetic resonance. *N Engl J Med* 1981; **304**: 1338–42.

[4] Gadian D, Radda G, Ross B, Hockaday J, Bore P, Taylor D, et al. Examination of a myopathy by phosphorus nuclear magnetic resonance. *Lancet* 1981; **2**: 774–5.

[5] Argov Z, Lofberg M, Arnold DL. Insights into muscle diseases gained by phosphorus magnetic resonance spectroscopy. *Muscle Nerve* 2000; **23**: 1316–34.

[6] Bore PJ, Chan L, Gadian DG, Radda GK, Ross BD, Styles P, et al. Noninvasive pHi measurements of human tissue using 31P-NMR. *Kroc Found Ser* 1981; **15**: 527–35.

[7] Arnold DL, Matthews PM, Radda GK. Metabolic recovery after exercise and the assessment of mitochondrial function in vivo in human skeletal muscle by means of ^{31}P NMR. *Magn Reson Med* 1984; **1**: 307–15.

[8] Taylor DJ, Styles P, Matthews PM, Arnold DA, Gadian DG, Bore P, et al. Energetics of human muscle: exercise-induced ATP depletion. *Magn Reson Med* 1986; **3**: 44–54.

[9] Arnold DL, Taylor DJ, Radda GK. Investigation of human mitochondrial myopathies by phosphorus magnetic resonance spectroscopy. *Ann Neurol* 1985; **18**: 189–96.

[10] Kemp GJ, Meyerspeer M, Moser E. Absolute quantification of phosphorus metabolite concentrations in human muscle in vivo by 31P MRS: A quantitative review. *NMR Biomed* 2007; **20**: 555–65.

[11] Chance B, Eleff S, Bank W, Leigh JS Jr, Warnell R. 31P NMR studies of control of mitochondrial function in phosphofructokinase-deficient human skeletal muscle. *Proc Natl Acad Sci USA* 1982; **79**: 7714–8.

[12] Argov Z, Bank WJ, Maris J, Chance B. Muscle energy metabolism in McArdle's syndrome by in vivo phosphorus magnetic resonance spectroscopy. *Neurology* 1987; **37**: 1720–4.

[13] Argov Z, Bank WJ, Maris J, Leigh JS Jr, Chance B. Muscle energy metabolism in human phosphofructokinase deficiency as recorded by 31P nuclear magnetic resonance spectroscopy. *Ann Neurol* 1987; **22**: 46–51.

[14] Argov Z, Bank WJ, Boden B, Ro YI, Chance B. Phosphorus magnetic resonance spectroscopy of partially blocked muscle glycolysis. An in vivo study of phosphoglycerate mutase deficiency. *Arch Neurol* 1987; **44**: 614–7.

[15] Vorgerd M, Grehl T, Jager M, Muller K, Freitag G, Patzold T, et al. Creatine therapy in myophosphorylase deficiency (McArdle disease): A placebo-controlled crossover trial. *Arch Neurol* 2000; **57**: 956–63.

[16] Martinuzzi A, Liava A, Trevisi E, Frare M, Tonon C, Malucelli E, et al. Randomized, placebo-controlled, double-blind pilot trial of ramipril in McArdle's disease. *Muscle Nerve* 2008; **37**: 350–7.

[17] DiMauro S, Moraes CT. Mitochondrial encephalomyopathies. [Review]. *Arch Neurol* 1993; **50**: 1197–208.

[18] Zeviani M, Di DS. Mitochondrial disorders. *Brain* 2004; **127**: 2153–72.

[19] Matthews PM, Allaire C, Shoubridge EA, Karpati G, Carpenter S, Arnold DL. In vivo muscle magnetic resonance spectroscopy in the clinical investigation of mitochondrial disease. *Neurology* 1991; **41**: 114–20.

[20] Argov Z, Arnold DL. MR spectroscopy and imaging in metabolic myopathies. *Neurol Clin* 2000; **18**: 35–52.

[21] Barbiroli B, Funicello R, Iotti S, Montagna P, Ferlini A, Zaniol P. 31P-NMR spectroscopy of skeletal muscle in Becker dystrophy and DMD/BMD carriers. Altered rate of phosphate transport. *J Neurol Sci* 1992; **109**: 188–95.

[22] Lodi R, Taylor DJ, Tabrizi SJ, Hilton-Jones D, Squier MV, Seller A, et al. Normal in vivo skeletal muscle oxidative metabolism in sporadic inclusion body myositis assessed by 31P-magnetic resonance spectroscopy. *Brain* 1998; **121**: 2119–26.

[23] Lodi R, Muntoni F, Taylor J, Kumar S, Sewry CA, Blamire A, et al. Correlative MR imaging and 31P-MR spectroscopy study in sarcoglycan deficient limb girdle muscular dystrophy. *Neuromusc Disord* 1997; **7**: 505–11.

[24] Argov Z, Taivassalo T, De SN, Genge A, Karpati G, Arnold DL. Intracellular phosphates in inclusion body myositis – a 31P magnetic resonance spectroscopy study. *Muscle Nerve* 1998; **21**: 1523–5.

[25] Argov Z, Bank WJ, Maris J, Peterson P, Chance B. Bioenergetic heterogeneity of human mitochondrial myopathies: Phosphorus magnetic resonance spectroscopy study. *Neurology* 1987; **37**: 257–62.

[26] Argov Z, De Stefano N, Taivassalo T, Chen J, Karpati G, Arnold DL. Abnormal oxidative metabolism in exercise intolerance of undetermined origin. *Neuromusc Disord* 1997; **7**: 99–104.

[27] Argov Z, De Stefano N, Arnold DL. ADP recovery after a brief ischemic exercise in normal and diseased human muscle – a 31P MRS study. *NMR Biomed* 1996; **9**: 165–72.

[28] De Stefano N, Matthews PM, Ford B, Genge A, Karpati G, Arnold DL. Short-term dichloroacetate treatment improves indices of cerebral metabolism in patients with mitochondrial disorders. *Neurology* 1995; **45**: 1193–8.

[29] Matthews PM, Ford B, Dandurand RJ, Eidelman DH, O'Conner D, Sherwin A, et al. Coenzyme Q_{10} with multiple vitamins is generally ineffective in treatment of mitochondrial disease. *Neurology* 1993; **43**: 884–90.

[30] Penn AM, Lee JW, Thuillier P, Wagner M, Maclure KM, Menard MR, et al. MELAS syndrome with mitochondrial tRNA$^{Leu(UUR)}$ mutation: Correlation of clinical state, nerve conduction, and muscle ^{31}P magnetic resonance spectroscopy during treatment with nicotinamide and riboflavin. *Neurology* 1992; **42**: 2147–52.

[31] Taivassalo T, De Stefano N, Argov Z, Matthews PM, Chen J, Genge A, et al. Effects of aerobic training in patients with mitochondrial myopathies. *Neurology* 1998; **50**: 1055–60.

[32] Taivassalo T, Chen J, Karpati G, Arnold DL, Argov Z. Short-term aerobic training response in chronic myopathies. *Muscle Nerve* 1999; **22**: 1239–43.

[33] Newman RJ, Bore PJ, Chan L, Gadian DG, Styles P, Taylor D, et al. Nuclear magnetic resonance studies of forearm muscle in Duchenne dystrophy. *BrMed J Clin Res* 1982; **284**: 1072–4.

[34] Younkin DP, Berman P, Sladky J, Chee C, Bank W, Chance B. 31P NMR studies in Duchenne muscular dystrophy: Age-related metabolic changes. *Neurology* 1987; **37**: 165–9.

[35] Kemp GJ, Taylor DJ, Dunn JF, Frostick SP, Radda GK. Cellular energetics of dystrophic muscle. *J Neurol Sci* 1993; **116**: 201–06.

[36] Barbiroli B, Funicello R, Ferlini A, Montagna P, Zaniol P. Muscle energy metabolism in female DMD/BMD carriers: A 31P-MR spectroscopy study. *Muscle Nerve* 1992; **15**: 344–8.

[37] Lodi R, Kemp GJ, Muntoni F, Thompson CH, Rae C, Taylor J, et al. Reduced cytosolic acidification during exercise suggests defective glycolytic activity in skeletal muscle of patients with Becker muscular dystrophy. An in vivo 31P magnetic resonance spectroscopy study. *Brain* 1999; **122**: 121–30.

[38] Cole MA, Rafael JA, Taylor DJ, Lodi R, Davies KE, Styles P. A quantitative study of bioenergetics in skeletal muscle lacking utrophin and dystrophin. *Neuromusc Disord* 2002; **12**: 247–57.

[39] Dalakas MC. Polymyositis, dermatomyositis and inclusion-body myositis. *N Engl J Med* 1991; **325**: 1487–98.

[40] Park JH, Vansant JP, Kumar NG, Gibbs SJ, Curvin MS, Price RR, et al. Dermatomyositis: Correlative MR imaging and P-31 MR spectroscopy for quantitative characterization of inflammatory disease. *Radiology* 1990; **177**: 473–9.

[41] Newman ED, Kurland RJ. P-31 magnetic resonance spectroscopy in polymyositis and dermatomyositis. Altered energy utilization during exercise. *Arthr Rheum* 1992; **35**: 199–203.

[42] Park JH, Niermann KJ, Olsen N. Evidence for metabolic abnormalities in the muscles of patients with fibromyalgia. *Curr Rheumatol Rep* 2000; **2**: 131–40.

[43] Park JH, Niermann KJ, Ryder NM, Nelson AE, Das A, Lawton AR, et al. Muscle abnormalities in juvenile dermatomyositis patients: P-31 magnetic resonance spectroscopy studies. *Arthr Rheum* 2000; **43**: 2359–67.

[44] Adams LB, Park JH, Olsen NJ, Gardner ES, Hernanz-Schulman M, King LE Jr. Quantitative evaluation of improvement in muscle weakness in a patient receiving extracorporeal photopheresis for scleroderma: Magnetic resonance imaging and magnetic resonance spectroscopy. *J Am Acad Dermatol* 1995; **33**: 519–22.

[45] King LE Jr, Park JH, Adams LB, Olsen NJ. Phosphorus 31 magnetic resonance spectroscopy for quantitative evaluation of therapeutic regimens in dermatomyositis. *Arch Dermatol* 1995; **131**: 522–4.

[46] Chariot P, Ruet E, Authier FJ, Labes D, Poron F, Gherardi R. Cytochrome c oxidase deficiencies in the muscle of patients with inflammatory myopathies. *Acta Neuropathol* 1996; **91**: 530–6.

[47] Blume G, Pestronk A, Frank B, Johns DR. Polymyositis with cytochrome oxidase negative muscle fibres. Early quadriceps weakness and poor response to immunosuppressive therapy. *Brain* 1997; **120**: 39–45.

[48] Oldfors A, Moslemi AR, Jonasson L, Ohlsson M, Kollberg G, Lindberg C. Mitochondrial abnormalities in inclusion-body myositis. *Neurology* 2006; **66**: S49–55.

[49] Matthews PM, Berkovic SF, Shoubridge EA, Andermann F, Karpati G, Carpenter S, et al. In vivo magnetic resonance spectroscopy of brain and muscle in a type of mitochondrial encephalomyopathy (MERRF). *Ann Neurol* 1991; **29**: 435–8.

[50] Lodi R, Kemp GJ, Montagna P, Pierangeli G, Cortelli P, Iotti S, et al. Quantitative analysis of skeletal muscle bioenergetics and proton efflux in migraine and cluster headache. *J Neurol Sci* 1997; **146**: 73–80.

[51] Lodi R, Montagna P, Soriani S, Iotti S, Arnaldi C, Cortelli P, et al. Deficit of brain and skeletal muscle bioenergetics and low brain magnesium in juvenile migraine: An in vivo 31P magnetic resonance spectroscopy interictal study. *Ped Res* 1997; **42**: 866–71.

[52] Penn AM, Roberts T, Hodder J, Allen PS, Zhu G, Martin WR. Generalized mitochondrial dysfunction in

Parkinson's disease detected by magnetic resonance spectroscopy of muscle. *Neurology* 1995; **45**: 2097–9.

[53] Taylor DJ, Krige D, Barnes PR, Kemp GJ, Carroll MT, Mann VM, et al. A 31P magnetic resonance spectroscopy study of mitochondrial function in skeletal muscle of patients with Parkinson's disease. *J Neurol Sci* 1994; **125**: 77–81.

[54] Arnold DL, Silver K, Andermann F. Evidence for mitochondrial dysfunction in patients with alternating hemiplegia of childhood. *Ann Neurol* 1993; **33**: 604–07.

[55] Vorgerd M, Schols L, Hardt C, Ristow M, Epplen JT, Zange J. Mitochondrial impairment of human muscle in Friedreich ataxia in vivo. *Neuromusc Disord* 2000; **10**: 430–5.

[56] Lodi R, Cooper JM, Bradley JL, Manners D, Styles P, Taylor DJ, et al. Deficit of in vivo mitochondrial ATP production in patients with Friedreich ataxia. *Proc Natl Acad Sci USA* 1999; **96**: 11492–5.

[57] Lodi R, Hart PE, Rajagopalan B, Taylor DJ, Crilley JG, Bradley JL, et al. Antioxidant treatment improves in vivo cardiac and skeletal muscle bioenergetics in patients with Friedreich's ataxia. *Ann Neurol* 2001; **49**: 590–6.

[58] Hart PE, Lodi R, Rajagopalan B, Bradley JL, Crilley JG, Turner C, et al. Antioxidant treatment of patients with Friedreich ataxia: Four-year follow-up. *Arch Neurol* 2005; **62**: 621–6.

[59] Schick F, Eismann B, Jung WI, Bongers H, Bunse M, Lutz O. Comparison of localized proton NMR signals of skeletal muscle and fat tissue in vivo: Two lipid compartments in muscle tissue. *Magn Reson Med* 1993; **29**: 158–67.

[60] Boesch C, Slotboom J, Hoppeler H, Kreis R. In vivo determination of intra-myocellular lipids in human muscle by means of localized 1H-MR-spectroscopy. *Magn Reson Med* 1997; **37**: 484–93.

[61] Boesch C, Kreis R. Observation of intramyocellular lipids by 1H-magnetic resonance spectroscopy. *Ann NY Acad Sci* 2000; **904**: 25–31.

[62] Perseghin G, Scifo P, De Cobelli F, Pagliato E, Battezzati A, Arcelloni C, et al. Intramyocellular triglyceride content is a determinant of in vivo insulin resistance in humans: A 1H-13C nuclear magnetic resonance spectroscopy assessment in offspring of type 2 diabetic parents. *Diabetes* 1999; **48**: 1600–06.

[63] Virkamaki A, Korsheninnikova E, Seppala-Lindroos A, Vehkavaara S, Goto T, Halavaara J, et al. Intramyocellular lipid is associated with resistance to in vivo insulin actions on glucose uptake, antilipolysis, and early insulin signaling pathways in human skeletal muscle. *Diabetes* 2001; **50**: 2337–43.

[64] Petersen KF, Shulman GI. Pathogenesis of skeletal muscle insulin resistance in type 2 diabetes mellitus. *Am J Cardiol* 2002; **90**: 11G–18G.

[65] Machann J, Haring H, Schick F, Stumvoll M. Intramyocellular lipids and insulin resistance. *Diabetes Obes Metab* 2004; **6**: 239–48.

[66] van Loon LJ. Intramyocellular triacylglycerol as a substrate source during exercise. *Proc Nutr Soc* 2004; **63**: 301–07.

[67] Fayad LM, Barker PB, Bluemke DA. Molecular characterization of musculoskeletal tumors by proton MR spectroscopy. *Semin Musculoskelet Radiol* 2007; **11**: 240–5.

Index

Locators in **bold** type denote major entries
Locators in *italic* denote figures/tables
Locators for main headings which have subheadings refer to general aspects of the topic

Abbreviated Injury Scale (AIS) 162
abscesses, brain 77, *83*, 110, **111–112**, *112*, *123*
acetate 111, *123*
acetone 14
acetylcholine 162
ACRIN study, prostate cancer 223
acute disseminated encephalomyelitis (ADEM) 11, **122**, *122*, *123*
AD (Alexander disease) **202–204**, *203*
ADC (apparent diffusion coefficient) 67
adenosine triphosphate *see* ATP
adrenoleukodystrophy (ALD) **201–202**, *202*
aging *see* brain development/aging
AIDS *see* HIV/AIDS
AIS (Abbreviated Injury Scale) 162
alanine 8, 14, 75–76, 111, *123*
ALD (adrenoleukodystrophy) **201–202**, *202*
Alexander disease (AD) **202–204**, *203*
ALS (amyotrophic lateral sclerosis) 11, **153–154**
alternating hemiplegia 250
Alzheimer's disease 13, **144–145**, 184
   differential diagnosis 148, *148*, *149*
   proton spectroscopy 145–147, *146*, *147*
amino acidopathy 181, **183–184**, 192, *192–193*
   hyperhomocysteinemias **194–195**, *195*
   maple syrup urine disease **184–185**
   urea cycle defects **192–194**, *194*
amino acids 111, *123*
ammonia 192
amygdala 161
amyotrophic lateral sclerosis (ALS) 11, **153–154**
angiogenesis 64
animal models 7, 10, **93–94**
apoptosis/apoptotic index 63, 162
apparent diffusion coefficient (ADC) 67
arabitol 14
arachnoid cysts 111
arginine 192
arterial spin labeling 174
ascorbic acid 27

aspartate 8, 14
aspartocyclase **185–186**
astrocytomas 62, *63*, 79; *see also* brain tumor
   apoptosis/apoptotic index 63
   histopathological classification of tumor cells 64
   pediatrics 83
   therapy planning/monitoring 82
   WHO grading/patient survival 66
ATP (adenosine triphosphate) 7
   mitochondrial disorders **248–249**
   muscular dystrophy **249–250**
   neonatal brain MRS 97
   ^{31}P spectroscopy – muscle **244–245**, *246–247*

bacterial infection *see* abscesses; intracranial infections
band reject filters 35–36, *35*
basal ganglia
   brain metabolic disorders 181
   HABC **200**
   human prion disease 153
   Huntington's disease 152
   neonatal brain MRS 98
   stroke 94, *95*
   traumatic brain injury 174
baseline corrections 37, *38*
baseline distortions 34
BASING sequences 25, *27*
BCKA (branched-chain α-keto acids) 184–185
Becker disease **249–250**
betaine 14
β-hydroxy-butyrate 14
bioenergetics 7
biopsy 111, 116, 152
birth, and hypoxic–ischemic encephalopathy *see* hypoxic–ischemic encephalopathy
BOLD (blood oxygenation level) effect 44
Boltzmann factor 2
brain biopsy 111, 116, 152
brain development/aging 51, **57–59**, *57*
   anatomical variations, young adults **51–55**, *52*, *53*, *54*, *55*, *56*, *57*

early brain development in children **55–56**, *58*, *59*
elderly brain **56–57**, *58*
key points 51
brain injury *see* traumatic brain injury
brain metabolic disorders 180, 182, *192–193*, **207**; *see also* amino acidopathy; leukoencephalopathy; mitochondrial disorders; organ acidopathy
   Canavan's disease 10, **185–186**, *186*
   classification of **180–181**
   creatine deficiency **182–183**, *183*
   key points **180**
   maple syrup urine disease **184–185**
   MRI scanning **181–182**
   N-acetyl aspartate (NAA) deficiency **183**
   nonketotic hyperglycinemia **187–188**
   phenylketonuria **183–184**
   proton (^{1}H-MRS) spectroscopy **182**, *192–193*, **207**
   proton (^{1}H-MRSI) spectroscopy 182
   sialicic acid storage disorders **186–187**, *187*
   succinate-dehydrogenase deficiency **184**, *185*
brain pH 14
brain stem 31, 173, 174
brain temperature 14
brain tumor 61, **85–86**
   choline 12
   conventional MR imaging **66**
   diagnostic accuracy 82
   differential diagnosis 65, **76–77**, *77*, **80–82**, *80*
   frequently asked questions 76–82
   grading/patient survival **65–66**, **77–80**, *78*, *80*
   histopathological classification **64–65**
   indications 76
   key points **61**
   lactate 13, 93
   and N-acetyl aspartate 11
   natural history of brain tumors **61–64**, *63*

Index

pediatrics **83–85**, *86*
protocol design 28, 45
proton (¹H-MRSI) spectroscopy
  66–76, *68–69*, *70*, *71*, *72*, *73*, *74*, *75*
single-voxel techniques 21
therapy planning/monitoring
  82–83, 85
branched-chain α-keto acids (BCKA)
  184–185
breast cancer **229–230**, *230*, **239–240**
  contrast agents 239
  field homogeneity **235**, *235*, *236*
  high magnetic field breast
    spectroscopy **238**, *238*
  key points 229
  magnetic resonance spectroscopy
    231–233, *232*, *233*, **234**
  monitoring treatment response
    233–234, *234*
  MRSI **236–237**, *237*
  protocol design 239
  screening and diagnosis **230–231**
  spectroscopic quantitation
    techniques 238
  water/lipid suppression **235–236**
breast tissue 14

CADASIL (cerebral autosomal
  dominant arteriopathy) 148
Canavan's disease 10, **185–186**, *186*
cancer *see* brain tumor; breast cancer;
  prostate cancer
carbon-13 spectroscopy 7, 243
Carr–Purcell method 4, *5*
caudate 43, 152
CBD (corticobasal degeneration) 150
CBF (cerebral blood flow) **93–94**
cell invasion, by tumor cells 64
central nervous system (CNS); *see also*
  brain
  disorders, ³¹P spectroscopy 250
  health, and *N*-acetyl aspartate 10
cerebellum 13, 31, 173, 181, **200**
cerebral atrophy 56
cerebral blood flow (CBF) **93–94**
cerebritis *see* abscesses
cerebrospinal fluid (CSF) 42
chemical shift 1, **4–5**, *6*, **45–46**, *46*, *47*;
  see also MRSI
chemistry, application of NMR to 1
chemotherapy 65, 82, 116, 234
CHESS pulses 25
children *see* pediatrics
choline 8, 9, **11–12**
  acute disseminated
    encephalomyelitis *122*, **124–122**
  adrenoleukodystrophy **201–202**

Alexander disease 204
Alzheimer's disease 146
amino acidopathy 194
anatomical variations in young
  adults **51–55**, *53*, *54*
brain development/aging *57*
brain metabolic disorders 182, 184
brain tumor 67, *70*, *71*, *72*, 111
breast cancer **229–230**, **231–234**,
  232, *233*, **235–236**, 238
childhood epilepsies 137
contrast agents 239
cryptococcosis **117**
dementia with Lewy bodies **150**
differential diagnosis 77, *77*, 80, 81
early brain development in children
  55–56, *58*
elderly brain **56–57**
epilepsy 133–134, *134*
Erdheim–Chester disease 77
frontal lobe epilepsy **137**
gliomas *80*
globoid cell leukodystrophy **201**
grading/patient survival *71*, *72*, **77–79**
hepatic encephalopathy 30, *30*
herpes simplex encephalitis **114**
high magnetic field breast
  spectroscopy **238**, *238*
HIV/AIDS **114–115**, 117
hyperhomocysteinemias 195
hypomyelination 199
infection/inflammation/
  demyelinating lesions *123*
leukoencephalopathy 198, *198*,
  199, 205
malformations of cortical
  development 137
mitochondrial disorders 189
multiple sclerosis 120–122
neonatal brain 98
oligodendroglioma 78
Parkinson's disease **151**
pediatrics 85, **164**, **165–166**
Pelizaeus–Merzbacher-like
  disorder **200**
progressive multifocal
  leukoencephalopathy 118
prostate cancer **215–216**, 217,
  219–220
reference signals 40
sialic acid storage disorders 187
stroke **93**, 94
subacute sclerosing panencephalitis
  118–119
therapy planning/monitoring 73, 83
traumatic brain injury (TBI) 163,
  167, **168–169**, **170**, **172–173**, 174

vegetative states **168–169**
volumetric/whole brain
  spectroscopy studies **173**
white matter rarefaction **204**
cholinergic system 162
cingulate gyrus 146
citrate/citrate metabolism **214–216**,
  214, *215*, **219–220**
CJD (Creutzfeldt–Jacob disease) 152
CNS *see* central nervous system
Cobalamin-responsive phenotypes 196
cognitive function related to
  metabolism **166**, 168
computed tomography 91, 162, 163
concussion/repeated brain injury **170**
continuous-wave (CW) signals 6
contrast agents **47–48**, *48*, 239
contrast-enhanced MR imaging 79
corpus callosum 56, 95, *96*, 173
cortex 163
cortical spreading depression (CSD) 93
corticobasal degeneration (CBD) 150
craniopharyngioma 83
creatine 8, 9, **13**; *see also*
  phosphocreatine
  adrenoleukodystrophy **201–202**
  Alexander disease 204
  amino acidopathy 194
  amyotrophic lateral sclerosis
    153–154
  anatomical variations in young
    adults **51–55**, *54*
  brain development/aging *57*
  brain metabolic disorders 182, 184,
    192–193
  brain tumors **67–69**, *70*, *71*,
    77–79, *80*
  concussion/repeated brain
    injury **170**
  deficiency **182–183**, *183*
  dementia with Lewy bodies **150**
  dietary supplements 248
  elderly brain **56–57**
  frontal lobe epilepsy **137**
  frontotemporal dementia 149
  globoid cell leukodystrophy
    (GCL) **201**
  herpes simplex encephalitis **114**
  hyperhomocysteinemias 195
  hypomyelination 199
  infection/inflammation/
    demyelinating lesions *123*
  leukoencephalopathy 198, *199*, 205
  mitochondrial disorders 189
  multiple sclerosis 120–122
  musculoskeletal disease 251
  Parkinson's disease **151**

creatine (cont.)
  pediatrics 85
  progressive multifocal leukoencephalopathy **118**
  prostate cancer **215–216, 219–220**
  reference signals 40
  sialic acid storage disorders 187
  stroke **93**
  temporal lobe epilepsy 134, 135–136
  traumatic brain injury **164, 165–166,** 167–168, **170,** 174
  vascular dementia 148
  vegetative states 168–169, *169*
  volumetric/whole-brain spectroscopy studies **173**
  white matter rarefaction **204**
Creutzfeldt–Jacob disease (CJD) 152
CSD (cortical spreading depression) 93
CSF (cerebrospinal fluid) 42
cryptococcosis **117,** *123*
CT (computed tomography) scanning 91, 162, 163
CW (continuous-wave) signals 6
cytomegalovirus 114

DAI (diffuse axonal injury) 162, 163
delayed radiation necrosis (DRN) *73, 83*
dementia *see* Alzheimer's disease; neurodegenerative disease; vascular dementia
dementia with Lewy bodies (DLB) **149–150**
demyelinating disorders 13, 93, *192–193,* 198, **200;** *see also* acute disseminated encephalomyelitis; multiple sclerosis
  adrenoleukodystrophy **201–202,** *202*
  Alexander disease **202–204,** *203*
  globoid cell leukodystrophy **201,** *201*
  inflammation **119,** *123,* **124**
  metachromatic leukodystrophy **200–201**
dendrite development 56
development, brain *see* brain development/aging
diabetes 252
diagnostic accuracy *see* differential diagnosis
diagnostic criteria, vascular dementia 148
diamagnetic effect 4
dietary supplements, creatine 248
differential diagnosis
  abscesses 111
  amino acidopathy 194
  brain tumors 65, **76–77,** *77,* **80–82,** *80*

leukoencephalopathy 181–182
Parkinson's disease-related disorders 150–151
vascular dementia/Alzheimer's disease *148,* **148,** *149*
diffuse astrocytoma *see* oligodendroglioma
diffuse axonal injury (DAI) 162, 163
diffusion, tumor cells 63–64
diffusion MRI **102–104,** *104*
diffusion tensor imaging 174
diffusion-weighted imaging (DWI) 97, 111
digital filters 34–36, *35*
digital rectal examination 212
Disability Rating Scale 163
DLB (dementia with Lewy bodies) **149–150**
Down's syndrome 184
DRN (delayed radiation necrosis) *73, 83*
Duchenne disease 249–250
DWI (diffusion-weighted imaging) 97, 111

ECD (Erdheim–Chester disease) *77*
echo-planar spectroscopic imaging (EPSI) 24, 52, *54*
eddy current correction 35
elderly brain **56–57,** *58*
electromyography (EMG) 153
EMCL (extramyocellular lipids) 251–252, *251*
encephalitis **114;** *see also* herpes simplex encephalitis
energy metabolism 243; *see also* P spectroscopy
entorhinal cortex 145
ependyoma 83
epilepsy **131–132,** *132,* **139;** *see also* temporal lobe epilepsy
  childhood epilepsies/Rasmussen's encephalitis 137, *138*
  frontal lobe epilepsy 137
  GABA (γ-amino butyric acid) **137–138**
  key points **131**
  malformations of cortical development 137
  MRS protocol design *10,* **12, 138–139**
  P spectroscopy **132–133,** *133*
  proton (¹H-MRSI) spectroscopy **133–134,** *134*
EPSI (echo-planar spectroscopic imaging) 24, 52, *54*
Epstein–Barr virus 114, 115

Erdheim–Chester disease (ECD) *77*
ethanol 14
ethanolamine 14
ethylmalonic aciduria **197**
ethylmalonic encephalopathy 190
exercise recovery paradigms **246–247**
external referencing 40–41

fast Fourier transformation (FFT) 6
fast MRSI techniques **24–25**
fast spin-echo (FSE) imaging 42, *42*
fatal familial insomnia (FFI) 152
fatty acid oxidation disorders **188–189**
Fermi filters 37
FFT (fast Fourier transformation) 6
field homogeneity/inhomogeneity
  breast cancer **235,** *235,* **236**
  MRSI artifacts **42–44,** *43*
field of view (FOV) 23
filters 34–36
fMRI (functional magnetic resonance imaging) 174
Fourier transform spectroscopy 1, **6–7,** *7,* 34–36, 37
free induction decay (FID) 3
frequency domain analysis **38,** *39*
frequency domain processing **36–37,** *36*
frequency suppression saturation pulses 26
Friedreich ataxia 250
frontal lobe epilepsy 137
frontotemporal dementia (FTD) *149,* **149**
FSE (fast spin-echo) imaging *42*
functional magnetic resonance imaging (fMRI) 174

GABA (γ-amino butyric acid) 8, 9, 14
  amino acidopathy 192
  chemical shift displacement 46
  epilepsy **137–138,** *139*
  MEGA-PRESS sequence **26–27,** *26*
  Parkinson's disease **151**
gadolinium-chelates **239**
gadolinium contrast agents **47–48,** *48*
galactitol 14
ganglioma 83
GBM tumors 64, 65, 66, 77
GCL (globoid cell leukodystrophy) *201,* **201**
genotype analysis of tumor cells 62
Gerstmann–Sträussler disease (GSD) 152
Gibbs Ringing patterns 23, *24*
Glasgow Coma Scale (GCS) 162, 163
Glasgow Outcome Scale (GOS) 163, 164

Index

glial tumors 62; *see also* brain tumor
gliomas 63–64; *see also* brain tumor
   anaplasia indicators 65
   grading/patient survival 77–79, 80
   histopathological classification of tumor cells 64
   MRSI 67
   therapy planning/monitoring 82
gliosis 28
globoid cell leukodystrophy (GCL) *201*, **201**
globus pallidus 43
glucose 14, 163
glutamate 8, 9, *10*, **13–14**
   amino acidopathy 192
   epilepsy 133–134, **139**
   Huntington's disease **152**
   infection/inflammation/demyelinating lesions *123*
   MEGA-PRESS sequence 27
   microdialysis studies **169–170**
   neonatal brain MRS 98, *99*
   short echo time spectroscopy 174
   traumatic brain injury 162, 163, **165**, *165*, 174
glutamine 8, 9, *10*, **13–14**, 194
   amino acidopathy 194
   hepatic encephalopathy *30*, **30**
   Huntington's disease **152**
   short echo time spectroscopy 174
glutaric aciduria type 1 **197**
glutathione 14, 27
glycerol **169–170**
glycerol 3-phosphoethanolamine (GPE) 67
glycine 13, 14, 188
glycogen 14
glycogenolytic/glycolytic disorders **247–248**, *248*
GOS (Glasgow Outcome Scale) 163, 164
grading, brain tumors 65–66, 79–80
gray matter
   acute disseminated encephalomyelitis 124
   Alzheimer's disease 146
   atrophy 145
   choline 12
   creatine 13
gray/white matter boundaries 161, 162
gray/white matter comparisons **51–56**, *53, 54*
   mild cognitive impairment 145
   multiple sclerosis 121
   neonatal brain MRS 98, *98*
   spiral-MRSI 24
   vascular dementia 148
GSD (Gerstmann–Sträussler disease) 152

HABC (hypomyelination and atrophy of basal ganglia and cerebellum) 181, **200**
Hamming filters 37
Hanning filters 37
HD (Huntington's disease) **151–152**
head injury *see* traumatic brain injury
head movements, MRSI artifacts **44**
hepatic encephalopathy 12, 13, 28, *30*, **30**
herpes simplex encephalitis 110, **114**, *123*
HIE *see* hypoxic–ischemic encephalopathy
high bandwidth radiofrequency slice-selective pulses 27
high magnetic field breast spectroscopy *238*, **238**
hippocampus 21, *22*, 51, 145, 161, 163
histidine 14
HIV/AIDS 11, 110, **114–115**, *123*; *see also* cryptococcosis; primary CNS lymphoma; progressive multifocal leukoencephalopathy; toxoplasmosis
homocysteine **194–195**, *195*
HSV (herpes simplex encephalitis) 110, **114**, *123*
human prion disease (HPD) **152–153**
Huntington's disease (HD) **151–152**
hydroxy-methylglutaryl coenzyme A lyase **197**
hyperhomocysteinemias **194–195**, *195*
hypomyelination 186, 198, **199–200**, *200*
hypomyelination and atrophy of basal ganglia and cerebellum (HABC) 181, **200**
hypothalamus 12, 51
hypoxia 8, 13
hypoxic–ischemic encephalopathy (HIE) **95–97**, *105*
   diffusion MRI **102–104**, *104*
   key points **91**
   MRS acquisition protocols **105**
   P spectroscopy 100, *101*
   proton (^{1}H) spectroscopy **100–102**, *101, 102, 103*

IMCL (intramyocellular lipids) **251–252**, *251*
infection 110, *123*, **124**; *see also* abscesses; intracranial infections
inferior frontal lobe 31
infiltration, tumor cells 63–64
inflammation 119, *123*, **124**; *see also* demyelinating disorders; infection
inflammatory myopathies **250**
information content, proton spectra **9–10**, *11*

inner-volume suppression (IVS) 46
inorganic phosphate 7, **247–248**
   inflammatory myopathies **250**
   mitochondrial disorders **248–249**
   muscular dystrophy **249–250**
   neonatal brain MRS 97
   P spectroscopy **244–245**, *246–247*
insular cortex 12, 51
insulin resistance 252
internal intensity reference 40
intracranial infections **110–111**, *123*, **124**
   abscesses 83, 110, **111–112**, *112*
   cryptococcosis **117**
   herpes simplex encephalitis **114**
   HIV-related infections **114–115**
   primary CNS lymphoma **115–117**, *116*
   progressive multifocal leukoencephalopathy **117–118**, *118, 119*
   subacute sclerosing panencephalitis **118–119**
   toxoplasmosis **115–117**, *116*
   tuberculoma **112–114**, *113*
inversion recovery 3
ischemia 13, 28, **93–94**; *see also* stroke
isoleucine 185
IVS (inner-volume suppression) 46

JC virus **117–118**, *118, 119*
J-coupling interactions **220–222**, *221*

Ki-67 labeling index 67
Krebs cycle 13

L-2-hydroxyglutaric aciduria **197–198**
lactate 8, **13**
   abscesses 111
   acquisition protocols/protocol design 28
   acute disseminated encephalomyelitis **122–124**
   adrenoleukodystrophy **201–202**
   amino acidopathy 194
   brain metabolic disorders 182, 184, *192–193*
   chemical shift displacement 46, *47*
   cryptococcosis **117**
   differential diagnosis 111
   epilepsy 133–134
   globoid cell leukodystrophy **201**
   herpes simplex encephalitis **114**
   HIV-related infections 117
   Huntington's disease **152**
   hypomyelination 199
   hypoxic–ischemic encephalopathy 97
   infection/inflammation/demyelinating lesions *123*

Index

lactate (cont.)
　ischemia 93
　leukoencephalopathy **190–192**, 205
　MEGA-PRESS sequence 27
　metabolic changes in early stage TBI **172–173**
　microdialysis studies **169–170**
　mitochondrial disorders 189–192, 249
　MRS acquisition protocols **105**
　MRSI in brain tumors 70, *74*
　multiple sclerosis 120–122
　neonatal brain MRS 98–99
　progressive multifocal leukoencephalopathy **118**
　prostate cancer **215–216**
　proton (^{1}H) spectroscopy **100–102**
　stroke **93**, 94, *95*, *96*
　therapy planning/monitoring 83
　traumatic brain injury **164**, 167, **169**, 174
　tuberculoma 114
Larmour frequency 2
LBSL (leukoencephalopathy with brainstem and spinal cord involvement and lactate elevation) **190–192**
LCModel (linear combination model) 27, 38, *39*
Leigh disease 189, *190*
lesion size/position, MRSI artefacts **45**
leucine 185
leukoaraiosis 148
leukodystrophy 10
leukoencephalopathy 181, **184**, **185–186**; *see also* demyelinating diseases; hypomyelination; white matter rarefaction
　adrenoleukodystrophy **201–202**, *202*
　Alexander disease **202–204**, *203*
　brain metabolic disorders **184**
　Canavan's disease **185–186**, *186*
　differential diagnosis 181–182
　globoid cell leukodystrophy *201*, **201**
　hereditary **198–199**, *198*, *199*
　megalencephalic leukoencephalopathy with subcortical cysts **205–206**, *206*
　metachromatic leukodystrophy **200–201**
　Pelizaeus–Merzbacher disease 186, **199–200**
　Pelizaeus–Merzbacher-like disorder *200*, **200**
　white matter rarefaction 198
leukoencephalopathy with brainstem and spinal cord involvement and lactate elevation (LBSL) **190–192**

leukoencephalopathy with vanishing white matter (VWM) 181, **204–205**, *205*
Lewy bodies 150; *see also* dementia with Lewy bodies
LGG (secondary) tumors 64, 67, *68–69*, 83
line excitation techniques 21
lineshapes 34
linewidth 27, 43–44, *43*, 99–100
lipid metabolism 243; *see also* proton (^{1}H-MRSI) spectroscopy
lipid suppression **25–26**, **44**, **235–236**
lipids 8, 9
　abscesses 111
　brain metabolic disorders 182
　brain tumors 69–70, 73
　differential diagnosis 111
　HIV-related infections 117
　hypoxic–ischemic encephalopathy 100
　infection/inflammation/demyelinating lesions *123*
　multiple sclerosis 120–122
　musculoskeletal disease 251–252
　prostate cancer **215–216**
　therapy planning/monitoring 83
　tuberculoma 114
liver disease 13
long echo time MRS 164, 174
loss of heterozygosity (LOH) 62
Lou Gehrig's disease *see* amyotrophic lateral sclerosis
lymphoma 85, *123*
lysosomal disorders 181, *192–193*

magnetic resonance imaging *see* MRI
magnetic resonance spectroscopy *see* MRS
magnetic resonance spectroscopy imaging *see* MRSI
malformations of cortical development (MCD) **137**
mammography 230–231
mannitol 14
maple syrup urine disease (MSUD) **184–185**, 194
MCA (middle cerebral artery) occlusion 93
McArdle's disease 247–248
MCI (mild cognitive impairment) 145
MD *see* mitochondrial disorders
measles virus **118–119**
medulloblastoma 83
MEGA-PRESS sequence **26–27**, *26*
MEGA sequences 25
megalencephalic leukoencephalopathy with subcortical cysts (MLC) 181, **205–206**, *206*

mental retardation 11, *12*
mesial temporal lobe 146
mesial temporal lobe sclerosis (MTS) 131
metachromatic leukodystrophy (MLD) **200–201**
methylmalonic aciduria (MMA) 195
methyl-sulfonyl-methane (MSM) 14
Michael–Levitt (MLEV) pulses *221*
microdialysis studies **169–170**
microscopic field inhomogeneity 44
middle cerebral artery (MCA) occlusion 93
migraine 250
mild cognitive impairment (MCI) 145
mitochondrial disorders 181, **188**, *192–193*
　acquisition protocols/protocol design 28
　fatty acid oxidation disorders **188–189**
　lactate 13, 93
　leukoencephalopathy with brainstem and spinal cord involvement and lactate elevation **190–192**
　MR spectroscopy **189–190**, *190*, *191*
　MRI scanning **189**
　and multiple sclerosis 121
　P spectroscopy **248–249**, 250
　pyruvate metabolism **188**
　respiratory chain disorders **188**
mixed oligoastrocytoma (MOA) 65, 66, 82
MLC (megalencephalic leukoencephalopathy with subcortical cysts) 181, **205–206**, *206*
MLD (metachromatic leukodystrophy) **200–201**
MLEV (Michael–Levitt) pulses 221, *221*
MMA (methylmalonic aciduria) 195
MOA (mixed oligoastrocytoma) 65, 66, 82
molecular genetics, tumor cells 62
MRI (magnetic resonance imaging) 1
　abscesses 111
　brain metabolic disorders 180, **181–182**
　brain tumors **66**
　cryptococcosis 117
　epilepsy 131–132, *135*
　human prion disease 153
　hypoxic–ischemic encephalopathy 96
　mitochondrial disorders **189**
　multiple sclerosis 120
　neurodegenerative disease 144
　progressive multifocal leukoencephalopathy 118

Index

prostate cancer **213–214**
stroke 91, 94
temporal lobe epilepsy **136–137**
traumatic brain injury **162–163**
MRS (magnetic resonance spectroscopy) **1–2**; *see also* P spectroscopy; proton spectroscopy; pulse sequences; single-voxel techniques
 basic theory **2**, *2*
 chemical shift 1, **4–5**, *6*
 Fourier transform spectroscopy **6–7**, *7*
 information content of proton spectra **9–10**, *10*, *11*
 key points **1**
 nuclei **9**, *9*
 prostate cancer **215–216**
 relaxation times **3–4**, *4*, *5*
 rotating frame **3**, *4*
 signal/signal-to-noise ratio **2–3**
 spin–spin coupling constants **5–6**, *6*
 in vivo **7–9**
MRSI artifacts; *see also* multi-slice MRSI
 chemical shift displacement **45–46**, *46*, *47*
 field inhomogeneity **42–44**
 Gd contrast agents **47–48**, *48*
 head movements **44**
 inadequate field homogeneity *43*
 lesion size/position **45**
 lipid suppression **44**
 out-of-voxel magnetization **45**, *46*
 protocol design **45**
 vertical scale **44**
 voxel/MRSI slice placement **44–45**
 water suppression **44**
MRSI techniques 19, **21–25**, *21*, *22*, *23*; *see also* multi-slice MRSI
 acquisition protocols/protocol design 19, **28–31**, **28–31**, *30*, *31*
 anatomical variations in young adults *52*, *53*, *54*
 data processing techniques **37–38**, *38*, *48*
 fast MRSI techniques **24–25**
 MRS at different field strengths **27**, *29*
 multiple RF receiver coils **27–28**
 PRESS MRSI **21–23**, *21*, *22*, *23*
 spatial resolution/scan time **23–24**, *24*
MSM (methyl-sulfonyl-methane) 14
MSUD (maple syrup urine disease) **184–185**, 194
MTS (mesial temporal lobe sclerosis) 131
multiple sclerosis (MS) 10, 11, 24, 110, **119–122**, *120*, *121*, *123*
multiple system degeneration (MSA) 150

multiplets **5–6**
multiple-voxel spectroscopic imaging *see* MRSI
multi-slice 2D MRSI **22–23**, *23*
multi-slice 3D MRSI
 acquisition protocols **30–31**, *31*
 brain metabolic disorders 180, 182
 PRESS MRSI 22
 prostate cancer 217
muscle tissue 15
muscular dystrophy **249–250**
musculoskeletal disease **243–244**, **252**
 energy metabolism *see* P spectroscopy
 key points **243**
 proton (^{1}H-MRSI) spectroscopy **251–252**, *252*
myelin 167, 198; *see also* demyelinating disorders
myo-inositol 13
 Alzheimer's disease **146–147**
 amino acidopathy 194
 amyotrophic lateral sclerosis **153–154**
 brain tumors **72–75**
 early brain development in children **55–56**, *58*
 epilepsy **133–134**
 hepatic encephalopathy 30, *30*
 HIV/AIDS **114–115**
 human prion disease 153
 infection/inflammation/demyelinating lesions *123*
 neonatal brain MRS 98
 short echo time spectroscopy 174
 stroke 94
 subacute sclerosing panencephalitis **118–119**
 traumatic brain injury **165**, *166*, 174
 vascular dementia 148
 young adults 52

N-acetyl aspartate 5, 8, 9, **10–11**, *10*, *12*
 acute disseminated encephalomyelitis **124–122**
 adrenoleukodystrophy **201–202**
 Alexander disease 204
 Alzheimer's disease **145–147**
 amino acidopathy 194
 amyotrophic lateral sclerosis **153–154**
 brain development/aging 57
 brain metabolic disorders 182, 184, *192–193*
 brain tumors 67, **71–72**, 75
 childhood epilepsies/Rasmussen's encephalitis 137
 concussion/repeated brain injury 170

cryptococcosis 117
deficiency **183**
dementia with Lewy bodies **150**
early brain development in children **55–56**, *58*
elderly brain **56–57**
epilepsy **133–134**, **139**
frontal lobe epilepsy **137**
frontotemporal dementia **149**
gliomas *80*
globoid cell leukodystrophy **201**
grading/patient survival *72*, **77–79**
herpes simplex encephalitis **114**
HIV/AIDS **114–115**, 117
human prion disease **153**
Huntington's disease **152**
hyperhomocysteinemias **195**
hypomyelination **199**
infection/inflammation/demyelinating lesions *123*
ischemia **93–94**
leukoencephalopathy **118**, **185–186**, 198, *198*, *199*, 205
malformations of cortical development **137**
metabolic changes in early stage TBI **172–173**
microdialysis studies **169–170**
mitochondrial disorders 189
MRS acquisition protocols **105**
multiple sclerosis **120–122**
neonatal brain 98, *98*
Parkinson's disease **151**
pediatrics 85, **164**, **165–166**
Pelizaeus–Merzbacher-like disorder **200**
proton (^{1}H) spectroscopy **100–102**
reference signals 40
sialic acid storage disorders **186–187**
stroke **91–92**, *92*, 94, *95*
subacute sclerosing panencephalitis **118–119**
temporal lobe epilepsy **29–30**, *29*, *30*, *134*, **135–136**
traumatic brain injury 163, **167–168**, **169**, *170*, 174
vascular dementia 148
vegetative states **168–169**, *169*
volumetric/whole brain spectroscopy studies 173
white matter rarefaction 204
young adults **51–55**, *52*
N-acetyl aspartyl glutamate 9, 10, *10*, 14, 27
N-acetyl-neuraminic acid 186, **186–187**

Index

necrosis *see also* delayed radiation necrosis
  abscesses 111
  grading/patient survival 79
  histopathological classification of tumor cells 64
  proton (^{1}H-MRSI) spectroscopy 67
  traumatic brain injury 162
  tumor cells 63
neonatal brain MRS **97–100**, *97*, 98, 99, *100*
neural stem cells 62, *63*, 65
neurodegenerative disease **144**, **154**
  amyotrophic lateral sclerosis **153–154**, *154*
  dementia **144–145**; *see also* Alzheimer's disease
  dementia with Lewy bodies **149–150**
  frontotemporal dementia *149*, **149**
  human prion disease **152–153**
  Huntington's disease **151–152**
  key points 144
  Parkinson's disease **150–151**, *151*
  vascular dementia **147–148**, *149*
neuronal marker substances 10–11; *see also* N-acetyl aspartate
neuron loss 56
neuropathy of unknown origin 184
Niemann–Pick type C disease 185
NINDS-AIREN criteria 148
ninety degree pulse 3
noise *see* signal-to-noise ratio
nominal spatial resolution 23
nonketotic hyperglycinemia **187–188**
nuclear magnetic resonance (NMR) 1
nuclei 9, *9*

oligoastrocytoma *see* oligodendroglioma
oligodendroglioma 62, *63*; *see also* brain tumor
  apoptosis/apoptotic index 63
  choline 78
  differential diagnosis 80–82, *80*
  histopathological classification of tumor cells 64–65
  therapy planning/monitoring 82
  WHO grading/patient survival 66
organ acidopathy 181, **195**
  ethylmalonic aciduria **197**
  glutaric aciduria type 1 **197**
  hydroxy-methylglutaryl coenzyme A lyase **197**
  L-2-hydroxyglutaric aciduria **197–198**
  methylmalonic aciduria 195, **196–197**
  propionic aciduria 195, **196**

outer-volume suppression (OVS) saturation bands 44
out-of-voxel magnetization **45**, *46*
oxidative stress 163

P spectroscopy 7
  epilepsy **132–133**, 133
  hypoxic–ischemic encephalopathy **100**, 101
  P spectroscopy of muscle tissue **244**, **252**
  basic theory **244**
  central nervous system disorders 250
  data analysis – assumptions/interpretations 247
  exercise recovery paradigms **246–247**
  glycogenolytic/glycolytic disorders **247–248**, *248*
  inflammatory myopathies 250
  mitochondrial disorders (MD) **248–249**
  muscle diseases – general 247
  muscular dystrophy **249–250**
  normal muscle **244–245**, *244*, *246*
PA (propionic aciduria) 195, **196**
papovavisur 114
parallel imaging methods 24–25
paramyxovirus 114
parietal lobe 146
Parkinson's disease (PD) **150–151**, *151*, 250
PDE (phosphodiesters) 97
pediatrics *see also* hypoxic–ischemic encephalopathy
  early brain development 55–56, *58*, *59*
  epilepsies **137**, *138*
  MRSI in brain tumors **83–85**, *86*
  neonatal brain MRS **97–100**
  traumatic brain injury 162, **164–166**
Pelizaeus–Merzbacher disease (PMD) 186, **199–200**, *200*
Pelizaeus–Merzbacher-like disorder (PMLD) 200, **200**, *192–193*
peroxisomal disorders 181, *192–193*
PET (positron emission tomography) scanning 132
pH, brain 14
phantom replacement technique 41, 238
phase correction 38
phased-arrays 27
phenylalanine 14
phenylketonuria **183–184**
phosphocholine 8, 67, **93**, 97
phosphocreatine 7, 8, 247–248; *see also* creatine
  P spectroscopy – muscle **244–245**, **246–247**

inflammatory myopathies 250
mitochondrial disorders (MD) **248–249**
muscular dystrophy **249–250**
neonatal brain MRS 97
phosphodiesters (PDE) 97
phosphomonoesters (PME) 97
phosphorus (^{31}P) spectroscopy *see* P spectroscopy
Pi *see* inorganic phosphate
pilocytic astrocytoma 83
Planck's constant 2
PMD (Pelizaeus–Merzbacher disease) 186, **199–200**
PML (progressive multifocal leukoencephalopathy) **117–118**, *119*, *123*
PMLD (Pelizaeus–Merzbacher-like disorder) 200, **200**
PNET (primitive neuroectodermal tumors) 83–85
point resolved spectroscopy *see* PRESS
point-spread function (PSF) 23, 37
polioencephalopathy 181
polyhydric alcohols 184
pons 51
positron emission tomography (PET) scanning 132
posterior cingulate 146, *146*
posterior fossa 31, *31*
prediction of illness course, traumatic brain injury **164**, *164*
prescan functions 28
PRESS (point resolved spectroscopy) sequence 3, 8, 19–20, *20*, 21–23
  lipid suppression 25
  prostate cancer 217, 221
primary CNS lymphoma **115–117**, *116*
primitive neuroectodermal tumors (PNET) 83–85
progressive multifocal leukoencephalopathy (PML) **117–118**, *119*, *123*
progressive supranuclear palsy (PSP) 150
proliferation capacity, tumor cells 62
propan-1,2-diol 14
propionic aciduria (PA) 195, **196**
prostate anatomy **213**, *213*
prostate cancer 212
  1.5/3.0 Tesla **222–223**, *222*
  citrate metabolism **214–215**, *214*, *215*
  incidence statistics
  J-coupling interactions **220–222**, *221*
  key points **212–213**
  magnetic resonance spectroscopy **215–216**

262

MRI scanning 213–214
MRSI 216–218, *217*, *218*
MRS/MRSI protocol 219, *219*
spectral interpretation 219–220, *220*
zinc 214–215, *214*
prostate tissue 14
prostate-specific antigen (PSA) test 212
protocol design
  breast cancer 239
  MRS acquisition protocols 19, **28–31**, *30*, *31*
  MRSI artifacts 45
  prostate cancer 219, **219**, **222–223**, *222*
proton (^{1}H) spectroscopy 7–9; *see also* MRS; MRSI
  abscesses 111–112, *112*
  acute disseminated encephalomyelitis *122*, **122–124**
  Alzheimer's disease **145–147**, *146*, *147*
  amyotrophic lateral sclerosis **153–154**, *154*
  brain metabolic disorders **182**, *192–193*, **207**
  brain tumors **66–76**, *68*, *70*, *71*, *72*, *73*, *74*, *75*
  cryptococcosis 117
  dementia with Lewy bodies 150
  epilepsy *133–134*
  frontotemporal dementia 149, **149**
  herpes simplex encephalitis 114
  HIV/AIDS **114–115**, **116–117**, *116*
  human prion disease 153
  Huntington's disease 152
  hypoxic–ischemic encephalopathy 97, **100–102**, *101*, *102*, *103*
  information content of proton spectra 9–10, *194*
  multiple sclerosis **120–122**, *120*, *121*
  musculoskeletal disease **251–252**, *251*, *252*
  neonatal brain MRS **97–100**, *97*, *98*, *99*, *100*
  Parkinson's disease *151*, **151**
  progressive multifocal leukoencephalopathy 118, *118*, *119*
  quantitation techniques summary *41*, *48*
  stroke 91
  subacute sclerosing panencephalitis **118–119**
  summary of compounds detected *11*
  tuberculoma **113–114**, *113*
PSA (prostate-specific antigen) test 212
PSF (point-spread function) 23, 37
PSP (progressive supranuclear palsy) 150

pulse sequences for spectroscopy 3, 4, 19
  fast MRSI techniques **24–25**
  key points 19
  lipid suppression **25–26**
  MRS at different field strengths 27, *29*
  multiple RF receiver coils **27–28**
  multiple-voxel techniques 19, **21–25**, *21*, *22*, *23*, **28–29**
  PRESS sequence 19–20
  shimming/prescan functions 28
  single-voxel techniques **19–21**, *20*, *28–29*; *see also* STEAM sequence
  spatial resolution/scan time **23–24**, *24*
  spectral editing techniques using selective pulses **26–27**, *26*
  STEAM (stimulated echo acquisition mode) sequence 19–20
  water suppression techniques 25
pulsed Fourier transform 6
purine nucleotides 14
putamen 43, 56
pyruvate 14, **169–170**, **188**

quality factor (Q) 41
quantitation **39–42**, *48*

radiofrequency (RF) coils 7
radiotherapy 82–83, *85*
Rasmussen's encephalitis **137**, *138*
reference signals 40–41
relaxation times 3–4, *4*, *5*
resonance overlap 34
resonant frequencies 1
respiratory chain disorders **188**; *see also* mitochondrial disorders
ribitol 14
rotating frame (RF) 3, *4*

saturation recovery 3
scalar coupling constants *see* spin–spin coupling constants
scyllo-inositol 14
SDH (succinate-dehydrogenase deficiency) **184**, *185*
selective pulses **26–27**, *26*
sensitivity-encoded (SENSE) MRSI 25, 28
shimming 28, 239
short echo time spectroscopy 174
sialic acid storage disorders **186–187**, *187*
signal-to-noise ratio (SNR) **2–3**, 20, 37, 232
signal-vs.-time curve 3, *4*
single-voxel techniques **19–21**, *20*; *see also* PRESS sequence; STEAM sequence

acquisition protocols/protocol design 28–29
anatomical variations in young adults *55*, *56*
brain metabolic disorders 180
slice profiles, STEAM/PRESS comparison 20
slice-selective spin-echo excitation techniques 8
SOD (superoxide dismutase) 153
SPECT imaging 116, 132
spectral analysis methods 34
  frequency domain processing **36–37**, *36*, **38**, *39*
  key points 34
  MRSI processing techniques **37–38**, *38*, *48*
  quantitation **39–42**, *41*
  spectral fitting **38–39**, *39*
  time domain processing *7*, **34–36**, *35*, *39*
spectral editing techniques 13, **26–27**, *26*
spectral fitting *35*, **38–39**
spectral–spatial pulses 25
spectroscopic quantitation techniques **238**
spin, nuclear 1
spin echoes 3
spin-echo depth pulse localization 8
spin-lattice relaxation time 3
spin–spin coupling constants **5–6**, *6*
spin–spin coupling effects 1
spiral-MRSI 24
splittings 5–6
STEAM (stimulated echo acquisition mode) sequence 3, 8, 19–20, *20*
  lipid suppression 25
  prostate cancer 217
  water suppression techniques 25
stroke 11, **91**, 105
  choline 93
  creatine 93
  key points 91
  lactate 93, *94*, *95*, *96*
  metabolic changes **94–95**, *95*, *96*
  MRS acquisition protocols 105
  N-acetyl aspartate **91–92**, *92*, *94*, 94, 95
subacute sclerosing panencephalitis (SSPE) **118–119**, *123*
substantia nigra 43
succinate 14, 111, *123*
succinate-dehydrogenase deficiency (SDH) **184**, *185*
superconducting magnets 1
superoxide dismutase (SOD) 153

Index

supra/infra-tentorial multi-slice MRSI 22, **30–31**, *31*
surface radiofrequency (RF) coils 7
surrogate markers 11, 152, 154
susceptibility-weighted imaging (SWI) 163, 174

taurine 14
TBI *see* traumatic brain injury
temperature, brain 14
temporal lobe epilepsy 11, **134–135**
   acquisition protocols/protocol design **29–30**, *29*, *30*
   diffuse metabolic abnormalities **135–136**, *136*
   MRS protocol design *10*, *12*, **138–139**
   use of MRS where MRI is normal **136–137**
thalamus 12, 51, 98, *98*, 174
threonine 14
time domain processing *7*, **34–36**, *35*, 39
time domain signal 6
tissue compartmentalization **41–42**, *42*
toxoplasmosis **115–117**, *123*
traumatic brain injury (TBI) **161–162**, 174
   adults **167–170**, *167*
   classification of head injury **162**
   cognitive function related to metabolism **166**, 168
   concussion/repeated brain injury, MRS studies **170**
   diffuse metabolic abnormalities **165–166**
   glutamate *165*, **165**
   incidence statistics 161
   key points **161**
   lactate **169**
   long-term symptoms 162
   magnetic resonance spectroscopy 163
   metabolic changes in early stage **172–173**, *172*, *173*
   microdialysis studies **169–170**
   MRI scanning **162–163**
   MRSI studies **170**, *171*
   myo-inositol **165**, *166*
   pediatric 162, **164–166**
   predictive value of MRS *164*, **164**
   vegetative states **168–169**, *169*
   volumetric/whole-brain spectroscopy studies **173**, 174
truncation artifacts 37
tuberculoma **112–114**, *123*
turbo-MRSI 24

ultrasound 231
units, tissue metabolite concentration **41–42**
unsuppressed tissue water signal 40
urea cycle defects **192–194**

valine 185
VAPOR scheme 25
   vascular dementia (VD) **147–148**, 149
vegetative states 168–169
vertical scale **44**
VHF (very high frequency) 2
visual cortex 51
vitamin C 27
volume of interest (VOI) 182
voxel shifting 37–38
voxel/MRSI slice placement **44–45**

water referencing 40
water suppression *7*, **25**, **44**, **235–236**
West Nile virus 114
WET scheme 25
white matter
   acute disseminated encephalomyelitis 124
   Alzheimer's disease 146
   choline 12
   creatine 13
   gray/white matter boundaries 161, 162
   gray/white matter comparisons **51–56**, *53*, *54*
   leukoencephalopathy 198, **204–205**, *205*
   multiple sclerosis 119
   neonatal brain MRS 98
   vascular dementia 148
   volumetric/whole-brain spectroscopy studies 173
white matter rarefaction 198, 204
   leukoencephalopathy with vanishing white matter **204–205**, *205*
   megalencephalic leukoencephalopathy with subcortical cysts **205–206**, *205*, *206*
World Health Organization (WHO) **64–66**

xenobiotics 14
X-ray mammography **230–231**

zero-filling *35*, 36
zinc **214–215**, *214*